IMMOBILIZED BIOSYSTEMS
Theory and Practical Applications

IMMOBILIZED BIOSYSTEMS
Theory and Practical Applications

Edited by

IVAN A. VELIKY

Tivelco International Inc., Ottawa, Canada

and

ROBERT J. C. McLEAN

Department of Biology,
Southwest Texas State University,
San Marcos, Texas, USA

BLACKIE ACADEMIC & PROFESSIONAL
An Imprint of Chapman & Hall
London · Glasgow · New York · Tokyo · Melbourne · Madras

Published by Blackie Academic & Professional, an imprint of Chapman & Hall, Wester Cleddens Road, Bishopbriggs, Glasgow G64 2NZ

Chapman & Hall, 2–6 Boundary Row, London SE1 8HN, UK

Blackie Academic & Professional, Wester Cleddens Road, Bishopbriggs, Glasgow G64 2NZ, UK

Chapman & Hall Inc., One Penn Plaza, 41st Floor, New York NY10119, USA

Chapman & Hall Japan, Thomson Publishing Japan, Hirakawacho Nemoto Building, 6F, 1-7-11 Hirakawa-cho, Chiyoda-ku, Tokyo 102, Japan

DA Book (Aust.) Pty Ltd., 648 Whitehorse Road, Mitcham 3132, Victoria, Australia

Chapman & Hall India, R. Seshadri, 32 Second Main Road, CIT East, Madras 600 035, India

First edition 1994

© 1994 Chapman & Hall

Typeset in 11/13pt Times by Variorum Publishing Limited, Rugby
Printed in Great Britain by Galliard (Printers) Ltd, Great Yarmouth

ISBN 0 7514 0175 7   1–85861–031–1

A catalogue record for this book is available from the British Library

Library of Congress Cataloging-in-Publication data

Immobilized biosystems: theory and practical applications / edited by Ivan A. Veliky and Robert J. C. McLean.
  p.   cm.
Includes bibliographical references and index.
ISBN 1–85861–031–1
  1. Immobilized cells.   2. Immobilized proteins.   3. Immobilized enzymes.   4. Immobilized microorganisms.   I. Veliky, Ivan A.
  II. McLean, Robert J. C.
TP248.25.I55144   1994
660'.6 — dc20   93-18949
                                                              CIP

♾ Printed on permanent acid-free text paper, manufactured in accordance with ANSI/NISO Z39.48-1992 (Permanence of Paper).

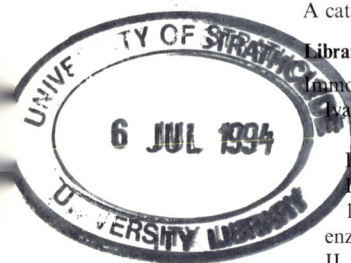

CONTENTS

PREFACE

This condensed view on immobilization of viable and non-viable cells, proteins, enzymes, active molecules and their interaction with the natural or synthetic carriers for performing biochemical and chemical reactions suggested a few thoughts for the present and future use of immobilized biosystems. It is essential that we understand the coordination of principles and theories of various scientific fields. Without understanding and using the basic functions in their broad integration many valuable applications and results of scientific research could be lost.

Even for Nature, it is difficult to combine all the necessary broad and detailed knowledge, imagination and creativity into one person's mind. However, there is a possibility of interlocking several minds and connecting the missing links between scientific fields and proceeding faster by transforming data and results into practice. Yet, one of the interlocking minds, the central coordinating mind, must have a powerful imagination and excellent background knowledge to be creative.

The above are generally applicable thoughts but are closely related and reflected in the topic of this book. Immobilized biosystems, as described by the authors, indicate examples of how the principles of integration work.

The first chapter describes a series of natural and synthetic carriers used for immobilization of viable cells and active organelles and molecules. The authors have shown the interaction and requirements of physico-chemical knowledge needed to interpret such a semi-synthetic biosystem.

The second chapter reaches even deeper into theory and indicates the potential of immobilized proteins for use as sensitive biosensors. The third chapter evolves the integration of physico-chemical principles into bioactive catalytic systems reflected by biochemical reactions. How to apply the preserved bioactivity of the immobilized viable cells and active proteins is demonstrated in Chapter 4. The authors suggested the use of glass fibers as support for the bioactive systems as inert, non-toxic and stable recoverable carriers.

The last chapter is a demonstration of another immobilized system in Nature. The authors describe a natural polymer produced by micro-organisms partially as a protection against environmental conditions. The polymer forming a biofilm inside the human body may carry and protect cells of micro-organisms difficult to fight with antimicrobial agents. In this case we have a natural immobilized system which we would like to 'dis-immobilize' in order to remove and cure the infection caused by the immobilized micro-organisms.

In this book, the described knowledge of immobilization techniques and characteristics of various natural and synthetic carriers for biochemical reactions is only a small indication of its great potential. The high concentrations of biocatalytic active molecules on the vast surfaces of many carriers offer their elevated interaction with the transforming substrate increasing the reaction and the yields. Among other benefits of the immobilized biosystems is the increased protection against physico-chemical environmental effects. Such batteries of biocatalytic units can perform in a number of cases with much higher capacities than the classical chemical technologies.

The present knowledge and applications of many immobilized biosystems are in their initial stages. Additional knowledge and understanding of many interactions between the active molecules and carriers are being gained. In the near future we shall see many new processes, medical treatments, transplants of active tissues and cells immobilized on inert carriers, detoxification, environmental-contamination detection and purification, all based on integrated knowledge of immobilization principles and their imaginative applications.

Ivan A. Veliky

LIST OF CONTRIBUTORS

S. Birnbaum
Process Analysis, Research and Development, Kabi Pharmacia Peptide Hormones, Kabi Pharmacia AB, S-11287, Stockholm, Sweden

R. Stephen Brown
Chemical Sensors Group, Department of Chemistry, Erindale Campus, University of Toronto, 3359 Mississauga Road North, Mississauga, Ontario, Canada L5L 1C6

D. E. Caldwell
Department of Applied Microbiology & Food Science, University of Saskatchewan, Saskatoon, Saskatchewan, Canada S7N 0W0

J. W. Costerton
Center for Biofilm Engineering, College of Engineering, 409 Cobleigh Hall, Montana State University, Bozeman, MT 59717-0007, USA

F. DiCosmo
Centre for Plant Biotechnology, Department of Botany, University of Toronto, 25 Willcocks Street, Toronto, Ontario, Canada M5S 3B2

P. Gemeiner
Institute of Chemistry, Slovak Academy of Sciences, Dúbravská cesta 9, CS-842 38 Bratislava, Slovak Republic

U. J. Krull
Chemical Sensors Group, Department of Chemistry, Erindale Campus, University of Toronto, 3359 Mississauga Road North, Mississauga, Ontario, Canada L5L 1C6

R. J. C. McLean
Department of Biology, Southwest Texas State University, San Marcos, TX 78666 USA

W. NEUMANN
Institute of Biomedical Engineering, University of Toronto, Toronto, Ontario, Canada M5S 3B2

O. NORRLÖW
Davison Product Line, W. R. Grace AB, S-251 06 Helsingborg, Sweden

L'. REXOVÁ-BENKOVÁ
Institute of Chemistry, Slovak Academy of Sciences, Dúbravská cesta 9, CS-842 38 Bratislava, Slovak Republic

F. ŠVEC
Institute of Macromolecular Chemistry, Czech Academy of Sciences, Heyrovského námestí 2, CS-162 06 Praha 6, Czech Republic

H. TANAKA
Institute of Applied Biochemistry, University of Tsukuba, Tsukuba, Ibaraki-ken, 305 Japan

E. T. VANDENBERG
Chemical Sensors Group, Department of Chemistry, Erindale Campus, University of Toronto, 3359 Mississauga Road North, Mississauga, Ontario, Canada L5L 1C6

I. A. VELIKY
Tivelco International Inc., 613 Fielding Drive, Ottawa, Ontario, Canada K1V 7G7

INTRODUCTION

IVAN A. VELIKY
Tivelco International Inc., Ottawa, Canada

'Creativity is a clone of knowledge and imagination.'

It is mankind's natural behaviour to challenge the unknown. Most of us are constantly fascinated by unexplored phenomena and complexity in Nature, our daily lives and the environment. Progress and scientific achievements increase our curiosity and imagination. The human brain has extremely high potential and energy. It also has a built-in, self-activating and stimulating, exceptionally complex mechanism as the main source of accumulated knowledge and inherited imagination and creativity.

Accumulated knowledge as a result of learning may not always be sufficient to produce progress. To use the knowledge one must have imagination. This combination will result in creativity and eventually will bring new discoveries and stimulate progress in science and technology. Imagination helps to combine the bits and pieces of gathered knowledge into a logical, useful product.

Such combination of knowledge and imagination is essential for all scientific fields and research areas. One of the most complex and, on knowledge and imagination, demanding fields is biology. Living systems and their interaction with the environment have created a well balanced biosphere harmony. There is a tendency to relate this harmony to interaction within the living systems only. However, there is a structure, a foundation needed for initiating, preserving and supporting the functions of living systems. Such a structure has an essential importance for the existence and functioning of living systems. Therefore, a better comprehension of the principles of coexistence in our environment can have a significant effect on creativity in science and can motivate new achievements much faster. Where to look for such a motivation? Nature is a source of infinite information. Nature is the World's largest data bank of facts.

There are many examples around us to indicate the importance and role of supporting structures or carriers of living systems. All biosystems are actually attached to one or another kind of a carrier or structure made of inorganic or organic materials. Relative to the carrier or structure, they are recognized as the natural immobilized biosystems. There are macro- and micro-systems. There are immobilized tissues, cells and active molecules. All of those individual cells survive and perform better in the immobilized state. Trees grow with roots attached to soil, algae grow attached to solid supports such as stones in streams and lakes. Some cells or tissues grow on supporting carriers made of inorganic or organic polymers. The catalytic systems of cells – enzymes – are part of a protein or other polymeric structure. One can continue selecting examples from our environment and biosystems to indicate the importance of a carrier for life performance. On the pages of this book the reader will find a description of old, natural and new synthetic carriers, most of them simulating Nature's structures or physico-chemical laws. The chapters of this book describe the theory, application and use of the carriers for immobilization of bioactive molecules, cells or tissue.

The forms of association between Nature and immobilization of bioactive systems are numerous. This book offers selected examples for the immobilization or cells and enzymes and indicate some of the infinite applications and potential at present and in the future.

Without a doubt, the biosystems represent the only example of the most complex of coordinated chemical reactions which also have a built-in memory bank of every one of them. The reactions can repeat not only at the same rate, but also with the same precision by reproducing the same biosystem as selected from the DNA memory bank of data. We also know that biosystems by some standards and expectations may not be perfect. For example, yeast cells convert sugar into ethanol and carbon dioxide. During the process the cells divide, using some of the sugar molecules as energy source and building material for the new cells and, obviously, they produce by-products from the nutrients necessary for their survival and reproduction. Some of the by-products in the wine industry are important for the taste and bouquet which determine the wine's quality. In industry which produces ethanol in large quantities as a pure product, the lower yields and by-products are simply economical losses. To improve this process and increase the production rates, it was attempted to immobilize viable yeast cells in various polymers. The cells immobilized by encapsulation or entrapment in those polymers formed high concentrations of active units in relatively small areas. To prevent the cells from dividing and multiplying, they were fed with the limited nutrient essential only for ethanol production. This nutrient control helped to increase the efficiency of producing ethanol of such immobilized biosystems when compared with the regular fermenters.

The viable cell immobilization approach is not always the best. Some polymers used for immobilization could allow leakage of cells and consequent clogging of bioreactors. Other polymers could slow down or prevent the transport of substrates or metabolites because of their molecular sizes. Such effects could lower the efficiency of a bioreactors. It is therefore important to select a suitable carrier.

In some simple cases a viable cell can be replaced by an immobilized non-viable cell form as an economical source of specific oxidases, hydrolyses, and other enzymes. A simple example would be the enzymatic hydrolysis of sucrose by invertase. The resulting products of this reaction are glucose and fructose. Many strains of Saccharomyces are known to produce enzyme invertase. To obtain non-viable cells as a source of invertase, *Saccharomyces cerevisiae* cells were homogenized to remove the viable cells. The homogenized mixture was immobilized by encapsulation in calcium alginate. A cylindrical bioreactor was filled with spherical beads containing encapsulated homogenate of non-viable cells with invertase activity. A continuous input of sucrose to the bioreactor resulted in continuous output of a mixture of fructose and glucose. The column was active for several weeks without replacing the immobilized homogenate. The advantage of such immobilized homogenate of cells (non-viable cells) or tissue is mainly in better stability of the enzyme in encapsulation. The enzyme is actually 'immobilized' on the non-viable structure of the cell or tissue and this complex is then encapsulated in natural or synthetic biologically inert polymer. Such 'double' immobilized systems could prevent leakage of enzymes for which the molecules are smaller than the pores of the gel used for encapsulation.

The example was completed by a demonstration of one potential practical application. A second bioreactor containing immobilized viable cells of another yeast strain known to utilize only glucose was fed the effluent of the first bioreactor containing glucose and fructose products from sucrose conversion by invertase. The glucose was utilized by the immobilized 'glucose-eating' yeast and converted to ethanol leaving fructose syrup as a by-product.

The demonstrated principle of immobilized biosystems based on enzymatic activities of non-viable and viable yeast cells could be used in practical applications such as fructose syrup production, fructose sweetened alcoholic beverages, dessert wines, etc. The process can be improved by replacing the immobilized non-viable yeast cells as the source of invertase by purified and obviously much more highly active pure enzyme immobilized on another carrier. A number of options and techniques are described in this book.

The authors of the chapters in this book directly or indirectly indicate the ways and areas of potential benefits of the immobilization of bioactive molecules and complexes for industry, technology and medicine.

One naturally occurring immobilized biosystem in the human body is worth mentioning specifically. Some of the problems of treating human and animal chronic infections may result from encapsulation of infectious micro-organisms in polymers formed and excreted by the micro-organism on the body's tissue, organs or walls of body fluid transporting passageways. Details of such cases are described in one of the chapters in this book. Polymers produced by the micro-organisms form a film on tissue and create a 'hiding' place for the micro-organisms. The antimicrobial molecules are effective and kill only the freely floating micro-organisms in the body fluids. However, they cannot penetrate and reach the micro-organisms embedded or encapsulated in the polymeric substance. The molecular sizes or physico-chemical reactions between the antimicrobial compound and the polymer produced by the micro-organisms can alter their transport mechanism. Therefore 'hidden' microorganisms in the polymer will survive and may become a source of chronic infection. This example demonstrates one of the principles of how naturally occurring immobilized biosystems could protect micro-organisms against environmental effects.

It was indicated in general terms what are immobilized biosystems, how they function and participate in biocatalytic reactions and simulation of complex biochemical processes performed in living systems. There are numerous books on the market at present and more will be coming out with details on immobilization of cells, enzymes and their combination for various applications. Although Nature is the best teacher and the source of infinite information, it is not necessarily perfect. It is Man's natural desire to learn, and continually improve his or her knowledge. The biosynthetic processes in Nature can produce healing products for many if not all diseases. Most of the biosynthetic processes are performed in plants as well as animals and microbes. Such biosynthetic processes are lengthy, multistep procedures. However, they are not necessarily the most efficient process. The biosynthesis of an active compound can sometimes be improved when performed outside the living system. It may be possible to dispose of a side reaction not essential for obtaining the final product, whereas in the living system the side reaction may be necessary as an intermediate protective step. Therefore, in a carefully designed artificial system, the biosynthetic process can run more efficiently and produce elevated yields and purity of the final product.

A good example of improved biosynthetic activity is the oxidases of certain cells. Plant cells isolated from carrot root perform bioconversion of cardiac glycosides (digitoxigenin, gitoxigenin) by introducing a hydroxyl group into the C_5 position. This reaction is difficult to perform by a classical chemical process. The biosynthetic process using viable and non-viable immobilized cells is efficient and provides a yield of over 80 per cent. This yield cannot be reached by present synthetic chemical methods. Also, the hydroxylated products of genins are most stable in the human body. Their therapeutic dose for heart patients is easier to determine than in the case of the non-modified digitalis drugs.

There are certainly many other reasons why and where the immobilized system could apply in practice. The immobilized cells or tissue have an important place in medical practice. The immobilized cells or tissue can replace cells, tissue or even glands damaged by illness or the aging process, or can replace non-developed or damaged brain cells, etc. Why immobilized cells or tissue? Why not a simple transplant of healthy tissue only? In some cases a simple transplant of tissue may be sufficient. The advantage of using immobilized cells for such repairs of damaged body organs is in preventing rejection and increasing stability and location of the transplanted cells or tissue. Before a transplant of the healthy cells is done, the cells are attached to an inert carrier and cultured in the medium gradually enriched with the body fluids of the recipient. The donor cells grow attached to the carrier and gradually adapt to the conditions of the recipient's body, hence reducing the rejection to a minimum. This cell conglomerate, an artificially created 'mini-organ' or 'mini-gland', is then used to replace the damaged human tissue.

Production of active medicines and drugs could be improved by using more efficient systems of immobilized active cells (microbial, plant or animal) in batch or continuous bioreactors. Fermentation processes could be controlled by using immobilized enzymes as biosensors for detecting metabolite concentration and the supply of substrates and nutrients. This book has a few innovative demonstrations in the construction and use of biosensors.

Immobilized viable cells in fermentation processes in the food industry are well recognized. The ancient way of vinegar production from wine or diluted alcohol is based on simple immobilization of bacteria on wood chips. There are other places in the food and fermentation industry, such as the production of fructose syrup, ethanol and other solvents, where immobilized enzymes or viable cells are used. The cheese-making industry is also using immobilized enzymes for processing milk into various cheese products. The process is continuous and preserves very well the flavour of the final product.

An open area for application of immobilized systems is the agriculture industry. Biological control of pests in agriculture is on the rise. Improvement of the propagation of predators has yielded good results by using nutrient saturated carriers.

The few abbreviated notes on the usefulness of immobilized biosystems in research and development indicates the need to challenge the unknown in this field. The applications of the findings and discoveries in the broad spectrum of the complexity of immobilized biosystems are yet to be recognized.

ERRATUM

The following section is missing from Chapter 1, p. 19.

Because the alignment of individual cellulose chains is frequently imperfect, irregularities in fibril structure occur. This is reflected in the structure of the whole fiber which consequently consists of polycrystalline aggregates separated by amorphous regions (Hon, 1988).

Cellulose was one of the first materials to be used as a matrix for covalent binding of enzymes. A survey of methods reported in the past 10 years (Kennedy, 1978; Sturgeon, 1982; White, 1985; Kennedy & Cabral, 1987a; Gemeiner *et al.*, 1989a) showed, however, that cellulosic materials are not as popular as other polysaccharide supports. The reasons are susceptibility to microbial degradation and nonspecific adsorption properties. Nevertheless, cellulose has a major advantage over other polysaccharide supports in that it has become available in many different physical forms, such as fibers, microgranules, micro-crystals, beads, gel particles, membranes, tubings, emulsions, etc. For industrial reactors, less pure cellulosic material can be used in the form of ropes, pulps, chippings, cloths or papers.

A variety of morphological properties of cellulose should be assessed with respect to its specific application prior to selection for evaluation. Sufficiently adequate for full characterization of carrier properties of cellulose are parameters such as surface area and pore diameter. Both of these parameters affect in turn the loading of cellulose by the enzyme. To take account of these parameters celluloses can be very roughly classified as low-porous and high-porous matrices based on their morphology.

1.3.1.1 *Low-porous Celluloses*
Low-porous celluloses available on the market are designed mainly as fibrous, powder, microgranular and microcrystalline. Cellulose powder is prepared either by mechanical grinding of fibrous cellulose or by its heterogeneous hydrolysis with mineral acids.

Chapter 1

NATURAL AND SYNTHETIC CARRIERS SUITABLE FOR IMMOBILIZATION OF VIABLE CELLS, ACTIVE ORGANELLES, AND MOLECULES

PETER GEMEINER,[1] L'UBOMÍRA REXOVÁ-BENKOVÁ,[1] FRANTIŠEK ŠVEC,[2] AND OLOF NORRLÖW[3]

[1]Institute of Chemistry, Slovak Academy of Sciences, Bratislava, Slovak Republic
[2]Institute of Macromolecular Chemistry, Czech Academy of Sciences, Prague, Czech Republic
[3]Davison Product Line, W.R. Grace AB, Helsingborg, Sweden

CONTENTS

1.1   INTRODUCTION

The first immobilization of a biological catalyst by sorption of invertase on the activated charcoal goes back to the 1916. However, the first impulse has soon been forgotten. A new push came in the mid-1950s starting a real immobilization rush. In the following years appeared thousands of papers and patents and many reviews and monographs (e.g. Zaborsky, 1973; Sundaram & Pye, 1974; Mosbach, 1976; Chibata, 1978; Wiseman, 1985).

A few basic immobilization techniques were discovered in the course of time classified as carrier-binding, crosslinking, and entrapping (Chibata, 1978).

A binding to an insoluble carrier may be due to a physical adsorption (hydrogen bonds, hydrophobic interactions, ionic bonds) or due to a covalent bond. Carrier-binding is widely used for enzyme immobilization. It takes advantage of carrier activity towards some groups localized in the protein molecule. The crosslinking of enzymes or whole cells employs polyfunctional reagents the glutaraldehyde being the most popular one (Marconi, 1989). The molecular (enzyme molecules) or granular (cells) entities also may be entrapped in a continuous polymer structure (network) or in a formation resembling a packing (microcapsules, liposomes, vesicles, hollow fibers, tubes, etc.). The literature also describes methods combining two of the basic ones, e.g. sorption of an enzyme on the surface of solid particle and subsequent crosslinking.

1.2 CLASSIFICATION OF CARRIERS

The success of any immobilization relies on the proper choice of the carrier. Some of them are developed specifically for a special type of immobilization technique (carrageenan or alginate for entrapping) while others are universal and may be used in all methods (agarose, polyacrylamide copolymers). Properties of any carrier can be reviewed in relation to the following criteria (Mosbach, 1976; Royer *et al.*, 1976):

1. Strength	5. Porosity
2. Form	6. Reactivity
3. Stability	7. Reusability
4. Hydrophilicity	8. Economy

An ideal carrier that possesses all optimal properties does not exist, because some of them are controversial. To choose the best support for the immobilization one has to find an optimal combination of the parameters. Any real carrier represents a compromise, anyway.

The criteria suggest additional categories for classification of polymers used for an immobilization regardless of whether the polymer is a natural or synthetic one. The former are particularly polysaccharides and inorganic compounds, while the synthetic polymers are more variable, their chemical structures and also their properties can better match the expected application.

In the following we shall try to show how the known supports may be further rated according to criteria and document the extent of types and properties available.

1.2.1 Mechanical Strength

Immobilization of any catalyst including the biological ones is driven by an idea of simple removal of the catalyst from the reaction mixture or vice versa, and its repeated use. Therefore the mechanical strength has to be evaluated in relation to the technology concerned in which the immobilized enzyme or cells will be used. Thus, the hollow fibers or tubes have to resist the pressure inside them under which the liquid is being pumped through, membranes should withstand the tangential force, and so on.

The most frequently used carriers are particulate, possessing a spherical shape. They are employed in both packed bed and stirred tank reactors.

To run a column reactor the substrate solution has to flow through and it is pumped under the pressure assuring the flow rate. The pressure acts on all the particles inside the column and tends to deform them in the flow direction. When the column operates in the downflow the force representing the weight of the support in the column itself should be added. The pressure in a packed bed reactor grows up with the length of the column and the flow rate. Therefore the soft carriers may be used in both batch and continuous stirred tank reactors where the immobilized catalyst is dispersed in the medium by gentle stirring. The catalyst particles need not resist a pressure vector, the mechanical stress is low and mechanical strength is thus less important.

An intermediate between packed bed and stirred tank is a fluidized bed reactor in which the immobilized catalyst is held in a fluid state by a stream of liquid or processed gas. Similarly to the previous case the cata-lyst has to resist abrasion caused during mutual collisions of the particles.

The previous issue suggests a division of insoluble supports into two extremes, e.g. soft (gel-like) and hard ones. The linking between them is, however, continuous.

The crosslinked polymers containing none or less than 10 per cent of crosslinking agent appear always in the dry state as a glass. They are transparent and exhibit almost no porosity when measured by a B.E.T. technique. The only pores are the distances between chains. This implies the traditional term microporous polymers. When immersed in a solvent they swell to the extent allowed by density of the network. Their volume may increase often even when the crosslinking density is low and they dissolve totally when not crosslinked at all. The apparent porosity, called swelling porosity, appears only when the polymer is swollen with solvent.

The slightly crosslinked carriers are used in the carrier binding and entrapping in the laboratory scale quite often (Mosbach, 1976). The group comprises natural polymers like dextran, agarose, alginate as well as the synthetic polymers like polyacrylamide, poly(vinyl pyrrolidon), poly(acrylic acid), etc.

The highly crosslinked non-porous hard particles do not swell at all. The accessible surface for immobilization is rather small and the amount of immobilized protein determining the specific activity of the composite is also low. The only advantage is absence of intraparticular diffusion inside particles typical for porous carriers. To increase the accessible surface extending it also into the particle, macroporous structures were developed.

The group of hard carriers contains inorganic porous materials based on silica, alumina, titania, zirconia, etc., metal particles, and excessively crosslinked polymers.

It should be kept in mind that the mechanical properties of some carriers may change depending on the medium in which they are located. A hard polymer may become soft when immersed in a surrounding fluid differing in type, pH value, or ionic strength. The typical example is a support bearing electrostically charged groups.

1.2.2 Physical Form

Any technology employing immobilized catalyst requires suitable shape of its body. Besides fibers, tubes, membranes, etc., those most often used for enzyme immobilization are carriers in the bead shape. The size of them depends on the process in which they operate and varies from a few nanometers up to some millimeters, i.e. within several orders of magnitude. The desired size of the support defines the method of its preparation.

1.2.2.1 Suspension Process

The most popular process of bead preparation proceeds in a suspension. The polymerization method has been known since 1909. Particle size of the produced beads is influenced by viscosity of the continuous phase, interfacial tension, ratio between the dispersed and continuous phase, and intensity of stirring that depends again on shape of both stirrer and reactor, and the stirring speed (Švec *et al.*, 1975; Horák *et al.*, 1981*a,b*).

Dispersion of continuous phase is a random process. It results in a system in which coalescence and redispersion reach a dynamic equilibrium which results in droplets of different sizes. The product of a suspension polymerization consists of particles having non-uniform size.

Any suspension process starts by dispersing one liquid in the other one in the form of small droplets. They are precursors of the final solid beads but they have to become insoluble during the process to become easily separable. The obvious method to do this is crosslinking. Two different ways are available to get a crosslinked carrier for immobilization: crosslinking of soluble polymeric molecules and crosslinking polymerization.

1.2.2.2 Subsequent Crosslinking
In the field of preparation of polymer-based carriers for enzyme immobilization, crosslinking of existing polymers is very common. The most famous representatives of this class of materials produced by Pharmacia Uppsala are based on natural polysaccharides dextran and agarose

(Sephadex, Sepharose) crosslinked to different degrees by epichlorohydrin. The products range from soft gels to rigid beads. Their hydrophilic surface has to be activated prior to immobilization (Mosbach, 1976).

A process similar to the subsequent chemical crosslinking by bifunctional reagents is sol-gel transformation. Some naturally occurring polysaccharides or proteins are insoluble in water at ambient temperature due to non-covalent interactions between their chains, e.g. hydrogen bonds, which successfully replace the chemical crosslinks. They readily dissolve at increased temperature. When the hot water solution is dispersed in an immiscible liquid and cooled, water swelled beads result. Typical examples of thermoreversible gels are agarose, starch, or gelatin (Hupkes & Tilburg, 1976; Kuu & Polack, 1983).

Cellulose is also a polysaccharide insoluble in any individual solvent even at high temperature due to the many hydrogen bonds and its crystalline structure. It becomes liquid after chemical transformation to xantogenate (viscose) by a reaction with CS_2. The viscose can be dispersed in an immiscible liquid and under higher temperature it decomposes back to the cellulose keeping the form of porous, mechanically strong beads (Baldrian *et al.*, 1978).

1.2.2.3 Crosslinking Polymerization
The crosslinking polymerization does not differ from any other polymerization with the exception that the monomer mixture has to contain a crosslinking agent, e.g. 1,4-divinylbenzene (I), methylene-bis-acrylamide (II), ethylene dimethacrylate (III), etc (Scheme 1.1).

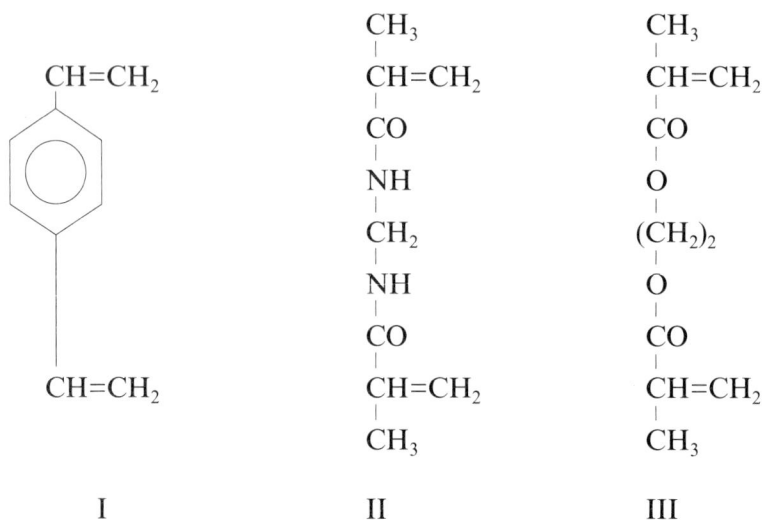

	CH_3	CH_3
	|	|
$CH=CH_2$	$CH=CH_2$	$CH=CH_2$
	|	|
	CO	CO
	|	|
	NH	O
	|	|
	CH_2	$(CH_2)_2$
	|	|
	NH	O
	|	|
	CO	CO
	|	|
$CH=CH_2$	$CH=CH_2$	$CH=CH_2$
	|	|
	CH_3	CH_3
I	II	III

Scheme 1.1

The product of crosslinking polymerization is an insoluble network that swells to an extent depending on the amount of crosslinking agent present in the mixture (Alfrey *et al.*, 1952; Kun & Kunin, 1968, Dušek, 1969; Popov & Schwachula, 1981).

1.2.2.4   Emulsion Polymerization

To get beads smaller than approximately 2 μm by a polymerization the emulsion process has to be employed. Although this type of polymerization has been used for the preparation of carriers for enzyme immobilization, examples of successful use are rare in the literature (Seitz & Pauly, 1979; Ohtsuka *et al.*, 1984; Bahadur *et al.*, 1985; Kawaguchi *et al.*, 1988).

The principle of emulsion polymerization is simple. Several molecules of a low molecular weight surfactant (a soap) when dissolved in water aggregate to colloidal moieties called micelles. A monomer added to the micellar system dissolves partly inside micelles where the polymerization takes place initiated by a water soluble free radical initiator. The stirring is less important than in the suspension polymerization. The product of emulsion polymerization is called latex and represents colloidally stabilized dispersion of a polymer in water. However, the polymer cannot be separated in the dry state without losing its particulate character (Poehlein, 1989).

1.2.2.5   Two-step Swelling Polymerization

The latex particles produced by an emulsion polymerization produce very uniform particles. These seeds can be enlarged by swelling with another monomer or monomer mixture to the expected size and polymerize again in order to stabilize the new size reaching up to several hundreds of μm. The enlargement does not spoil the uniformity (Ugelstad *et al.*, 1985).

The uniformity of particles is advantageous. Pressure drop in any filled column is directly proportional to the particle size distribution (Chibata, 1978). The uniformly sized carriers may be packed in higher columns or a less powerful pump may be used to get equal flow rate.

1.2.2.6   Spraying of Polymer Solution

This less common method was used to get beads from thermoreversible polymer solutions or from acrylamide monomer by polymerization (Kostner & Mandel, 1976; Woodward *et al.*, 1982). The method consists of solidification of droplets running from a capillary or set of capillaries into an immiscible liquid. The solidification results during the falling down and the product is collected on the bottom. The method allows preparation of entrapped enzymes or cells under mild conditions.

1.2.3   Chemical, Thermal, and Biological Stability

The immobilized enzymes or cells are used in various media differing in acidity (basicity), ionic strength, containing various substrates. The carrier should be stable to withstand the conditions without deteriorating.

All natural and many synthetic polymers include heteroatoms in the main chain or in the crosslinks that can be subject of a chemical attack and cleavage. When the extent of breaks is high the originally insoluble carriers dissolve. An example of hydrolytically weak bond is the amide bond –CO–NH– which is a part of polyamide main chain or methylene-bis-acrylamide (II) crosslinks, ester bond –CO–O– of polyesters or diester crosslinking agents (III). The hydrocarbon –C–C– or ether bonds –C–O–, are very stable and do not decompose at all.

The working pH range of the majority of enzymatic reactions is between pH 3 and 10. In this range the vast majority of organic carriers are stable despite the presence of bonds mentioned above. However, aqueous solution with high pH is lethal for some inorganic supports, e.g., silica dissolves at pH exceeding 8.

Attention should be devoted not only to the carrier itself but also to the linking between the carrier and the immobilized enzyme. It also can break up under the action of chemical agents. A typical example is two weak imine bonds produced between a glutaraldehyde activated carrier bearing amino groups and an amino group of the enzyme : $|–CH_2–N=CO–(CH_2)_3–CO=N–ENZYME$. To improve the stability of conjugates prepared by the popular activation method the imino group is reduced by sodium borohydride to an amino group. At low pH the silane bond between the residue of activation reagent (often γ-aminopropyltrimethoxysilane) and the matrix is not stable and the immobilized enzyme leaks.

The alginate gel often specially used for entrapping whole cells dissolves when the bivalent calcium ion causing crosslinking is exchanged by a monovalent metal ion or replaced by a proton.

The thermoreversible gels must not be used at temperatures at which they dissolve and lose their shape.

Carriers based on natural polysaccharides or proteins are good nutrients for microbes. When the carrier is digested the immobilized species is released into solution. The synthetic polymers and inorganic supports resist microbial attack very well. Also, in this case, the enzyme itself is threatened. Some microbes produce extracellular proteinases that destroy the immobilized enzyme and render the conjugate inactive (Messing, 1975*a*).

1.2.4   Hydrophilicity

Almost all biological reactions take place in aqueous media. This implies that the immobilized biocatalyst should also be located in a hydrophilic environment one part of which is the support. Moreover, the hydrophilic porous carrier allows penetration of water soluble substrates by diffusion to the immobilized active entity which catalyzes the transformation. The enzyme molecules keep the correct conformation in water and express the highest activity.

The hydrophilic carriers also prevent any non-specific interactions of a hydrophobic character which may block just the essential groups of an enzyme or distort its higher structure and thus decrease or even kill its activity.

The hydrophilicity of a carrier is caused by the presence of hydrophilic groups like ether $-C-O-C-$, amide $-CO-NH-$, or hydroxyl groups $-OH$, on all its surface. Also the electrostatically charged groups, e.g., carboxyl $-COO^-$, different amino groups $-NH_2$, $-NHR$, $-NR_2$, $-N^+R_3$, sulfo groups $-SO_4^{2-}$, etc., are very hydrophilic when ionized. The electrostatic interaction with the opposite charged groups of enzymes is employed for immobilization but it also can induce non-specific interactions with the consequences shown above.

1.2.5   Porosity

Mass transfer plays an important role in any heterogeneous system in which a reaction proceeds. In the case related to our topic the substrate is transported from the bulk liquid (or gaseous) phase into the solid where the immobilized catalyst is located. Product of the reaction is transported back to the bulk phase. The stirring and diffusion are considered to be the most influential effects.

1.2.5.1   Non-porous Carriers

The simplest situation arises when the catalyst is immobilized just on the outer surface of a non-porous solid. The reaction rate (enzymatic activity) is governed only by mass transfer in the bulk, i.e. out of the particles. The easily available surface of a non-porous carrier is, however, small and the area suited for accommodation of enzyme molecules or even whole cells is thus limited. The immobilized amounts of the catalyst are low and also the overall activity. An example of such a carrier is a tube activated on the inner wall only.

Because of evident drawbacks of non-porous materials the porous carriers are preferred.

1.2.5.2   Gel-type Carriers

The slightly crosslinked polymers are not porous in the dry state, but when swelled in water the structure 'opens' and all parts of the carrier are accessible for penetration of even large molecules. The extent of crosslinking determines the 'mesh size' of the support. When a crosslinking proceeds in the presence of enzyme molecules or cells, they remain entrapped in the matrix and their leakage is widely prevented. The network allows the substrate and the product to penetrate freely through the 'screen' of the carrier while the catalyst cannot leave the 'cage'. The advantage of entrapping is simplicity of the immobilization method and low cost, possibility of immobilization of more than one enzyme at any level of purity or a mixture of cells, high yields of immobilization and easy recovery of reaction products which, however, is typical for any heterogeneous catalyst. The method is successful in the industrial processes (Marconi, 1989).

The most often described polymer for entrapping is possibly polyacrylamide gel first used by Bernfeld (Bernfeld & Wan, 1963; Tori *et al.*, 1972; Ohmiya *et al.*, 1977; Couderc & Baratti, 1980: Morikawa *et al.*, 1980; Freeman & Aharonowitz, 1981; Škodová *et al.*, 1981; Bang *et al.*, 1983; Furusaki & Asai, 1983*a*; Furusaki *et al.*, 1983*b*). Scheme 1.2 schematically depicts the conjugate.

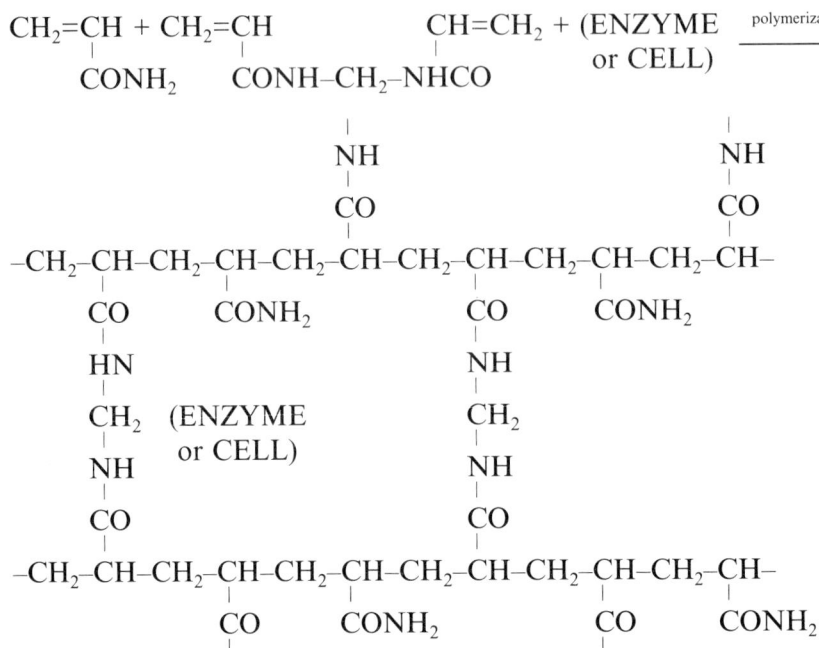

$$CH_2=CH + CH_2=CH \qquad\qquad CH=CH_2 + (ENZYME \xrightarrow{\text{polymerization}}$$
$$\quad\;\; | \qquad\qquad\;\; | \qquad\qquad\qquad | \qquad\qquad or\ CELL)$$
$$\;\; CONH_2 \quad\; CONH-CH_2-NHCO$$

$$| \qquad\qquad\qquad\qquad |$$
$$NH \qquad\qquad\qquad\quad NH$$
$$| \qquad\qquad\qquad\qquad |$$
$$CO \qquad\qquad\qquad\quad CO$$
$$| \qquad\qquad\qquad\qquad |$$
$$-CH_2-CH-CH_2-CH-CH_2-CH-CH_2-CH-CH_2-CH-CH_2-CH-$$
$$\quad\;\; | \qquad\qquad | \qquad\qquad\qquad\; | \qquad\qquad |$$
$$\;\; CO \qquad\; CONH_2 \qquad\qquad CO \qquad\; CONH_2$$
$$\;\; | \qquad\qquad\qquad\qquad\qquad | $$
$$\;\; HN \qquad\qquad\qquad\qquad\quad NH$$
$$\;\; | \qquad\qquad\qquad\qquad\qquad |$$
$$\;\; CH_2 \quad (ENZYME \qquad\quad CH_2$$
$$\;\; | \qquad\;\; or\ CELL \qquad\qquad |$$
$$\;\; NH \qquad\qquad\qquad\qquad\quad NH$$
$$\;\; | \qquad\qquad\qquad\qquad\qquad |$$
$$\;\; CO \qquad\qquad\qquad\qquad\quad CO$$
$$\;\; | \qquad\qquad\qquad\qquad\qquad |$$
$$-CH_2-CH-CH_2-CH-CH_2-CH-CH_2-CH-CH_2-CH-CH_2-CH-$$
$$\qquad\quad | \qquad\qquad | \qquad\qquad\quad | \qquad\qquad |$$
$$\qquad\;\; CO \qquad\; CONH_2 \qquad\qquad CO \qquad\; CONH_2$$
$$\qquad\quad | \qquad\qquad\qquad\qquad\quad |$$

Scheme 1.2

Besides crosslinked polyacrylamide many other polymers were used, e.g., poly(acryl acid), poly(vinyl alcohol), poly(vinyl pyrrolidon), poly(2-hydroxyethyl methacrylate) (Chibata, 1978; Manecke & Beier, 1982, 1983; Kamamura & Kaetsu, 1983, 1984; Cantarella *et al.*, 1988). In the last decade the use of natural polymers like carrageenan, alginate, or agar for the entrapping procedure, grew.

The immobilization by entrapping is expected to be mild but one should keep in mind the negative effect of free radical and reaction heat released during polymerization especially when large scale batches are produced. The general drawback of the entrapped catalysts is poor mechanical strength, and their easy deformation that limits their use in columns.

A special category of entrapping is encapsulation. The enzyme is entrapped inside a polymeric cover with a wall permeable for low molecular weight compounds. The final catalyst may have different size and shape starting from liposomes and vesicles (50 nm) up to fibers, tubes, and membranes the size of which is not limited at all (Chibata, 1978; Inloes *et al.*, 1983).

1.2.5.3 Macroporous Polymers

Highly crosslinked systems and inorganic carriers are not supposed to swell. The only way to increase the surface available for immobilization is to make them porous. Extensive studies of crosslinking polymerization revealed that the presence of an inert solvent or solution of soluble polymer can cause rigidity and porosity of the final polymer. Polymers which are porous even in the dry state are called macroporous and exhibit a specific surface area over 5 m^2/g (Seidl *et al.*, 1967: Jacobelli *et al.*, 1979).

Copolymerization of mono- and divinylic monomer leading to a macroporous polymer is a special type of heterogeneous crosslinking polymerization and its theory has been developed (Seidl *et al.*, 1967; Heitz & Platt, 1969; Dušek, 1971; Guyot, 1988).

1.2.5.4 Morphology of Macroporous Polymers

The unifying peculiarity of all macroporous polymers is the spherical entities inside beads that reach a size up to a few hundreds of nm called globules (Kun & Kunin, 1968; Pelzbauer *et al.*, 1979). They are most important morphological features of macroporous polymers. The schema of a globular structure is shown in Fig. 1.1. The definition of surface area and porosity follows from the illustration. The former roughly represents the sum of surfaces of all globules, while the latter is the free space among them. The globules may be arranged very regularly in an array,

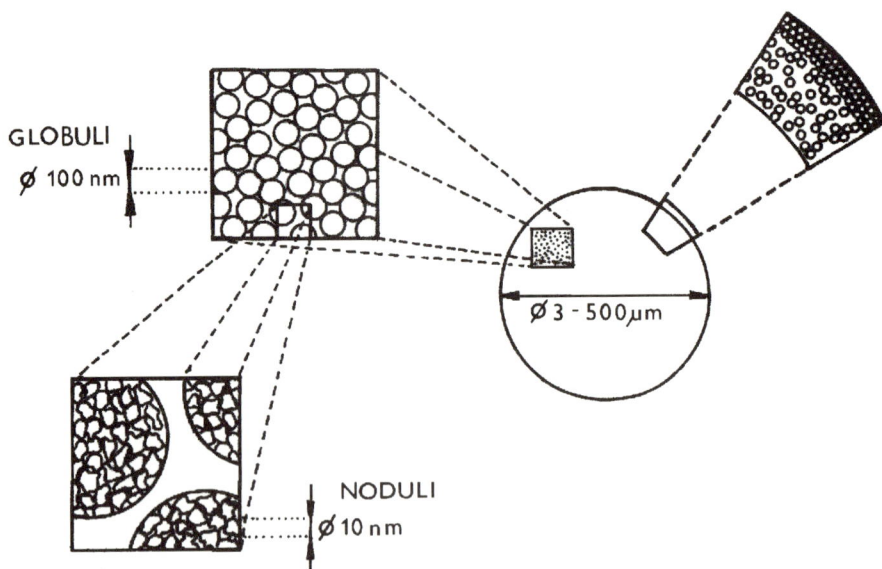

Fig. 1.1 Morphological features of a macroporous polymeric bead.

the interstitial vacancies are small and their size is more uniform. When, on the other hand, the pattern is not regular the pores become larger and their size distribution broader. Large pores can better accommodate the bulky enzyme molecules leaving enough space for diffusion of substrate. The pore size should exceed the size of immobilized enzyme at least three times. The catalyst is attached to the surface of the globules inside a bead.

Some authors found that macroporous beads are covered with a shell which is less porous than the inside part of the bead (Fig. 1.1). The presence of the shell may influence the diffusion (mass transfer) of substrate from the bulk phase to the immobilized enzyme inside the bead and products of enzymatic reaction (Heitz, 1970; Horák *et al.*, 1981*a*).

1.2.5.5   *Inorganic Carriers*

The porosity and morphology of the porous particles depend on the method of their preparation and it is affected by the chemical nature of the material used. The porous particles prepared from soluble inorganic polymers like silica, titania, zirconia, etc. exhibit morphological features similar to that of macroporous polymers described in the previous section. The mechanism of creation of the structure is closely related.

Glass carriers with controlled pore size are somewhat different but still

Table 1.1
Comparison of Enzyme Carriers

Morphological feature	Advantage	Disadvantage

Non-porous	Low diffusional limitations	Low surface area
		Low enzymes loading
		Use of fine particles:
		  Difficult to remove
		  Difficult for continuous process
		  High pressure drop
		  Low flow rate

Porous	High surface area High enzyme loading Protection from external environment	Diffusional limitations High cost (CPG) Micropores

very popular. The narrow pore size distribution provides the glass with a great potential in the field of immobilized enzymes. The preparation is based on the fact that certain borosilicate glass compositions exist which after heat treatment (500–700°C) can be leached to form a porous glass framework. During the heating the base glass separates into two inter-mingled and continuous glassy phases. One of them, rich in boric acid, is soluble in acid, the other one is high in silica and stable towards acid. In contrast to the globular morphology of traditional macroporous carriers the morphology of controlled pore glass is different and resembles tortu-ous capillaries. The pore size distribution is very narrow, the porosity low (up to 30 per cent) and specific surface area reaches up to 300 m²/g. A similar method was developed also for preparation of controlled pore ceramics (silica, titania, etc.) (Messing, 1978).

Table 1.1 compares both advantages and disadvantages of carriers relating to their porosity.

1.2.6 Reactivity

Reactivity of a carrier means its ability to interact with an enzyme molecule which is to be immobilized. The reactivity is important when a carrier binding technique is employed. The solid carrier dominates in enzyme immobilization but it is less suitable for immobilization of cells. The cells are immobilized only in a thin layer on the outer surface of the carrier (Ghommidh & Navarro, 1982; Messing, 1982; Žurková *et al.*,

1983; Black *et al.*, 1984) and the ratio of weight of active cells to the volume or weight of the support is low. The main part of the conjugate represents the carrier.

The immobilization is simple when the enzyme is bound only by a physical interaction to the carrier which can be an ion-exchange resin (Chibata, 1978), organic polymer (Marconi, 1989), ceramics (Filbert, 1975), or metal particles (Charles *et al.*, 1974). The first industrial process exploiting immobilized enzyme (separation of D- and L-amino acids) was based on ion-exchange interaction between DEAE Sephadex and enzyme (Chibata, 1978). The physical binding is advantageous in regeneration of the carrier.

The covalent linking of an enzyme to the carrier is the most usual in the literature but not in industry (Marconi, 1989). For covalent immobilization the chemical reaction by which the enzyme is attached to the carrier is the most important. The monographs and reviews describing the chemical reactions are exhaustive, but much less attention has been paid to the carrier itself (Messing, 1975*a*, 1978).

All carriers for enzyme immobilization can be divided into three categories according to their reactivity.

The first level polymers are represented by soluble macromolecules which are, in fact, not suitable for an immobilization. To this category belong not only the majority of native or modified natural polymers, e.g., starch, dextran, agarose, cellulose and its derivatives, gelatin, but also some synthetic and inorganic polymers.

The carriers of second level are available either by processing of the polymers of the first level (down direction), e.g. by subsequent crosslinking or by a polymerization or copolymerization of a monomer or monomers that are not explicitly reactive like 2-hydroxyethyl methacrylate, vinylpyrrolidone, maleic anhydride, acrylamide and its derivatives, styrene (Specht & Brendel, 1977; Gerig & Loehr, 1980; Schulte & Horser, 1982; Handa *et al.*, 1983; Kolarz *et al.*, 1989; Marconi, 1989). These carriers possess a bead shape and may be directly used only for a simple type of immobilization, e.g., for physical adsorption. For covalent immobilization they cannot be used directly.

Only the carriers of the highest third level exhibit appropriate reactivity to be able to react with an enzyme molecule yielding a covalent bond. The third level carriers are similar to the second level, available again by two ways. The chemical modification (activation) of second level carriers may represent several consecutive reaction steps leading to an active group which converts the polymer to a real carrier for covalent immobilization. As typical examples we can mention an activation of the agarose gels by bromcyan, diazotation of amino groups linked to a polymer, reaction of copolymer of 2-hydroxyethyl methacrylate with epichloro-hydrin, reaction of silica with γ-aminopropyltrimethoxysilane followed by activation with glutaraldehyde, etc. (Mosbach, 1976).

A corresponding product is accessible also directly by direct polymer-ization of an active monomer in only one step. An illustration is the copolymerization of 4-fluorostyrene or 3-fluoromethacrylanilide (Manecke & Pohl, 1978*a*; Manecke & Vogt, 1978*b*), 2,3-epoxypropyl methacrylate (Švec *et al.*, 1975; Kramer *et al.*, 1975), 4-iodobutyl methacrylate (Brown & Joyeau, 1973), vanilin methacrylate (Brown & Joyeau, 1974), acrolein (Tarhan & Pekin, 1983), etc.

It is obvious that carriers prepared by polymerization of an activated monomer require more sophisticated preparation of the monomers and development of specific polymerization conditions for all of them. The advantage of the third level horizontal approach is the chemical homo-geneity, the beads contain only the requested functionalities. This is, however, outweighed by the large consumption of the activated monomer most of which is buried inside the matrix. The groups are not exposed on the surface where the reaction between the carrier and the enzyme can only take place.

The vertical approach directing to chemical modification of polymer of lower level results in higher chemical heterogeneity. The beads contain not only the expected reactive groups but also the starting groups as well as various groups, products of intermediate or parallel reactions. None of them can be removed because they all are connected to one polymer network. The chemical heterogeneity can suppress an optimization of the reaction conditions but never fully prevent it. In spite of drawbacks the approach is most frequently used. It produces a variety of activated poly-meric carriers from a single basic polymer without changing their physi-cal structure, e.g. porosity, specific surface area, pore size and distribution, particle size and distribution, etc. Moreover, the modifica-tion may proceed under conditions where the major part of groups will be located only on the surface, either on the outer surface of non-porous beads or on the inner surface too, in porous ones.

From Scheme 1.2 it also follows that for obtaining any carrier at least one crosslinking polymerization or one crosslinking of soluble polymer has to be made.

The chemical nature of the carrier is evidently important. The counter-part – an enzyme or a cell – also takes place in the reaction and its chem-ical composition is not less important. The protein molecules are composed from 21 amino acid residues with different pendant chains bearing various chemical groups. For the reactions the choice is limited to several functional groups localized in side chains, ends of protein molecules and on the surface of cells or their parts, particularly, the amino groups of lysine and arginine and the N terminal amino group, carboxyls of aspartic and glutamic acid and the C terminal carboxyl, phenolic hydroxyl of tyrosine and aliphatic hydroxyl of serine and threonine, imidazole group of histidine and indole group of tryptophane. The most frequent are the hydroxyls (mean content of serine in proteins amounts 7·8 per cent and threonine 6·5 per cent) and primary amino groups (mean content of lysine is 7·0 per cent) (Means & Feeney, 1971). Despite high content the nucleophilicity of hydroxyl is poor and for polymerization reactions is not used very often. On the other hand, the amino groups are very popular. Of course, in some immobilization reactions more than one type of group participates.

1.2.7   Reusability

Although the stability of an enzyme is often reported to be enhanced by the immobilization the activity of an immobilized enzyme declines gradually when stored or during operation (Chibata, 1978). Factors influencing the decay of enzyme activity are denaturation, bacterial contamination, leakage of enzyme, incorrect operation, etc.

The easiest way to reuse a carrier is when the enzyme is immobilized by weak interaction only. The hydrophobic or the ion–ion interactions are broken under the change of ionic strength or pH of the surrounding medium. After rinsing with a buffer solution a new portion of fresh enzyme will be immobilized. The simplicity of repeated use is compensated by high leakage of enzyme during operation and storage.

The other extreme represents an enzyme covalently bound to an organic carrier. The bond between the enzyme and the carrier is strong and it cannot be broken without damage to the carrier. The disposable organic carriers present an environmental problem.

Most inorganic carriers are readily regenerated by a pyrolyzing process. They can be treated in this fashion because of their dimensional stability at high temperature. The immobilized enzyme is burned off in a furnace at a temperature above 400°C in the presence of air or oxygen. After cooling, the carrier is generally ready for a new activation and reuse for immobilization of a fresh enzyme (Messing, 1975a). The inorganic carriers are the best from the point of view discussed but their instability out of an optimal pH range limits their use.

1.2.8 Economy

The industrial use of immobilized enzyme and cells is an evident confirmation that the costs of production are not higher than in conventional processes (Chibata, 1978; Marconi, 1989). The economy of any process depends on many different factors, e.g. cost of labor, energy, inputs, overheads, etc., and to discuss all of them would be beyond the scope of this chapter. The economic problem is why despite speculations on the great potential of immobilized enzymes and cells and a lot of enthusiasm, only a few industrial processes were generated.

Let us focus on only one of the inputs, the carrier itself. From the previous discussion it seems to be clear that a universal carrier possessing exclusively positive features definitely does not exist and a compromise carrier has to be found relating to the use. It is also obvious that for a highly sophisticated sensor containing a small amount of a delicate enzyme the cost of the carrier may be neglected. On the other hand, immobilized glucose isomerase for large-scale production of high fructose syrups or immobilized penicillin acylase for production of 6-aminopenicillanic acid must compete with traditional well established technologies based on soluble enzymes or cells. Here, the cost of the carrier has to be compensated by other advantages, e.g. lower labor and production costs.

Supports produced in large quantities will be definitely cheaper than the same carriers prepared in the laboratory scale. The overall cost of enzyme immobilization can be substantially reduced by employing a regenerable carrier but the such carriers are limited in their use.

Economy continues to be the crucial point in the deciding whether the immobilized enzyme or cells will be inserted in the process.

1.2.9 Summary of the Various Carriers in Present Practice

The natural and synthetic polymeric carriers are rather important in the field of immobilization of biological catalysts. The major part of all immobilization literature is devoted just to the use of that class of compounds. The synthetic polymers exhibit a wide variety of physical forms and chemical structures thus matching many demands on an ideal support. The natural polymeric carriers have some advantages over the synthetic polymers. Actually, the polysaccharides are more 'physiological' just because both support and species to be immobilized are of natural origin and even in nature they appear very often together. The natural supports come from renewable sources growing all over the world. Finally, the natural polymers are also easily biodegradable thus less contaminating the environment. Unfortunately, the choice of their forms and structures is limited and prevents even more extensive use, though some of them are now used in large-scale production.

Despite the large number of papers dealing with the immobilization on organic polymer carriers the number of applications did not reach its upper limit. Now, use is concentrated mainly in the field of preparation of fine and speciality chemicals where the manifold reaction possibilities are unlimited. To this category belong industrial processes using immobilized enzymes, such as synthesis of L-amino acids, 6-aminopenicillanic acid or high fructose corn syrups, hydrolysis of lactose in milk and whey, and immobilized cells, such as synthesis of asparagic and other amino acids, glucose isomerization, *cis*-1,2-dihydroxycyclohexa-3,5-diene, etc. Also popular in analysis are immobilized active molecules, organelles and cells. Many very sensitive diagnostic methods in medicine are based on enzyme activity employed as a chemical amplification device. The demand on construction of biosensors grows steadily. They are able to monitor continuously the concentrations of hormones, proteins or drugs in the body liquids thus enabling proper care. Another spectacular development is the enzyme electrode combining the selectivity and sensitivity of enzymatic methods with the speed and simplicity of ion-selective electrode measurements (Marconi, 1989). The enzymes also make it possible to measure concentration of various compounds while monitoring the excess of heat produced during the enzymatic transformation of the measured substrate.

The further development will be closely connected with progress in molecular biology and genetic engineering (Walker & Gingold, 1990). They will supply the enzymes with desired specificity in large enough quantities and at reasonable cost. Then the enzymatic processes will be able to compete successfully with the traditional homogeneous and heterogeneous catalytic reactions. The advantage of enzymatic catalysis is low energy demand, no risk of environmental contamination with hazardous compounds and enhanced specificity increasing yields and thus simplifying separation and isolation.

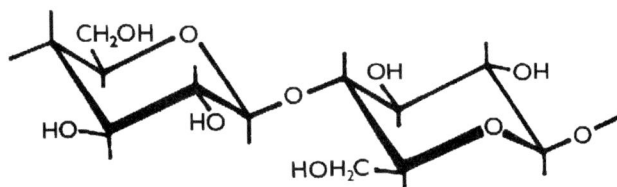

Fig. 1.2 Repeating structure of a cellulose polymer based on 1,4-linked β-D-glucose residues.

The field of immobilized biologically active species has still a great potential for further development. The enzymes, wherever they are located and immobilized, may in the near future open new ways to the products that are either tediously or not at all available now, to the more simple analytical and diagnostic tests, to the living-body-like systems fully substituting defective organs.

1.3 POLYSACCHARIDES

1.3.1 Cellulose and Derivatives

Cellulose is a particularly important natural polymer because it is the most abundant renewable organic resource. According to diverse estimations approximately 10×10^{11} tons of cellulose are yearly biosynthesized and destroyed in nature. Commercial cellulose from higher plants supplies annual world consumption of about 150 million tons of fibrous raw material. Of this amount, 7 million tons, mostly from cotton, represents chemical-grade cellulose (Hon, 1988). In the period 1983–85 almost 0·5 million tons of cellulose and cellulose derivatives were consumed by the US market only and of this amount more than 75 per cent was cellulose acetate (Yalpani & Sanford, 1987).

Cellulose is a polydisperse, linear syndiotactic polymer of plant origin. Its basic monomeric unit is D-glucose. The latter links successively via glucosidic bonds (in the β-configuration) between carbons 1 and 4 of adjacent units to form long chain 1,4-β-glucans. Figure 1.2 represents a structural diagram of a part of a cellulose chain.

The size of the naturally occurring cellulose molecule is indicated by its degree of polymerization (DP) or chain length, and it is heavily dependent on its source. In some cases it may exceed a DP of 10000. Aggregation of these long cellulose molecules through inter- and intramolecular hydrogen bridges between three hydroxyl groups forms the ribbon-like strands called microfibrils (< 6 nm). The finest agglomeration within a microfibril is called an elementary fibril (3·5 nm in width). Single fibrils then associate to form thicker and longer macrofibrils, which in turn aggregate giving the cellulose fiber.* Mechanical stress due to compression and/or shear forces is sufficient to change the topochemistry and reactivity of the polymer, and to cause homolytic chain scission. The stress-induced reaction results in disaggregation of the fiber bundle and shortening of fiber length. This is reflected in a loss in the fiber DP value, increment in reducing end groups and their accessibility, and in the degree of fiber crystallinity (Hon, 1985).

*Refer to p. xvii of prelims.

Acidic hydrolysis of fibrous cellulose yields (as insoluble portion) micro-crystalline cellulose with a DP value of 25–350 depending on both the properties of the original material and the conditions of hydrolysis. Micro-crystalline cellulose represents the chemically purest cellulose preparation containing 0·05 per cent of ash and an approximately similar amount of extractable compounds. Thus, microcrystalline cellulose is not soluble either in water or in diluted acids or organic solvents. Subsequent treatment of cellulose by the hydronium ion (HCl) and then by mechanical energy causes the individual unhinged microcrystals to disperse into a liquid medium as individual colloidal particles. The microcrystals of cellulose form aqueous suspensoids with unique functional properties similar to those of other members of the microcrystal polymer product family (Battista, 1975).

Besides its extreme insolubility in usual aqueous media (e.g. salt solutions) cellulose itself is not a gelling biopolymer, but its diverse derivatives may form a variety of thickened solutions and gels. Thus, esterification with acids in the presence of dehydrating agents, or reaction with acid chlorides, or etherification by treatment of solutions of alkali cellulose with alkyl halides, may produce a number of interesting cellulose derivatives capable of network formation in aqueous media (Clark & Ross-Murphy, 1987).

Crosslinking of microcrystalline cellulose with epichlorohydrin or formaldehyde yields cellulose gels consisting of rod/shaped gel particles. The latter are again insoluble in alkaline solutions. Crosslinking of cellulose decreases the permeability of high-molecular-mass substances but, at the same time, improves the separation of low-molecular-mass substances. The latter effect is documented in Table 1.2 (Luby *et al.*, 1971). In addition, via crosslinking with epichlorohydrin the amount of accessible hydroxyl groups increases the reactivity of cellulose in activation reactions (Gemeiner *et al.*, 1980; Gemeiner & Zemek, 1981*b*) .

Table 1.2
The Differences in Reduced Elution Volumes $(V_{red})^a$ of Substances with Molecular Weight = 100 and 1000 as Found for Various Gels

Gel	$(V_{red})^{100}$
Sephadex G-15	0·182
Starch crosslinked with epichlorohydrin	0·188
Cellulose crosslinked with epichlorohydrin	0·141
Cellulose crosslinked with formaldehyde	0·138
Non-modified powered cellulose	0·076

$^aV_{red} = V_e/V_t$ where V_e is elution volume and V_t the total volume.
All the data were taken from Luby *et al.* (1971).

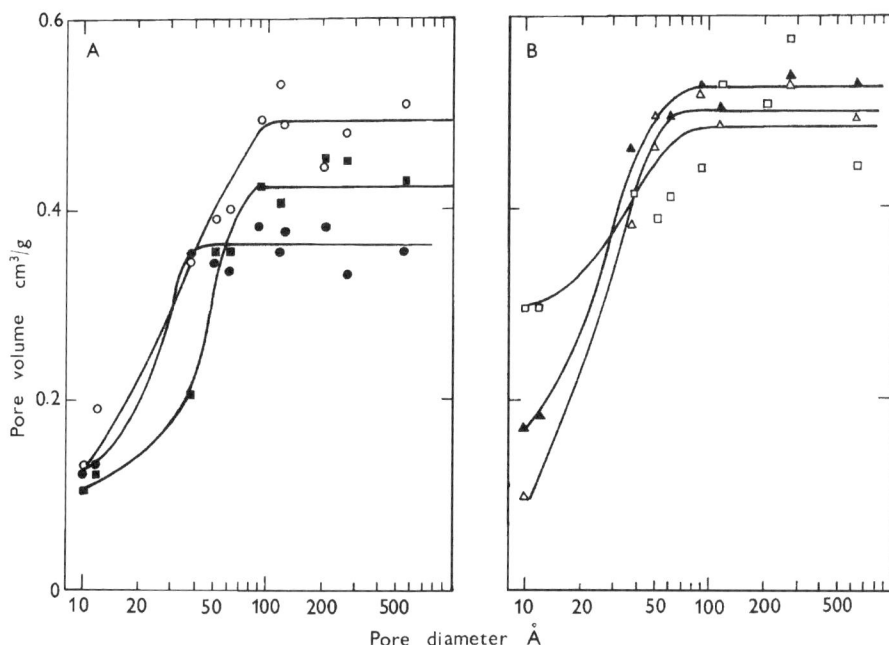

Fig. 1.3 Pore volume distribution of commercial cellulose as determined by solute exclusion techniques. (A) α-cellulose represented by a milled cotton (Sigma) (●), microcrystalline cellulose Sigmacell 20 (○) and Sigmacell 50 (■) (Sigma) with nominal fiber lengths at 20 and 50 μm; (B) noncrystalline cellulose Sigmacell 100 (□) (Sigma) with a nominal fibre length of 100 μm, regenerated celluloses Solka Floc SW-40 (△) and BW 100 (▲) (Brown & Co.) (Weimer & Weston, 1985).

Pore volume distribution data for different commercially available celluloses are summarized in Fig. 1.3. The celluloses vary in both fiber saturation point (range = 0·36–0·53 cm^{-3} g^{-1} dry cellulose) and maximum pore size (range = 45–100 Å) and display some differences in distribution of pore volumes at submaximal pore sizes (Weimer & Weston, 1985).

Water has a profound effect on the structure of cellulose. The specific surface area (SSA) of natural cellulose is known to increase drastically upon wetting. In addition, water is known to cause an increase in crystallinity due to a recrystallization effect. Consequently, the structural characteristics of vacuum-dried cellulose are different from those of water-swollen (or solvent-dried) cellulose (Table 1.3) (Lee *et al.*, 1983).

Swelling of cellulose matrix in water involves hydration and consequently a certain degree of constrained dissolution of the cellulose chains in their amorphous regions. This is of special importance because it is in these regions of low three-dimensional molecular order that chemical

Table 1.3
Structural Parameters of Untreated and Treated Celluloses

Cellulose[a]	Pretreatment method	Vacuum-dried		Solvent dried		
		CrI (%)	SSA (m² g⁻¹)	CrI (%)	SSA (m² g⁻¹)	DP
Untreated Solka Floc SW-40	Untreated	76·7	1·89	77·0	5·08[b]	1210
Milled Solka Floc SW-40 (Sweco 270)	Sweco milled	22·0	2·95	37·0	24·2[c]	380
Mercerized Solka Floc SW-40	5N NaOH	72·9	—	74·5	19·8	1140
Regenerated Solka Floc SW-40	85% H_3PO_4	2·4	26·2	4·0	239	1080

[a]Cellulose Solka Floc SW-40 was provided by Brown Co. (Berlin, NH, USA).
[b]Specific surface area upon soaking for 30 min.
[c]CrI – Crystallinity index; SSA – specific surface area; DP – degree of polymerization.
All the data were taken from Lee *et al.* (1983).

reactions most readily occur. Conjugation with biological molecules, particularly macromolecules, occurs more likely within these accessible gelatinous sites than in the highly ordered crystalline regions.

Although the hydroxyl groups of celluloses are not reactive enough to form covalent bonds between the enzyme and the support without previous activation, cellulose undergoes all the reactions associated with polyhydric alcohols and a wide range of active celluloses may be prepared in this way. The hydroxyl groups of cellulose may be activated directly by introduction of an electrophilic group, reactive towards the enzyme, into the matrix. However, the nucleophilic character of cellulose supports is so weak that pendant functional groups, such as aliphatic or aromatic amino groups, carboxyl or thiol groups have to be introduced as activators for coupling or before activation (indirect coupling) (Weliky & Weetall, 1965; Crook *et al.*, 1970; Goldman *et al.*, 1971; Zaborsky, 1973; Lilly, 1976; Chibata, 1978; Kennedy & Cabral, 1987a). There are also known examples of cellulose activation by radical reactions. These are usually applied in graft copolymerization of cellulose with a bifunctional monomer, e.g. glycidylmethacrylate, in the presence of an enzyme. Copolymerization can be initiated by use of cellulose peroxide (Focher *et al.*, 1975), Fe^{2+}–H_2O_2 system (D'Angiuro *et al.*, 1978, 1980a,b,c, 1982) or irradiation (Beddows *et al.*, 1984, 1986). In the latter case enzyme is additionally immobilized onto graft cellulose copolymer (Beddows *et al.*, 1984, 1986).

In practice, low-porous celluloses are complicated carriers. One might be tempted to regard cellulose as a trihydric alcohol similar to sugars in the type of its reactions. This, however, has been definitely shown not to be the case when reactions with cellulose take place. Then all the properties of cellulose as a fiber-forming, high polymeric substance become fully expressed. Considering the two-phase, crystal line-amorphous structural concept and physical and chemical reactivity of the hydroxyl groups the main reactions of low-porous celluloses may be classified into four classes:

(i) Reaction takes place exclusively with one (the accessible) of the two types of hydroxyl groups (either primary or secondary), and its nature is topochemical (micellar/heterogeneous). The duality of submicroscopic structure determines the rate, e.g. periodic acid oxidation.

(ii) Reaction takes place preferentially with one of the two types of hydroxyl groups and it is of permutoid (quasi-homogeneous) character. The different reactivity of the hydroxyls determines the reaction rate, e.g. tritylation.

(iii) Reaction takes place equally with hydroxyls of both types, and it is topochemical. The duality of submicroscopic structure determines the reaction rate, e.g., the sorption of water.

(iv) Reaction takes place equally with both types of hydroxyl groups, and it is permutoid. The reaction rate is uniform throughout. A typical example is nitration (Hebeish & Guthrie, 1981).

Permeability, the surface area available for enzyme attachment, as well as the reactivity of cellulose depend largely on the following factors: degree of crystallinity, the nature and size of the compound to be bound, and the swelling induced capacity of the activation medium (Kennedy & Cabral, 1987a).

Several types of chemically modified celluloses are commercially available (Table 1.4) and some of them were originally used as ion-exchangers.

Table 1.4
Commercially Available Modified Celluloses

4-Aminobenzyl-
Aminoethyl- (AE-)
Diethylaminoethyl- (DEAE-)
Carboxymethyl- (CM-)
Epichlorohydrin triethanolamine- (ECTEOLA-)
Oxy-
Phospho-
Sulfoethyl-
Triethylaminoethyl- (TEAE-)

Table 1.5
Cellulose Derivatives for Covalent Coupling of Enzymes

Triazinyl-cellulose
Bromacetyl-cellulose
Cellulose trans-2,3-carbonate
Cellulose imidocarbonate
Cellulose hydrazide
Cellulose azide
Cellulose carbonyl
Diazo-cellulose
Isothiocyanato-cellulose

Modified celluloses were used to immobilize enzymes directly by ionic binding but they also permitted a wide range of covalent binding by which the enzyme interacts mainly via its amino groups.

Different active derivatives that have been used for covalent coupling of enzymes (Weliky & Weetall, 1965; Crook *et al.*, 1970; Goldman *et al.*, 1971; Zaborsky, 1973; Lilly, 1976; Chibata, 1978; Kennedy, 1978; Kennedy & Cabral, 1987a; Scouten, 1987) are listed in Table 1.5. The major advantage of cellulose derivatives for immobilization of enzymes is that they are equipped with residual hydroxyl groups which provide a hydrophilic character protecting the attached enzyme.

1.3.1.2 High-porous Celluloses

Traditional microgranular forms of the commercially available cellulose were distinguished by unsuitable physical structure (low porosity) and unsatisfactory geometrical shape of their individual particles. Furthermore, extensive microcrystalline areas within the matrix exacerbate the problem. These disadvantages were eliminated by developing macroporous cellulose in the form of spherical beads (Štamberg, 1988) or gel particles of spherical and irregular shape (Štamberg *et al.*, 1982). Cellulose in this form exhibited good permeation, improved mechanical properties, and higher binding capacities. Moreover, bead cellulose usually exhibits better a chemical reactivity than its original forms. Commercially available bead cellulose is packed in wet, solvent-equilibrated form, i.e. ready to use.

Traditional porous cellulose materials also include membranes of both common geometries: planar (flat-sheet) and tubular (including hollow fibers) with porosity properties well described in business documentations.

(a) *Methods of preparation.* Various procedures have been described (Štamberg *et al.*, 1982; Štamberg, 1988) for preparation of beads or gel particles with spherical form. All of them involve the following three basic steps:

(1) Liquefaction of the cellulose polymer;
(2) Dispersion of the cellulose phase in a non-miscible medium;
(3) Solidification of liquid droplets and final modification of beads.

In step 1 the solutions of cellulose or its derivatives are prepared by procedures resembling those known for example in the production of cellulose fibers, sheets and films. The starting material may exceptionally also be a molten derivative, e.g. cellulose acetate.

In the second step, droplets of the liquid raw material are formed in an inert and non-miscible medium. Dispersion is carried out mostly by stirring, exceptionally by dropping or spraying. Particle size may be controlled by passing the fluid through a nozzle, by the efficiency of mixing during dispersion (Štamberg, 1988) or by addition of surface active compounds (Štamberg & Peška, 1980; Štamberg *et al.*, 1982; Dean *et al.*, 1985; Štamberg, 1988).

In the third step the conditions are set for sol/gel transition in drops by means of various mechanisms, such as chemical crosslinking, diminishing the solubility by chemical and physical effects or by change of state with the aid of cooling, etc. The procedures are carried out in a way avoiding the deformation of spherical shape and adhesion of individual particles to yield agglomerates. Finally, the pure product is isolated. The required beads are selected and in some cases their chemical and physical structure is modified.

Bead cellulose produced by 'TSGT' (thermal-sol-gel-transition) process is the preparation most frequently described. It is produced from technical viscose (i.e. from aqueous solution of cellulose xanthate) by dispersion with stirring in an organic solvent and subsequent thermal solidification without crosslinking through covalent bonds. In the course of solidification, the xanthate groups are gradually decomposed, and the solubility in water disappears. Finally, the decomposition is completed in an alkaline medium. After washing, the product is spherical, porous, regenerated cellulose (Štamberg, 1988). The TSGT process may be applied also for composite formation of bead cellulose and magnetite, i.e. for preparation of magnetic bead cellulose (Gemeiner *et al.*, 1989*a*).

The TSGT process or other three step procedures give rise to beads or gel particles of spherical or irregular shape, with dimensions (diameter or width $\times$ length) 20×1000 μm (Kuga, 1980*a*; Štamberg, 1988; Gemeiner *et al.*, 1989*a*). Triacetylcellulose microspheres with an average diameter less than 2 μm may be prepared in this way (Ikada & Tabata, 1988), however, the procedure for their preparation slightly differs from that used in the universal scheme. First, triacetyl cellulose microspheres are created by means of gradual evaporation of the solvent and by centrifugation. Then, pure cellulose microspheres are regenerated by alkaline saponification of the triacetyl cellulose microspheres (Ikada & Tabata, 1988).

Cellulose microspheres may be prepared also in a simpler way, i.e. the precipitation at the interface of aqueous solutions of suitable oppositely charged polyelectrolytes. The course of precipitation may easily be kept under control so that spherical capsules are formed during spontaneous polyelectrolyte complex formation. The basic condition is an appropriate water-solubility of both oppositely charged polyelectrolytes with carefully controlled molecular characteristics. Among the most advantageous cellulose derivatives is sulfate (polyanion, PA) in combination with poly(dimethyldiallylammonium chloride) (polycation, PC) (Dautzenberg *et al.*, 1985*a,b*). This method of preparing cellulose capsules exhibits several advantages over the other methods. The main advantage is the possibility of maintaining physiological conditions during encapsulation and thus securing non-toxicity and biocompatibility of the resulting material. Other advantages are simplicity of handling and variability of properties. Dautzenberg *et al.* (1985*a,b*) have shown that with varying the polyanion/polycation combinations the capsules' properties such as mechanical strength, elasticity and deformability, capsule wall properties such as morphology, permeability and transparency, can be varied in a wide range by changing the parameters of the polyelectrolyte precursors as well as the conditions of capsule preparation. Both homogeneous (isotropic) and heterogeneous (anisotropic) capsules may be obtained by changing the membrane morphology.

Spherical shape is typical also for semipermeable microcapsules prepared from cellulose nitrate by emulsification followed by spherical ultrathin membrane formation. The next step is represented by secondary emulsification. The principle of secondary emulsification has been used for microencapsulation of biocatalysts into the cellulose polymer material. In the latter case, an aqueous enzyme containing solution was emulsified in cellulose nitrate and the second emulsification proceeded in an aqueous solution (Chang, 1976, 1985, 1987). The cellulose polymer might be allowed to solidify by removal of the organic solvent.

The wet-spinning procedure for manufacturing man-made fibers was adjusted to produce porous fibers also from cellulose: an organic solvent of a fiber-forming polymer is emulsified with an aqueous solution of the enzyme, and the resulting emulsion is then extruded, through the holes of a spinneret, into a coagulation bath whereby cellulose polymer is precipitated in filamentous form. At the end of the process, a bundle containing several parallel continuous individual filaments consisted of a macroscopically homogeneous dispersion of small droplets of the material entrapped in the porous polymer gel. The entrapment itself is rendered possible by microcavities inside the fibers (Marconi & Morisi, 1979) which transform the nearly isotropic cellulose fibers into hollow ones (Dinelli *et al.*, 1976). The fibers are made from cellulose and its derivatives such as di- and triacetate, nitrate and ethylcellulose. The physical properties as well as suitability of some of these materials for immobilization of biocatalysts are well documented (Chambers *et al.*, 1976).

Among the above discussed porous celluloses suitable for immobilization of biocatalysts the most widely used is the bead cellulose which will receive the most attention in the following sections.

(b) *Structure and properties.* The chemical structure of bead celluloses does not differ from that of analogous non-spherical preparations. On the other hand, the difference in physical structure, particularly in porosity, accessibility, and shape of individual particles is essential.

The advantage of bead cellulose prepared by the TSGT process is its chemical purity and the corresponding hydrophilicity. Natural celluloses contain hardly removable lipids which make them hydrophobic and are responsible for their non-specific adsorption in interactions with biochemical substrates. Spherical, regenerated cellulose contains a substantially lower amount of lipids. TSGT bead cellulose has, like other cellulose preparations, low content of carboxyl groups (<0.02 mmol g^{-1}), negligible content of sulfur ($<0.03\%$), and its non-combustible portion decreases upon washing with diluted mineral acids (from $<0.15\%$ to $<0.01\%$). In the course of the TSGT process cellulose is partially decomposed. The starting raw material (cellulose pulp) exhibits DP values of 600–700, the regenerated spherical product 300–500 (Baldrian *et al.*, 1978). Bead cellulose consists of particles with perfectly spherical shape. Examination in a microscope revealed a standard deviation of approximately 1 per cent from the circular shape. The beads were transparent, or slightly turbid, and practically free of aggregates (Baldrian *et al.*, 1978).

Porosity of never-dried TSGT bead cellulose, i.e. the percentage (v/v) of water in the swollen mass, may be as high as 90 per cent. Additional adjustments, especially drying in various media or partial drying and reswelling, gives rise to products with porosity values between 50 and 90 per cent. Drying with nonpolar solvents (solvent-exchange drying, i.e. the s.e.d. method) yields dry, highly porous materials with large internal specific surface area (SSA). The values of SSA calculated from nitrogen adsorption isotherm were within the range of 135–190 m^2 g^{-1} and were considerably greater than those of the samples of microcrystalline cellulose and cotton cellulose fibers (5·4 and 2·5 m^2 g^{-1} respectively) (Lee et al., 1982). An investigation of the X-ray diffraction pattern of s.e.d. samples indicated that the substance was partially crystalline; the crystallinity was about 30 per cent (Baldrian et al., 1978).

The porous structure of TSGT bead cellulose is typically heterogeneous, resembling a macroreticular structure rather than soft-gel polymer carriers. The insolubility of the material may be ascribed to the existence of hydrogen bonds crosslinking the polymer chains. It leads to the formation of rigid, oriented regions of quaternary structures, alternating with highly porous amorphous skeleton in accordance with the classical model of the cellulose structure. This results in a wide distribution of pore sizes, but also in relatively high mechanical strength, i.e. low deformability of particles. The cumulative pore size distribution resembles more that of the cellulose gel (Fig. 1.4) than that of the swollen cellophane and cotton linter (Peška et al., 1976; Kuga, 1980b).

Table 1.6 shows that swollen TSGT bead cellulose exhibits rigidity modulus values of the same order of magnitude as does the swollen car-

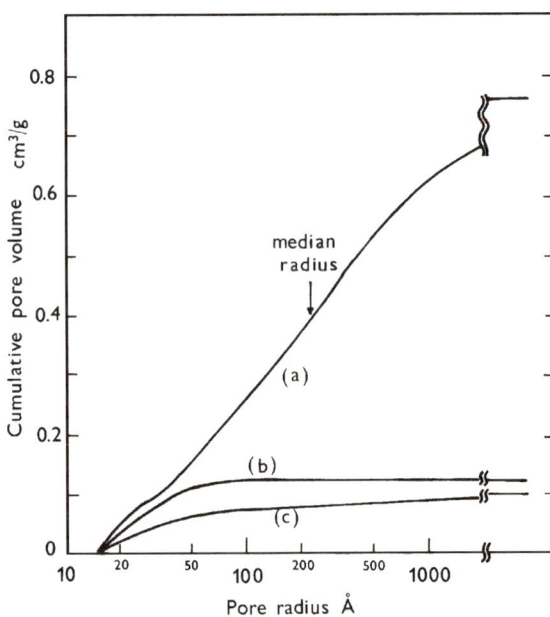

Fig. 1.4 Cumulative pore size distribution curves of solvent-exchange dried celluloses. (a) cellulose gel; (b) cellophane film; (c) cotton linter. Capillary condensation analysis was carried out by the method of C. Pierce (Kuga, 1980b).

Table 1.6
Mechanical Characteristics of Bead Materials

Swollen bead cellulose			Dry bead cellulose			Swollen commercial ion exchangers		
No.	Radius, r (mm)	Penetration modulus, A (MPa)	No.	Radius, r (mm)	Penetration modulus, A (MPa)	Name	Radius, r (mm)	Penetration modulus, A(MPa)
6[a]	0·533	2·80	8	0·347	24·73	Zerolit[c]	0·535	1·44
	0·503	2·56		0·332	25·84		0·605	1·50
	0·500	2·46	9	0·419	17·21		0·700	1·35
	0·483	2·63		0·408	20·74		0·638	1·30
7[b]	0·610	0·44				Zerolit[d]	0·692	0·28
	0·605	0·44					0·834	0·27
	0·615	0·34				Amberlite[c]	0·510	27·62
	0·590	0·41					0·473	26·92
	0·595	0·59						

[a] Bed volume 4·1 ml g^{-1}.
[b] Bed volume 8·45 ml g^{-1}.
[c] Carboxylic cation exchanger Zerolit 226 (H$^+$).
[d] Zerolit 226 in equilibrium with 0·01 N NaOH.
[e] Strongly acid styrene sulphonic cation exchanger Amberlite IR-120 (H$^+$).
All the data were taken from Peška *et al.* (1976).

boxylic cation exchanger; the values found for dry bead cellulose are comparable with those given for swollen styrene ion exchangers, represented here by Amberlite IR-120 (Peška *et al.*, 1976). Like cellulose gel particles TSGT bead cellulose (Fig. 1.5) (Motozato & Hirayama, 1984) has essentially higher mechanical strength than dextran or agarose gels and during flow-through in a column exhibits much lower pressure drop (Peška *et al.*, 1976). Straight lines in Fig. 1.5 indicate that both cellulose gel particles and cellulose beads are very rigid and suitable as packing material for HPLC. Never-dried TSGT bead cellulose is accessible to dextrans with molecular masses up to $0·5 \times 10^6$ (Kuga, 1980*a*; Gemeiner *et al.*, 1989*a*) but there have been reported also bead celluloses accessible to proteins with molecular masses of several millions (Fig. 1.6) (Kuga, 1980*a*) having maximum pore size 50 nm. Irregularly shaped gels show lower capacity ratios and poorer resolution than spherical ones. The latter exhibit a performance as good as do commercial agarose or dextran gels (Table 1.7) (Kuga, 1980*a*).

Bead cellulose is available from several suppliers: Amicon Grace Co., Chisso Co., Pierce Chemical Co. (USA), Daicel Chemical Industries (Japan), and North-Bohemian Chemical Works Secheza, Lovosice (Czechoslovakia). Amicon-Grace Co. and Chisso Co. recommend their

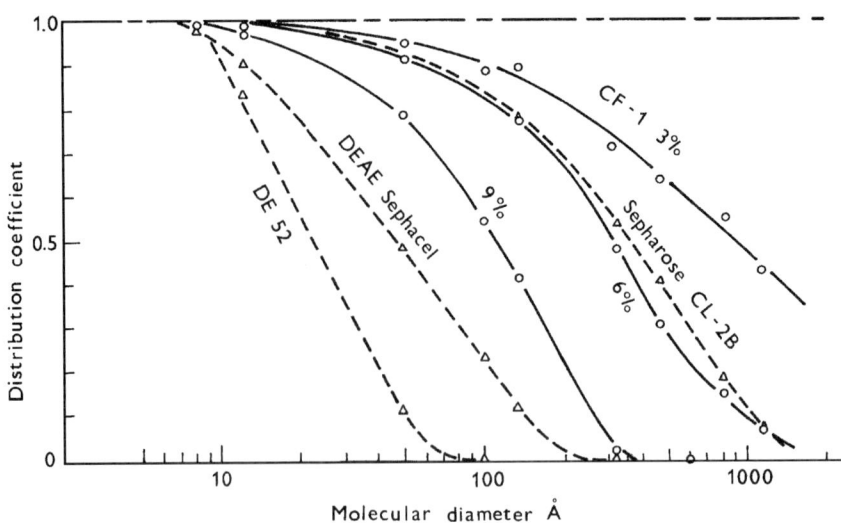

Fig. 1.5 Calibration curves for cellulose gel particles and commercial cellulose and agarose gels. The points above 300 Å correspond to the poly(ethylene oxide) fractions; the other to the dextran fractions. The curves can be regarded as normalized pore size distributions of the gels (Kuga, 1980*a*).

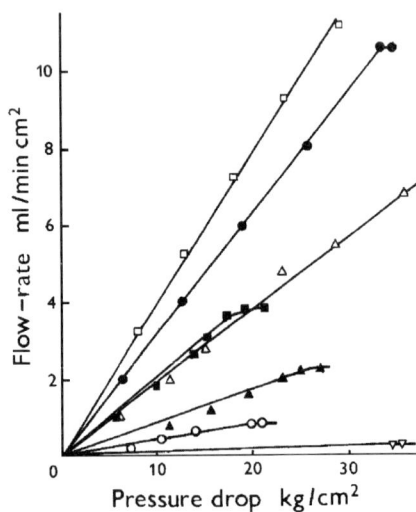

Fig. 1.6 Relationship between flow-rate and pressure drop. All measurements were made by the use of a 150×8 mm I.D. metal column packed with gels ranging from 44 to 105 μm in diameter (Motozato & Hirayama, 1984).

Table 1.7

Gel Filtration Performances of Cellulose and Other Gel Packing

Gel	Shape and size (μm)[a]	HETP[b] (mm)	Flow rate (ml/min)	Capacity ratio, V_i/V_0[c]
CF-1: 3%	Spherical, 74–210	0·69	0·164	1·86
6%	Irregular, 149–297	1·13	0·169	0·69
6%	Spherical, 74–210	0·30	0·162	1·27
9%	Irregular, 105–210	0·46	0·242	0·82
Cotton linter: 1·0%	Spherical, 44–210	0·31	0·224	1·22
1·5%	Spherical, 44–149	0·12	0·229	1·94
3·0%	Spherical, 105–210	0·23	0·227	1·00
DEAE-Sephacel	Spherical, 40–130[b]	0·44	0·157	1·94
DE-52	Fibrous, 30–60 × 100–300[d]	1·04	0·157	1·00
Sepharose CL-2B	Spherical, 60–250[d]	0·23	0·155	1·78

[a]Nominal value from sieve opening.
[b]HETP = height equivalent to a theoretical plate.
[c]V_i = internal volume gel; V_0 = void volume.
[d]Obtained by microscopic observation. For DE-52, width × length.
All the data were taken from Kuga (1980a).

Matrex Cellufine and Cellulofine media as medium-pressure gels ideally suited for a wide range of chromatographic separations.

They demonstrate on individual examples the outstanding flow rates in several column sizes securing fast separations, high resolution due to small particle sizes, high capacity allowing large sample loads, and high throughput due to fast flow capabilities. Besides that, they report minimal fines generation even with rough handling, minimal changes in swelling and no shrinking during operational changes, compatibility with all commonly used chromatographic solvents and buffers as well, autoclavability, and low cost per unit separation capacity. Pierce Chemical Co. supplies crosslinked bead cellulose, the so-called Excellulose, having molecular mass exclusion limit of 5×10^3. Desalting of macromolecules and fractionation of small molecules ($400 < M < 5000$) have been successful with the material. Daicel Chemical Industries provides Cellulose Spherical Particles characterized by molecular mass exclusion limits ranging within 10^4–10^6 and differing in compressibility by the eluent flow (Gemeiner *et al.*, 1989a).

(c) *Methods of functionalization.* Porous bead celluloses probably consist of semi-crystalline cellulose microfibrils instead of single molecular chains or aggregates of several molecular chains as they may be seen in dextran or agarose gels. For this reason, the imbibed water can be replaced by any polar or non-polar liquid without significant contraction of the gel. In addition, the gel is thermally stable and contains densely distributed hydroxyl groups. These features are highly advantageous for chemical modification (Kuga, 1980a).

Cellulose derivatives in insoluble bead form may be prepared in two fundamental ways. Either the respective derivative is prepared in soluble form, converted into the bead form, and fixed in a suitable way or, on the contrary, first the bead cellulose carrier skeleton of respective porosity is prepared and then functionalized by reactions known with other polymers. Only the second method is used in practice (Gemeiner et al., 1988b). The same starting material may be functionalized by numerous experimental procedures applied for preparation of cellulose derivatives in general (Engelkirchen, 1987; Schnabelrauch et al., 1990) and for preparation of cellulose adsorbents (Weliky & Weetall, 1965; Kennedy, 1974a) in particular.

When the porosity of the starting cellulose material, which is stabilized by hydrogen bonds and disperse forces only, is to be preserved, it is necessary to devote appropriate attention to proper selection of reaction conditions. Substantial decrease in porosity may be obtained by direct evaporation of water or other polar solvents from macropores, but also by the action of alkali. Alkali is often present as reaction component in alkylation or acylation reactions and it is also used in activation (mercerization) prior to reactions. It is, of course, necessary to avoid conditions at which both the starting cellulose and the product of functionalization are soluble. The degree of substitution (DS) value plays an important role here. Above its critical value (varying from 0·3 to 0·6 according to the substituent) the product becomes soluble. Solubilization may be prevented by crosslinking or, sometimes, also by a mere change of conditions that, in heterogeneous reaction, lead to different distribution of functional groups on the cellulose chain, although the average DS value remains (Gemeiner et al., 1988a).

When modifying the bead cellulose in non-aqueous media, the evaporation of water from the swollen starting material is avoided by successive solvent exchange according to the WAN (water–alcohol–nonpolar solvent) method. The procedure has no unfavourable effect on porosity (Gemeiner et al., 1989a). To direct the activation reaction into the intra-particle domain it may be appropriate to fill the inter-particle volume with an inert medium such as a hydrocarbon.

An important factor is also the proper selection of the mode of biocatalyst binding for retaining the structure and geometry of pores of bead cellulose. Because of their ability to react directly with unmodified bead cellulose bifunctional reagents are preferred for this process. In most cases satisfactory results have been obtained with bead cellulose of low DS value, e.g. with chlorotriazine derivatives in the range of $0.001 \leq DS \leq 0.021$. As long as the structure and geometry of pores of the bead cellulose derivative are retained, the catalytic effect is a result of combination of size-exclusion properties of bead cellulose and of catalytic properties of the immobilized enzyme (Gemeiner *et al.*, 1989*a*).

There are universal procedures for activation of natural and synthetic polymers (diazotization/coupling, formation of amides and hydrazides) as well as procedures typical of activation of natural polysaccharides (reactions of vicinal diols with cyanogen bromide and chloroformates, etherification with cyanuric chloride). Cellulose, agarose (Sepharose), and crosslinked dextran (Sephadex) were the carriers most often activated by the aforementioned methods of activation of natural polysaccharides. Activation with cyanogen bromide is still, because of the considerable popularity of agaroses and dextrans, probably the most often used method. High reaction rate at moderate conditions and high substitution degree are the main advantages of this simple activation procedure.

Of course, in activation of bead cellulose the procedures elaborated for activation of traditional cellulose materials have mostly been applied, such as powdery and microcrystalline celluloses. One of these procedures is oxidation of bead cellulose with periodate. The indisputable advantage of this method over activation with cyanogen bromide is nontoxicity of the reagent. However, it is necessary to control the degree of oxidation in order to prevent destruction of beads. The other two universal procedures (activation with epichlorohydrin and 4-toluenesulfonyl chloride), often used in preparation of bead cellulose adsorbents, have not been used so often in immobilization of enzymes. Activation with cyanuric chloride is more promising because it provides a product (cellulose chlorotriazine) which is, like the product of periodate oxidation (cellulose 2,3-dialdehyde), sufficiently reactive for enzyme immobilization via covalent bonds (Gemeiner *et al.*, 1989*a*).

Series of methods for activation of bead cellulose and procedures for binding of biocatalysts with other hydroxyl-carriers were modified or newly developed. Regarding the stability of bonds and easy performability of the reaction, diazotization of arylamino derivative, activation with 2,4,6-trichlorotriazine, oxidation with periodate and quinone, and some variants of the glutaraldehyde method are considered to be most suitable. The classical activation with cyanogen bromide or esters of chloroformic acid (ethyl, nitrophenyl, N-oxysuccinimidyl), which then bind biocatalysts as derivatives of carbonic acid, are less suitable, due to the lower hydrolytic stability of these bonds. For binding of biocatalysts with amino groups (combined with $NaBH_4$ reduction), reactive aldehyde groups were introduced into bead cellulose by oxidation with periodate and also by treatment with 4-isothiocyanatobenzaldehyde, bromoacetaldehyde diethyl acetal, or by a series of consecutive reactions of 3-chloro-2-hydroxypropyl derivative (Gemeiner et al., 1989a).

The availability and reactivity of hydroxyl groups of the basic bead cellulose itself is not sufficient for some modifications and interaction of cellulose products. Introduction of hydroxyethyl, or rather $-(CH_2-CH_2O)_n-H$, $n = 1$–3, or $-NH_2$ for the preparation of a beaded $-NH_2$ derivative and/or $-SH$ groups into the material considerably improves its properties. Hydroxyethylation and crosslinking also facilitate continuous change from macroporous- to gel-type porosity. The reactions of alkylating cellulose derivatives (C–2, C–6) with ammonia or diamines are used for the preparation of a beaded $-NH_2$ derivative.

Toluenesulfate, chlorohydroxypropyl (Figs 1.7 and 1.8), chlorodeoxy, nitrate and nitrite derivatives are utilized as most dominant alkylating derivatives of bead cellulose. Their preparation and reactions proceed with a higher conversion rate and at lower temperature in the case of derivatives based on bead hydroxyethyl-cellulose. Alkylation reactions serve also for preparation of bead thiol celluloses by using either thiourea, xanthate or thiosulfate, and a subsequent hydrolysis of their intermediates. The reaction of thiol-cellulose with disulfide (e.g. 2,2'-dipyridyl disulfide) or the above-mentioned alkylation of thiosulfate results in reactive cellulose disulfides (Gemeiner et al., 1988b, 1989a). Upon alkylating phenol and ω-phenyl-α-alcohols a series of isohomologous derivatives of bead cellulose was prepared (Gemeiner et al., 1988b, 1989a).

Among bead cellulose derivatives, the ion-exchange derivatives occupy the most important place. They are prepared by treatment either with classical (for CM, DEAE, P, DEAHP) or, in the case of bead cellulose,

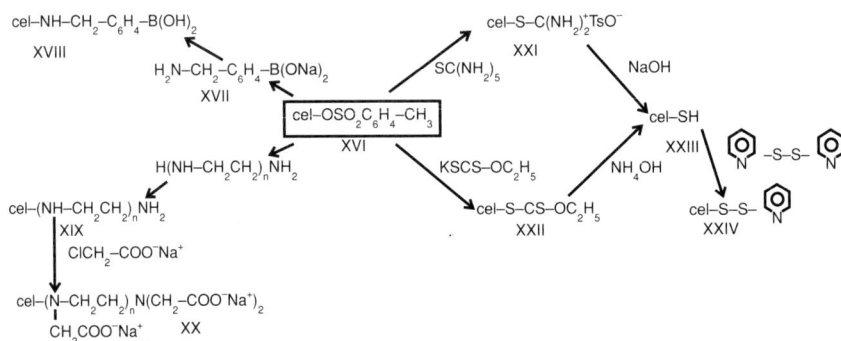

Fig. 1.7 Consecutive reactions on bead cellulose 4-toluenesulfonate (Gemeiner et al., 1988b).

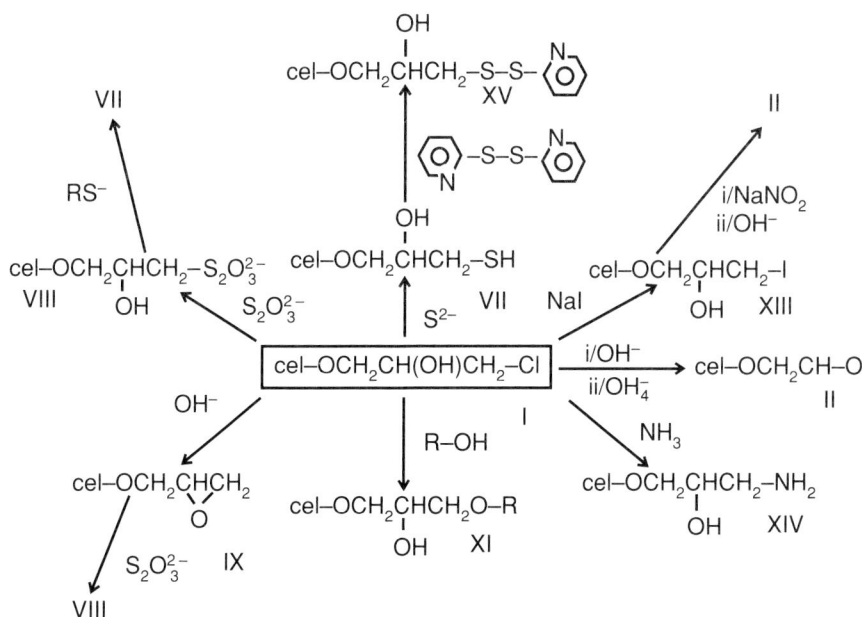

Fig. 1.8 Consecutive reactions on bead 3-chloro-2-hydroxypropyl cellulose (Gemeiner *et al.*, 1988*b*).

nontraditional (for SHP, TMAHP) alkylation reagents. At present, seven basic types of ion-exchangers in bead form are available on the market (Table 1.4). Besides the materials listed in Table 1.4 others such as ECTEOLA (product of the reaction of cellulose with epichlorohydrin and triethanolamine), amino-ethyl, and sulfoethyl, are also described.

The morphology of the final product depends on the method of preparation. The first study on carboxymethylation of bead cellulose in an aqueous-acetone medium already pointed to changes in the content of carboxy-methylated glucose units in both amorphous and crystalline portions as well as in the position on the ring. In the case of weak basic anion-exchangers (DEAE and DEAHP) the quality of the functional group is influenced by the method of preparation. Besides the generally given $-CH_2-CH_2-N(C_2H_5)_2$ groups, varying amounts (as high as 57 per cent) of strongly basic quaternary groups are also present. As follows from the titration curves, their structure is probably more complex (Gemeiner, 1989*a*).

Surface properties of positively-charged microcarriers may be decisive in microcarrier technology. Hirtenstein & Clark (1983) reported on the influence of DS in DEAE-substituted Sephadex on the density of anchored cells. Many positively charged bead celluloses are also used as microcarriers for surface-dependent cells (Klein & Kressdorf, 1989).

Morphology and intrinsic structure of these bead celluloses might represent further factors which may exert a determining influence on the interaction between the microcarrier and the cell (Klein & Kressdorf, 1989).

1.3.1.3   Immobilized Technology

Up to now cellulose was used in all four main *support-binding* methods.

Physical adsorption. If the binding of the enzyme onto the surface is not specific enough, competition with further biocatalysts and/or proteins occurs and the enzyme is released from the surface. Introduction of a recognition element to the carrier represents an effective way to avoid the leakage of enzymes. Hydrophobic interactions occurring during the physical adsorption of enzymes onto hydrophobized cellulose derivatives may be used as an example of the previous procedure. Hence, an enzyme immobilized in this way is less liable to leakage from the support in a wider range of aqueous solutions than are enzymes bound by van der Waals or ionic forces only (Kennedy & Cabral, 1987a).

Hydrophobic adsorption of an enzyme is dependent on well-known experimental variables such as pH, the nature of the solvents, ionic strength and temperature (Kennedy & Cabral, 1987a). Non-covalent adsorption of diverse enzymes and/or proteins on non-ionic cellulose derivatives (Table 1.8) may be considered strong in spite of the fact that it involves only interactions with solely hydrophobic character. Non-ionic detergents and in some cases ethylene glycol in high concentrations cause desorption. On the other hand, high concentrations of salts increase the strength of binding (Butler, 1975).

Owing to the relatively weak binding forces between the protein and the adsorbent, close control of additional variables is required for optimal adsorption and retention of activity. These variables or better intrinsic properties are: hydrophobicity of the adsorbent and the adsorptive, e.g. enzyme.

Hydrophobicity of non-ionic hydrophobic derivatives of cellulose is usually regulated by the degree of substitution (DS) and/or by the length of the hydrophobic chain (Butler, 1975; Dixon *et al.*, 1979; Gemeiner *et al.*, 1989b, 1990). The hydrophobic region of an enzyme may be modulated by changing its primary structure either by chemical modification (Schmid, 1979) or by site-directed mutagenesis (Regnier & Maszaroff, 1987). The intensity and/or strength of hydrophobic interaction usually increases with increasing hydrophobicity of both the adsorbent and the adsorptive. To reach the conditions optimal for immobilization by hydrophobic adsorption it is recommended to adapt the hydrophobicity of the adsorbent, e.g. cellulose derivative (Gemeiner *et al.*, 1989a,b) to that of the enzyme.

Table 1.8

List of Non-ionic Hydrophobized Cellulose Applied to Immobilization of Enzymes by Hydrophobic Adsorption

Cellulose physical form	Hydrophobic moiety	Number of immobilized enzymes	References
powdery, fibers paper, cloth, string	acetyl, lauryl, benzoyl, 4-phenyl-butyryl, phenoxyacetyl	10	Butler (1971)
powdery	hexanoyl, palmitoyl, benzoyl, phenoxyacetyl, phenylacetyl, phenylpropionyl	7	Dixon *et al.* (1979)
powdery	tannic acid	9	Watanabe *et al.* (1979)
cloth	naphthoxypropyl	3	Sharma & Yamazaki (1984)
microspheres	hexyl	5	Ikada & Tabata (1988)
bead	phenoxyhydroxypropyl	3	Gemeiner *et al.* (1989*b*)

It was supposed that some non-ionic cellulose esters of carboxylic acids, namely phenoxyacetyl, may also act as substrates for certain enzymes such as subtilisin, chymotrypsin, lipase, etc. Nevertheless, the fact that their catalytic activity remained unaltered points to the adsorption process occurring away from the active site (Butler, 1975).

Enzymes can also be immobilized by physical adsorption on to affinity matrices. Cellulose and its derivatives may be simultaneously both adsorbents and insoluble substrates. This may occur if glycoside hydrolases but mainly cellulases are adsorbed onto the cellulose. In such cases substrate is the adsorbent and preferential adsorption of the cellulase occurs leaving most of the other enzymes in solution. This type of adsorption is a combination of affinity binding, hydrogen bonding, hydrophobic forces, and permeation effects (Reese, 1982). The cellulase complex 'immobilized' on cellulose is perfectly stable. Practically a total recovery of the activity may be obtained even after desorption by the above eluents (Reese, 1982).

Concanavalin A, covalently bonded to cellulose, binds different glycoenzymes (glucose oxidase, invertase) as it must be admitted, *biospecifically*; additional stabilization (Iqbal & Saleemuddin, 1983) can be omitted in some cases (Dočolomanský *et al.*, 1993).

Ionic binding. Like physical adsorption, the immobilization of an enzyme by ionic bonds may be carried out easily by the same procedures, which may be considered as mild in comparison with those required in most methods of covalent coupling.

The use of ion exchange derivatives of cellulose in the ionic binding method does not need special comments. Indeed, there is an extensive literature concerning the use of traditional forms of ion-exchange cellulose derivatives, e.g. standard fibrous, powdery, microgranular, etc. Development and production of these cellulose materials in beaded form have become a matter of course also for renowned companies in the world. Despite that, the development of methods of preparation, characterization of new properties, and broadening the range of ion-exchange cellulose in beaded form have continued. One of the interesting results is, for example, the increase in cation- and anion-exchange capacity achieved by cross-linking of bead cellulose and the simultaneous manifold increase in porosity. Table 1.4 presents examples of use of ion-exchange bead celluloses the preparation and properties of which have been recently described in the literature (Gemeiner *et al.*, 1989*a*).

In contrast to physical adsorption ionic binding exhibits a stoichiometric character. The mass action law was found to rule ionic (and covalent) binding and also physical adsorption caused by biospecific interaction between an insoluble matrix and the enzyme. On the other hand, the physical adsorption of enzymes onto a certain class of hydrophobized matrices is regulated predominantly by the partition law. This points to the non-stoichiometric character of physical adsorption. The above contention is also supported by the finding that the partition of the enzyme from a solution on a hydrophobized solid–liquid interface may be simulated by its liquid–liquid partitioning. Thus, the primary recognition of the type of binding can be made by determining whether the respective adsorption process is regulated by the mass action law or by the partition law (Breier *et al.*, 1987; Gemeiner *et al.*, 1989*a*).

Metal-linking method. This is based on chelating properties of the transition metals, which can be employed to couple enzymes. Interaction of the transition metal compounds with cellulose is demonstrated in Fig. 1.9 on an example of a simple system consisting of titanium(IV) chloride and cellulose (Kennedy & Cabral, 1987*a,b*).

In titanium(IV) chloride solution, titanium ions are coordinated with molecules or species of ions that are essentially ligands of the complex ion. Nucleophilic groups, such as hydroxy-, amino-, sulfhydryl-, etc., are effective ligands for the transition metal ions. Hence, it may be expected that transition metal ions may form complexes with both supports and enzymes. Supports such as cellulose and silica have hydroxyl groups which act as new ligands replacing other ligands For example, cellulose contains vicinal diol groups not involved in glycosidic linkages between the residues. These diol groups are amenable to chelation by transition metal ions, the chelate being produced by replacement of two of the titanium ion ligands by polysaccharide hydroxyl groups. Thus, in chelated form of the cellulose there are many titanium centers with residual exchangeable water and/or chloride ligands. This exchangeable nature of original residual ligands imparts their reactivity to the derivatized cellulose and the insolubility of the cellulose provides a matrix suitable for immobilization of liquid-soluble molecules by chelation (Kennedy & Cabral, 1987*a,b*).

Enzymes have groups that can act as ligands such as carboxyl groups, phenolic and alcoholic hydroxyl groups, sulfhydryl and amino groups. Formation of metal-linking bonds between the titanium-treated polymers and the enzymes, yielding enzymatically active derivatives, depends on very similar requirements (Kennedy & Cabral, 1987*a,b*) as in the case of the covalent-binding method.

A relatively high specific activity retention (30–80%) may be achieved by the metal-linking method. Immobilized enzymes prepared by this method have also variable operational stability. Particularly low stabilities were obtained with high-molecular-mass substrates. The metal activator, e.g. titanium itself, also seems to influence the stability of the immobilized enzyme preparation, probably due to inhibition of the enzyme by the metal activator (Kennedy & Cabral, 1987*a*).

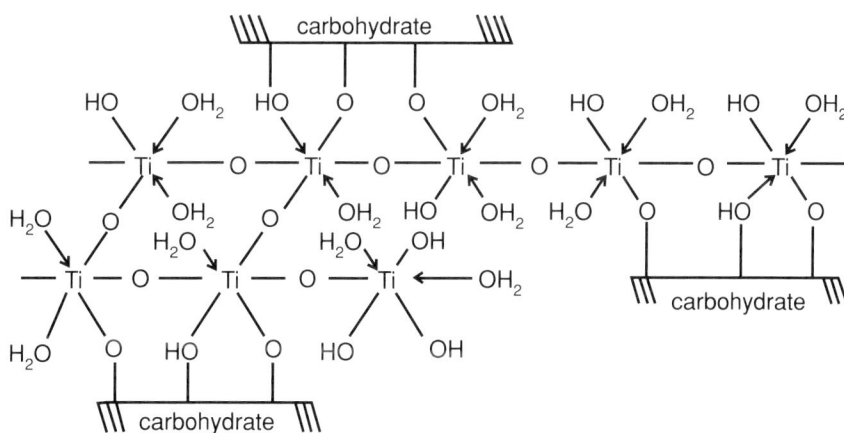

Fig. 1.9 Titanium-treated cellulose chelated and/or complexed along its chains with various aquo-, chloroaquo-, and chloro-complexes of titanium species prevalent in the solution (Kennedy & Cabral, 1987*b*).

P. Gemeiner, L. Rexová-Benková, F. Švec & O. Norrlöw

Table 1.9
Review of Enzymes and Other Proteins Immobilized Covalently on Activated Bead Cellulose via Method 1a

Derivative	Reagent used for activation	Site of activation	Activated bead cellulose	Proteins immobilized on activated cellulose	Reference
1	2	3	4	5	6
Ia	Ethyl chloroformate	OH	trans-2,3-Cyclic carbonate	Chymotrypsin A	Kennedy & Rosevear (1974), Kennedy et al. (1973), Kennedy (1974b)
Ib	4-Nitrophenyl chloroformate	OH	trans-2,3-Cyclic carbonate	Trypsin	Drobník et al. (1982)
Ic	N-Hydroxysuccinimidil chloroformate	OH	O-Ethoxycarbonyl	Trypsin	Drobník et al. (1982)
Id	5-Norbornene-2,3-dicarboximido carbonochloridate	OH	Active carbonate	Concanavalin A, Ovomucoid, Immunoglobulin G, Protein A, etc.	Büttner et al. (1989)
II	Cyanogen bromide	OH	Imidocarbonate	Trypsin, Glucoamylase, Methemoglobin, Immunoglobulin G	Drobník et al. (1982), Chen et al. (1981), Pommerening et al. (1979), Dean et al. (1985), Chandler & Johnson (1979)
III	2,4-Diisocyanatotoluene	OH	Isocyanatophenyl	Glucoamylase, Glucose isomerase, Invertase	Chen & Tsao (1977), Chen et al. (1981), Chen & Tsao (1977), Chen & Tsao (1977)

	Reagent	Group	Reactive form	Enzyme	References
IV	Sodium nitrite/HCl	NH$_2$(aromatic)	Diazonium salt	Glucoamylase Trypsin Invertase Lipase	Chen & Tsao (1977) Gemeiner et al. (1986, 1987) Kéry et al. (1990)
V	Glutardialdehyde	NH$_2$	Aldehyde	Glucoamylase Glucose oxidase Invertase β-Galactosidase	Chen & Tsao (1977) Valentová et al. (1981) Štefuca et al. (1990) Kmínková et al. (1988)
VI	Periodate, sodium	OH	2,3-Dialdehyde	Trypsin Glucose oxidase Chymotrypsin Lysozyme Ovalbumin Serum albumin Immunoglobulin G	Gemeiner & Breier (1982) Drobník et al. (1982) Gemeiner et al. (1986) Gemeiner et al. (1987) Gemeiner & Pašteka (1983) Valentová et al. (1981) Turková et al. (1979) Turková et al. (1980) Gemeiner & Pašteka (1983) Gemeiner & Pašteka (1983) Gemeiner & Pašteka (1983) Dean et al. (1985) Chandler & Johnson (1979)
VII	Benzoquinone	OH	Benzoquinone	β-Galactosidase Chymotrypsin Subtilisin Protease, neutral	Chun et al. (1980) Chun & Sernetz (1980) Hong et al. (1982) Škachová & Kučera (1983) Škachová & Kučera (1983) Škachová & Kučera (1983)
VIIIa	Epichlorhydrin + α,ω-diaminoalkane	OH	Alkylamine	Glucose oxidase	Valentová et al. (1981)

Table 1.9 — *contd.*

Derivative	Reagent used for activation	Site of activation	Activated bead cellulose	Proteins immobilized on activated cellulose	Reference
1	2	3	4	5	6
VIIIb	Epichlorhydrin + benzidine	OH	Arylamine	Glucose oxidase	Valentová et al. (1981)
IX	Carbodiimide	COOH	Activated ester	Glucose oxidase	Valentová et al. (1981)
X	4-Isothiocyanato-benzaldehyde	NH_2	Formylphenyl	Trypsin	Gemeiner & Breier (1982)
XI	Bromoacetaldehyde	OH	Formylmethyl	Trypsin	Gemeiner et al. (1986) Gemeiner et al. (1987)
	Diethyl acetal/HCl			Invertase	Štefuca et al. (1990)
XII	Cyanuric chloride	OH	Chlorotriazine	Invertase	Štefuca et al. (1988) Štefuca et al. (1990)
				Lipase	Kéry et al. (1990)
				Immunoglobulin G	Peng et al. (1987)
XIII	Tosyl chloride	OH	Tosyl	Immunoglobulin G	Peng et al. (1987)
XIV	Dimethyl sulfoxide -carbodiimide	OH	Aldehyde	Immunoglobulin G	Peng et al. (1987)
XV	Chromium trioxide	OH	Carbonyl	Immunoglobulin G	Peng et al. (1986) Peng et al. (1987)

Covalent-binding method. This is the most frequent method of immobilization of enzymes on celluloses. All three typical procedures may be found in the following examples:

1a Reaction of the enzymes with the activated carrier (Table 1.9);
1b Reaction of the enzyme with the carrier using a homobifunctional reagent (Table 1.10);
1c Reaction of the enzyme with a heterobifunctional reagent and subsequent reaction of the activated enzyme with the carrier (Table 1.11).

In examples listed in Tables 1.12–1.14 cellulose in beaded form was used predominantly.

One of the methods most frequently applied to covalent immobilization of biocatalysts (1a) on to bead cellulose represents reductive alkylation. It is a simple, cheap, and rapid procedure that can be applied to many enzymes since it is known that ϵ-amino groups of lysine residues utilized for binding are not essential for catalytic activity of numerous enzymes. This was proved in experiments with immobilization of chymotrypsin (Turková *et al.*, 1979, 1980; Šťovíčková *et al.*, 1988), glucose oxidase (Valentová *et al.*, 1981), and trypsin (Drobník *et al.*, 1982; Gemeiner *et al.*, 1986, 1987) on different aldehyde derivatives, 2,3-dialdehyde inclusive, where highly active preparations were obtained. The storage stability of dry chymotrypsin-bead cellulose conjugates prepared via reductive alkylation with 2,3-dialdehyde was observed to be excellent (Šťovíčková *et al.*, 1988). Drying was performed by lyophilization and in air at 25°C to 75°C.

In method 1c there are also included examples in which the enzyme, structurally a glycoprotein, was activated by periodate oxidation (Valentová *et al.*, 1981; Štefuca *et al.*, 1988). This example has an analogue in the procedure in which the glycoenzyme was prepared synthetically and the neoglycoenzyme was then activated by periodate oxidation (Gemeiner & Viskupič, 1981*a*). Covalent immobilization by procedure 1c is one of the ways to achieve oriented binding of enzymes into multienzyme complexes (Gemeiner & Zemek, 1988*c*). Glycosylation of the enzyme and periodate oxidation of the neoglycoenzyme might be the reaction steps of a universal procedure, used for this purpose.

Immobilization of enzymes by ionic bond on derivatives of bead celluloses exemplified in Table 1.12.

Bead cellulose is a porous carrier. The size and geometry of pores and the size and conformation of the enzyme determine the distribution of the enzyme on the carrier, on its surface and along the pores. In the study of kinetics of 'porous' immobilized enzymes, except mass transport limitations, the size-exclusion principle is a dominant factor. It is con-

Table 1.10

Review of Enzymes Immobilized on Bead Cellulose Derivatives Covalently via Method 1b using Homobifunctional Reagents

Derivative	Homobifunctional reagent	Site of activation	Activated bead cellulose	Enzymes immobilized	Reference
XVI	Glutardialdehyde	Amino, aromatic	Aldehyde	Glucoamylase	Chen & Tsao (1977)
XVII	2,4-Diisocyanatotoluene/water	Hydroxy	Amino	Glucoamylase	Chen & Tsao (1977)
XVIII	3-Aminopropylethoxysilane/ glutardialdehyde	Hydroxy	Aldehyde	Glucoamylase	Chen & Tsao (1977)
XIX	Polyethylenimine/ glutardialdehyde	Unknown	Aldehyde	Invertase	Štefuca et al. (1988)

Table 1.11

Review of Enzymes and Other Proteins Immobilized on Bead Cellulose Derivatives Covalently via Method 1c using Activated Enzymes

Derivative	Functional group of cellulose	Functional group of activated proteins	Immobilized proteins	Reference
XX	Hydrazide	2,3,-Dialdehyde	Invertase	Štefuca et al. (1988)
XXI	Hydrazide, hydrazinodeoxy	2,3-Dialdehydoglucose -galactose, -cellobiose -lactose	Serum albumin	Gemeiner & Viskupič (1981a)
XXII	Alkylamine	2,3-Dialdehyde imine	Glucose oxidase	Valentová et al. (1981)
XXIII	Alkylamine/cyclohexyl isocyanide	2,3-Dialdehyde imine	Invertase	Marek et al. (1984)

Table 1.12
Review of Enzymes Immobilized on Bead Cellulose Derivatives with Ionic Bonds

Derivative	Method of preparation and reagents used	Derivative	Immobilized enzyme	Reference
XXIVa	Cross-linking followed by 2-chlorotriethylamine	DEAE	Glucoamylase	Chen & Tsao (1976) Chen *et al.* (1981)
			Glucose isomerase	Chen & Tsao (1976) Chen *et al.* (1981)
XXIVb	2-chlorotriethylamine followed by crosslinking	DEAE	Glucoamylase Glucose isomerase	Chen & Tsao (1977) Maeda *et al.* (1979)
XXIVc	Cross-linking followed by 2-chlorotriethylamine and repeating crosslinking	DEAE	Glucoamylase	Dickensheets *et al.* (1979)
XXV	Diisocyanato, water, O-methylisourea	Guanidino	Glucoamylase	Chen & Tsao (1977)
			Glucose isomerase Invertase	Chen & Tsao (1977) Dickensheets *et al.* (1977)
XXVI	Epichlorhydrine, diethylamine	DEAHP	Invertase	Hradil & Švec (1982*a*) Hradil & Švec (1982*b*)
		DEAHP, DEAE	β-Galactosidase	Kmínková *et al.* (1989)

Table 1.13
Intrinsic Kinetics Constants of Immobilized Invertase Measured in Packed Bed Reactors

Bead cellulose prepared from	Bead cellulose derivative	Immobilization	Constants		Reference
			K_m (mmol dm^{-3})	K_i (mmol dm^{-3})	
Cellulose acetate	Guanidino	Ionic (2)	46·9	4·884	Dickensheets *et al.* (1977)
Cellulose xanthate	Chlorotriazine	Covalent (1a)	33·0	—	Štefuca *et al.* (1988)
Cellulose xanthate	Chlorotriazine	Covalent (1a)	48·5	1752·0	Štefuca *et al.* (1991)
Cellulose xanthate	Chlorotriazine	Covalent (1a)	47·6	1748·7	Štefuca *et al.* (1991)

Table 1.14
Proteins Covalently Immobilized on Bead Cellulose (Perloza®), Produced in Czecho-Slovakia

Protein	Method of immobilization	Activated bead cellulose	Reference
Chymotrypsin	1a	2,3-Dialdehyde	Turková *et al.* (1979)
			Turková *et al.* (1980)
		Benzoquinone	Škachová & Kučera (1983)
Trypsin	1a	2,3-Dialdehyde	Gemeiner & Breier (1982)
			Drobník *et al.* (1982)
			Gemeiner *et al.* (1986)
			Gemeiner & Pašteka (1983)
		Formylphenyl	Gemeiner & Breier (1982)
		Formylmethyl	Gemeiner *et al.* (1986)
			Gemeiner *et al.* (1987)
		2-(4-Aminophenylsulfonyl)ethyl[a]	Gemeiner *et al.* (1986)
			Gemeiner *et al.* (1987)
Subtilisin	1a	Benzoquinone	Škachová & Kučera (1983)
Protease, neutral	1a	Benzoquinone	Škachová & Kučera (1983)
Invertase	1a	Chlorotriazine	Štefuca *et al.* (1988)
			Štefuca *et al.* (1991)
		2-(4-Aminophenylsulfonyl)ethyl[a]	Štefuca *et al.* (1988)
			Štefuca *et al.* (1991)
		Formylmethyl	Štefuca *et al.* (1991)
	1b	PEI-cellulose/aldehyde	Štefuca *et al.* (1988)
	1c	Hydrazide	Štefuca *et al.* (1988)
	2	DEAHP	Hradil & Švec (1982*a*)
			Hradil & Švec (1982*b*)
β-Galactosidase	1a	Benzoquinone	Chun *et al.* (1980)
			Chun & Sernetz (1980)
Glucose oxidase	1a	Aldehyde	Valentová *et al.* (1981)
	1c	2,3-Dialdehyde	Valentová *et al.* (1981)
Hydroxysteroid Dehydrogenase + diaphorase	1a	Imidocarbonate	Bovara *et al.* (1981)
Polynucleotide Phosphorylase	1a	2-(4-Aminophenylsulfonyl)ethyl[a]	Kéry *et al.* (1988)
Serum albumin	1a	2,3-Dialdehyde	Gemeiner & Pašteka (1983)
	1c	Hydrazides, hydrazinodeoxy	Gemeiner & Viskupič (1981*a*)

[a]Commercially available as Ostsorb AV (United Chemical and Metallurgical Works, Ústí n/L).

firmed by empirical quantitative relations, respecting the relative catalytic activity of trypsin immobilized on various derivatives of bead cellulose (Perloza) vs molecular mass (Gemeiner & Breier, 1982), available distribution coefficient (Gemeiner *et al.*, 1986), and hydrodynamic diameter of the substrate (Gemeiner *et al.*, 1987). The constants of these equations were tested as parameters of quantitative relationships between the activity of the immobilized enzyme and the structure of the bead cellulose-trypsin conjugates (Gemeiner & Breier, 1982; Gemeiner *et al.*, 1986, 1987). In preparation of conjugates attention was paid to reductive alkylation, i.e. immobilization procedure utilizing aldehyde derivatives of bead cellulose, 2,3-dialdehyde inclusive.

For characterization of suitability of the enzyme carrier numerous criteria are used: availability (price), easy activation, resistance to chemical and microbial attack, hydrodynamic properties (e.g. drop pressure), etc. It is customary to compare bead cellulose as enzyme carrier with other specific biopolymers, mainly with agaroses (Sepharoses) and crosslinked dextrans (Sephadexes). As follows from data published recently (Gemeiner *et al.*, 1989*a*) bead cellulose is equal to traditional biopolymers in all criteria mentioned above, moreover, in price, hydrodynamic properties, resistance to chemical attack, and easy activation it even surpasses them.

Another criterion for suitability of the carrier are the properties of the enzyme after immobilization. In this respect the first source of information are the kinetic constants of immobilized and free enzyme and the effect of various factors on their values. Such information was obtained from packed bed reactors (PBR), packed with immobilized invertase (Dickensheets *et al.*, 1977; Hradil & Švec, 1982*b*; Štefuca *et al.*, 1988, 1990) and glucoamylase (Dickensheets *et al.*, 1977) and from stirred tank reactors (STR), packed with immobilized invertase (Hradil & Švec, 1982*a*; Štefuca *et al.*, 1988). Serious confrontation of kinetic data is impeded by diverse methods in immobilization (Tables 1.9, 1.10, 1.12), different procedures of preparation of bead cellulose (Štamberg, 1988; Gemeiner *et al.*, 1989*a*) as well as by the origin and quality of free enzyme. It follows also from the differences in values of kinetic constants of invertase from *Candida albicans* (Dickensheets *et al.*, 1977) and *S. cerevisae* (Štefuca *et al.*, 1988, 1990), immobilized in two different ways on two different bead celluloses (Table 1.13). Availability of the Czecho-Slovak bead cellulose both at inland and foreign workplaces at present offers the possibility of unifying the properties of the carrier. The enzymes/proteins and the methods used for their immobilization on the Czecho-Slovak bead cellulose are reviewed in Table 1.14.

Most recently developed mathematical models predicting kinetics of STR and PBR were extended by considering the role of cellulose bead geometry (Štefuca *et al.*, 1988). In the PBR, owing to the Sherwood number kept below 2 (flow rates 0·5–4 cm^3 min^{-1}), deformation of the ideal spherical shape of beads was observed. The spherical shape of cellulose beads was partly destroyed in STR, particularly when the number of revolutions exceeded 200 min^{-1}. When preservation of the original shape of beads is required, other types of reactor or reinforcement of beads is recommended (Štefuca *et al.*, 1988). Invertase was used as the model enzyme.

The same model (invertase immobilized on derivatives of bead cellulose) was used when the intrinsic kinetic constants of hydrolysis of saccharose were estimated using both direct (thermal assay probe) and indirect (differential reactor system) methods (Štefuca *et al.*, 1990). The effect of heat losses and flow nonideality on the values of intrinsic kinetic constants (K_m, K_i, V_m) was eliminated, but, the effect of flow rate on the thermistor sensor signal was evident. The obtained kinetic parameters were superverified by means of an independent differential reactor system. Of several Perloza derivatives, the chlorotriazine derivative was used successfully in all studies (Štefuca *et al.*, 1988, 1990).

When the cells are to be cultured *in vitro*, it is advantageous to utilize microcarriers. As such, also bead DEAE-, DEAHP-, and TMAHP-celluloses with the exchange capacity of 0·3 to 1·5 mmol g^{-1} and particle size 70 to 300 μm may be used. Veber *et al.* (1984) employed this method and succeeded in preparing cells L, HeLa, BHK, and others, suitable for diagnostic purposes or for industrial production of various biologically active compounds, e.g. human and animal virus vaccines, interferons, hormones, enzymes, etc.

Less frequently applied are those procedures in which biocatalysts are immobilized by *entrapping*. All main types of cellulose carriers, e.g. gel, fiber, (micro)capsule, membrane, were applied to the immobilization of biocatalysts by entrapping methods.

Entrapment in gel. Although a number of bead cellulose derivatives have been employed for immobilization of various enzymes, it soon became clear that the traditional methods for cellulose solubilization (Štamberg, 1988; Gemeiner *et al.*, 1989*a*) resulted in rapid enzyme inactivation. Linko *et al.* (1977, 1979, 1980, 1982, 1987) succeeded in developing such solvent systems for α-cellulose that allowed the regeneration of the biocatalyst of high enzyme loading as beads from water. They entrapped inside cellulose beads numerous bacteria and yeasts instead of enzymes; the solvent system consisted of N-ethylpyridinium chloride, dimethylformamide, and dimethyl sulfoxide. Catalytic activity recoveries of glucose isomerase, invertase, β-galactosidase, and urease ranged from 22 to 85 per cent (Linko, 1982). Similar techniques were also developed to prepare bead-shaped biocatalysts using cellulose di- and triacetates as carriers (Linko *et al.*, 1980; Linko, 1982). Activity yields of entrapped cells, e.g. invertase (*S. cerevisae*), up to 85 per cent were obtained (Linko *et al.*, 1980).

Marconi and coworkers (1987, Giovenco *et al.*, 1987) found that cellulose diacetate having a degree of polymerization around 300 can be advantageously used as a bead shaped matrix for cell entrapment. Acetone was used as the solvent, and the necessary aggregation of cells was achieved by addition of a polycation (polyethyleneimine). In such a way bacteria and yeasts were entrapped; catalytic activity of their glucose isomerase (Giovenco *et al.*, 1987; Marconi *et al.*, 1987), penicillin acylase, and β-galactosidase (Giovenco *et al.*, 1987) was used for industrial applications.

Entrapment and immobilization in fibers. A broad assortment of cellulose materials suitable for entrapment/immobilization of enzymes is available in the form of hollow-fiber membranes and/or hollow-fiber devices. Here belong predominantly the membranes available commercially from regenerated cellulose or cellulose acetate in isotropic form. These materials differ in nominal cutoff, surface area, inside diameter and geometry.

Several different techniques have been employed for entrapment/immobilization of soluble enzymes. Most simple among these is that in which the enzyme is physically entrapped by hollow fibers (Chambers *et al.*, 1976), continuous filaments, short stable fibers, woven fabrics or felts (Marconi & Morisi, 1979). This type of entrapment may normally be provided by means of a continuous-flow system in which the enzyme is recirculated through the hollow fiber and also through a small reservoir (Chambers *et al.*, 1976). A great number of enzymes, belonging to any class, have been entrapped with good retained activity (Marconi & Morisi, 1979).

In order to improve the half-life of soluble enzymes, a number of additional procedures have been developed. Prominent among these are the covalent coupling of soluble macromolecules (e.g. dextran) to the enzyme (O'Neill *et al.*, 1971). An alternative technique for stabilization of enzymes for use with hollow fibers is immobilization of the enzyme on particles small enough (DEAE-cellulose, 1-3 μm polyacrylamide beads) to be circulated through or placed within the hollow fiber lumen (O'Neill *et al.*, 1971; Chambers *et al.*, 1976).

The use of hollow-fiber immobilization techniques facilitates the construction of multienzyme systems. The low-molecular-weight cutoff of the device permitted the complete retention of all enzymes and co-factor (NAD), while allowing dialytic permeation of the substrates (Chambers *et al.*, 1974; Marconi & Morisi, 1979). Hollow-fibers, which are permeable to cofactors yet retain enzymes, have been also utilized, investigating multiple cycling of NAD. Further unique set-ups utilizing hollow-fiber immobilized cofactor-requiring enzymes have been reviewed by Chambers *et al.* (1976), Mansson & Mosbach (1987), etc.

Cellulose hollow fibers represent suitable materials also for covalent immobilization of enzymes. For the latter technique investigators applied procedures commonly used for activation of hydroxyl groups, e.g. their activation by cyanogen bromide (Giordano *et al.*, 1981) or tresyl chloride (Comfort *et al.*, 1989). It seems, that covalent immobilization of enzymes in cellulose hollow fibers will become a viable route for the use of immobilized enzymes in clinical medicine.

Entrapment in (micro)capsules. This method was made particularly attractive by Chang (1976, 1985, 1987) who applied it to construction of 'artificial cells', i.e. encapsulated multienzyme systems with co-factor regeneration. Artificial cells containing enzymes and proteins have been used in a number of therapeutic conditions. Semipermeable microcapsules were prepared from cellulose nitrate by emulsification followed by spherical ultrathin membrane formation. A typical artificial cell of mean diameter 20 μm has an ultrathin membrane of 20 μm thickness and equivalent pore radius of about 1·8 μm. Numerous applications require artificial cells meeting cellular dimensions. This stresses the importance of procedures enabling a permanent control of mean diameter and size distribution microcapsules during their formation (Poncelet de Smet *et al.*, 1989).

Functional properties of enveloping membranes prepared by emulsification depend on physico-chemical properties of cellulose and cellulose derivatives applied (Kamide, 1987). From derivatives of cellulose applied as precursors of capsules, besides the nitrates the acetates of cellulose (Pusch, 1987) and ethylcellulose (Kato *et al.*, 1985) are attractive. Unlike nitrate and acetate derivatives of cellulose ethylcellulose is used mainly for production of microcapsules designed for selective drug delivery (Kato *et al.*, 1985).

Application of emulsification was successfully avoided by Pommerening *et al.* (1983) who utilized for encapsulation of enzymes polyelectrolyte complex (PEC) formed using sodium cellulose sulfate (SCS) and poly(dimethyldiallyl ammonium chloride) (PDMDAAC) as polyelectrolyte components. With this polyanion/polycation combination capsule properties such as mechanical strength, elasticity and deformation, and capsule wall properties such as morphology, permeability and transparency, can be widely varied by changing the parameters of the polyelectrolyte precursors as well as the conditions of capsule preparation (Dautzenberg *et al.*, 1985). The pore width of the capsule membrane can be adjusted at the right order of magnitude to secure both complete retaining of the enzyme inside the capsule and, at the same time, free passage substrates and products. Another typical feature for microcapsules obtained by PEC formation is their full compatibility with aqueous solutions buffered with salts. Ionic strength of aqueous environment is qualified to be a factor controlling properties of PEC capsules (Dautzenberg *et al.*, 1985).

The PEC membrane formed by SCS and PDMDAAC was used to encapsulate lactate dehydrogenase (Štefuca *et al.*, 1991). The capsules obtained were checked for their functionality in a stirred batch reactor following the kinetics of NADH oxidation. The data were fitted with an isotropic kinetic model including competitive product inhibition phenomenon. The results of mathematical modelling demonstrated that the anisotropic system, like PEC capsules, could be satisfactorily described by the isotropic model.

1.3.2 Starch and Co-polymers

Starches are used as components and/or processing aids in the manufacture of products such as adhesives, textiles, paper, food, pharmaceuticals, and building materials. The use of this natural polymeric material is based on its thickening, gelling, adhesive, and film-forming properties, as well as its low cost, controlled quality, and ready availability. Like other polysaccharides, starch may be modified in its characteristics by chemical, physical and/or enzyme treatment. The latter will enhance or repress the intrinsic properties of starch or even impart new ones (Rutenberg, 1980).

Nevertheless, starch is the least suitable common polysaccharide for biocatalyst immobilization. When compared with other polysaccharides it is rarely employed as a carrier. Moreover, contradictory evaluations are available concerning the application of starch as a carrier.

The first property usually indicated as negative is its susceptibility to microbial attack, i.e. its biodegradability. Biodegradability and non-toxicity (biocompatibility) of starch as a material were, however, utilized, e.g., in encapsulation of insecticides into starch capsules (Riley, 1983; Trimnell & Shasha, 1988) and in blood substitution with hydroxyethylstarch-hemoglobin conjugates (Cerny *et al.*, 1982). Moreover, starch is sought as encapsulation material for preparation of 'biodegradable microspheres' which may transport targeted enzymes/proteins and drugs in the organism (Artursson *et al.*, 1984). Several methods were developed for the preparation of 'polyacryl starch microparticles' from hydroxyethylstarch or maltodextrin. These particles fulfill many requirements of drug carriers for application *in vivo*. This concerns particularly those requirements which are related to biocompatibility, non-toxicity and biodegradability (Laakso *et al.*, 1986; Stjärnkvist *et al.*, 1989).

A practical example of how to overcome the biodegradability of starch was demonstrated in the application of partially acid hydrolyzed starch-acrylonitrile graft copolymer for immobilization of glucoamylase (Slininger *et al.*, 1988). When evaluating starch as a carrier it is often claimed that its chemistry is well understood. However, the projection of this putative extensive knowledge into the development of immobilization methods (Kennedy & Cabral, 1983*a*; Rutenberg & Solarek, 1984; Kennedy & Cabral, 1987*a*; Scouten, 1987; Koch & Roper, 1988) is not particularly conspicuous.

Starch is a complex polysaccharide substance. As raw material extracted from a plant source starch normally exhibits a granular structure with particle size 2–100 μm. Granules involve a complex packing arrangement of starch polymers, and also contain other bio-molecular components such as lipids and proteins. The term starch, unlike cellulose, often applies to a mixture of two molecular entities, namely the branched amylopectin and the linear amylose.

Amylopectin and amylose are composed of repeated glucose residues the amylose being an $\alpha(1{\rightarrow}4)$-linked-D-glucose polymer. In amylopectin the D-glucose residues are bound with both $\alpha(1{\rightarrow}4)$ and $\alpha(1{\rightarrow}6)$ linkages (Fig. 1.10). Amylose is normally the minor constituent by weight, accounting for only 25 per cent of the total starch. In the granule, amylose and amylopectin are packed together in a way that is not yet fully understood. That the granules are partly crystalline has long been known (Clark & Ross-Murphy, 1987).

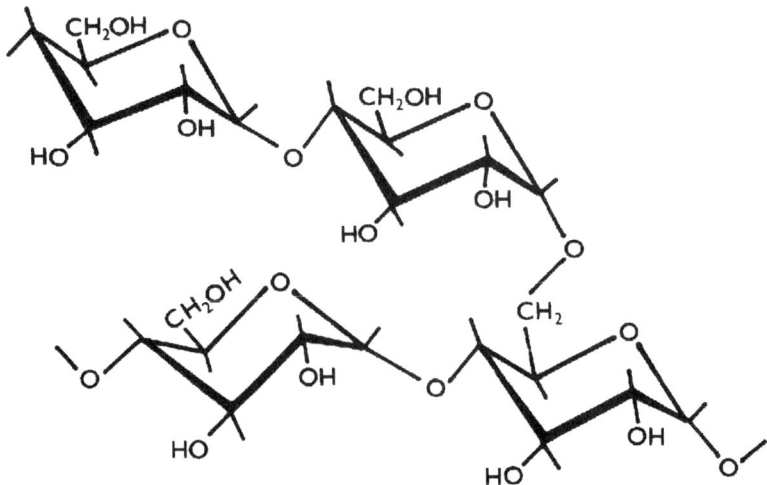

Fig. 1.10 Branched structure of the amylopectin component of starch based on 1,4- and 1,6-linked α-D-glucose residues. Note that amylose, the other polysaccharide component of starch, involves only 1,4-linkages and is a linear polymer (Clark & Ross-Murphy, 1987).

When heated in water above a specific temperature known as the gela-tinisation temperature starch granules in suspension swell suddenly and irreversibly, and the amylose component leaks into the solution. If the starch concentration is high enough this mixture of swollen granules and free polymer behaves as a viscoelastic paste. The resulting gels have been described as composite materials consisting of amylose gel matrices 'filled' with swollen granules (Clark & Ross-Murphy, 1987).

Aggregation and gelation characteristics of both the amylose and amylo-pectin fractions of starch have also received attention. Anyone who has worked with aqueous solutions of these molecules is familiar with the unstable nature of aqueous amylose and the stability of amylopectin in dilute solutions. It is generally well recognized that, compared to amylo-pectin, amylose molecules tend to associate and crystallize. In part, this contrasting behavior has made it difficult to accept the fact that amylo-pectin does indeed crystallize. Crystallization is evident through X-ray diffraction of waxy maize granules; it is less evident for precipitates from solution. Amylopectin generally requires to be highly concentrated (30 per cent or more) for crystallization. This is attained upon aging of gels or upon dehydration during film formation. Molecular association and/or crystallization occurring in gels gives rise to hydrogen-bonded physical, rather than chemical crosslinks (Zobel, 1988).

Amylose gels are stiffened due to involvement of the linear amylose molecule in more than one crystallite; amylopectin gels are softer due to fewer intermolecular interactions owing to the shape of the cluster. For example, at a concentration of 2 per cent, amylose forms a reasonably rigid, free-standing gel that sets up in a few minutes. At the same con-centration, amylopectin forms a freeflowing solution and remains so. Once formed, amylose gels generally require autoclave temperatures (110–160°C) for reversal (Zobel, 1988). In the case of amylopectin gels, the molecular chain branch points interfere with regular and extended association between molecular segments. Consequently, temperatures required to reverse association and solubilize amylopectin gels may vary from room temperature to 95°C, depending on the degree of molecular association or crystallization that has developed. Amylose is noted for its capacity to form coherent, free-standing films that are reasonably strong and flexible when plasticized. Amylopectin films are very brittle, gener-ally non-continuous, and have the appearance of mud-flats (Zobel, 1988).

Like cellulose, starch offers an interesting variety of physical forms in which it may be applied as a carrier for immobilization. The most typical physical forms of starch are summarized in Table 1.15. The number of examples given for application of various physical forms of starch is,

Table 1.15

Overview of Procedures Utilizing Different Physical Forms of Starch as the Matrix/Carrier for Immobilization

Components	Method of matrix preparation	Physical form of matrix	Method of immobilization	Reference
starch	gelatinization	gel	entrapment	Baumann et al. (1965)
starch	gelatinization	dried gel	entrapment	Baumann et al. (1967)
starch	gelatinization	microbeads	occlusion	Trimnell & Shasha (1988) Dunkle & Shasha (1988)
starch	gelatinization	beads	n.d.	Hanus et al. (1974)
starch	crystallization	microspheres	entrapment	Schröder (1985)
hydroxyethyl-starch-, maltodextrin-acroylate	graft copolymerization	microspheres/microparticles	occlusion/microencapsulation	Riley (1983) Laakso et al. (1987b) Laakso & Sjöholm (1987a)
starch, vinyl-enzyme	graft copolymerization/chemical modification	particles	covalent coupling	D'Angiurio et al. (1980b) D'Angiurio et al. (1982)
starch-(maleic anhydride styrene)	graft copolymerization	microparticles	n.d.	Beddows et al. (1986)
starch-acrylonitril	graft copolymerization acid hydrolysis	particles	affinity binding	Slininger et al. (1988)
starch derivatives	chemical modification	microparticles/powder	covalent coupling	Goldstein et al. (1970)
starch derivatives	chemical modification	microparticles	covalent coupling	Mehltretter & Weakly (1972)
alkali saponified starch-poly (acrylonitril)	precipitation	membrane	entrapment/encapsulation	Weaver et al. (1976)

n.d. = not determined.

however, small. Modest also is the available information about characterization of physical properties of these materials.

A typical property of starch is its capability to form gels. Enzymes are simply entrapped in starch gels and very little enzyme is lost during preparation. The basic procedure is as follows: Starch is placed into a boiling mixture of buffer and glycerine (or non-ionic surfactant). Subsequently, the resulting mixture is boiled further until a clear solution is obtained. This is covered and allowed to cool to 47°C. In another beaker, the enzyme is dissolved in the same buffer, and this solution is poured into the starch solution at 47°C. Finally, the enzyme starch solution is poured into an open urethane foam cell and cut to appropriate physical form (Bauman *et al.*, 1965). However, the entrapped enzymes were too weakly held in the starch gel, and much of the enzyme activity becomes lost due to washout (Guilbault & Das, 1970).

Starch gel particles in a porous, spongelike form were prepared by a procedure that involved heating the starch in water, freezing and thawing (Hanus *et al.*, 1974; Kučera, 1983). This procedure yields particles in both an irregular shape (Hanus *et al.*, 1974) and a beaded form (Kučera, 1983). Bead shaped starch, amylose, and amylopectin were prepared also by suspending an aqueous solution of corn starch, amylose or amylopectin in an oil medium and afterwards crosslinking it with epichlorohydrin (Matsumoto *et al.*, 1982). According to an alternative procedure beads of starch triacetate, amylose triacetate, and amylopectin triacetate were prepared by suspending solutions of the corresponding acetates in dichloromethane. Saponification of triacetate beads, followed by crosslinking with epichlorohydrin in kerosene containing a surfactant, yielded gel particles with a diameter of 10–500 μm (Matsumoto *et al.*, 1982).

It is proposed that, similar to porous cellulose beads, starch gel beads will meet the needs for immobilization by a covalent bond rather than for entrapment of the biocatalysts. Contrary to bead cellulose, however, procedures of activation of porous starch gels, suitable for covalent immobilization of biocatalysts were not elaborated. Examples of activation of granular starch and subsequent immobilization of biocatalysts are also not abundant (Goldstein *et al.*, 1970; Mehltretter & Weakly, 1972).

Like cellulose, starch is also used as a carbohydrate backbone with synthetic side chains which may be manufactured by grafting of synthetic monomers. For starch, the usual method is the grafting of monomers, such as maleic anhydride (Beddows *et al.*, 1986), acrylonitrile, styrene (Slininger *et al.*, 1988), and acrylic acid (Koch & Roper, 1988; Slininger *et al.*, 1988) onto granular starch by means of irradiation (Beddows *et al.*, 1986) or radical initiators like, Ce^{4+} compounds (Koch & Roper, 1988; Slininger *et al.*, 1988). The disadvantage of this method is the limited grafting yield since a considerable part of synthetic monomers is used in homopolymer formation. This drawback may be avoided by formation of primary radicals on the starch molecule which favours the co-polymerization on starch backbone as opposed to homopolymer formation (Koch & Roper, 1988).

Poly(maleic anhydride/styrene) and poly(maleic anhydride/acrylonitrile) graft copolymers of starch were also used as carriers for enzymes without further activation. This mode of immobilization, evidently an adsorption one, proved to be unsuitable (Beddows *et al.*, 1986). More promising results were provided by the procedure in which the poly(acrylonitrile) graft copolymer of starch was treated with hydrochloric acid. A series of carriers with increasing starch content was prepared by decreasing the extent of acid hydrolysis (Slininger *et al.*, 1988). The results suggest that both the poly(acrylonitrile) and the residual starch attached to it participate in enzyme binding. The authors assume that the binding of enzyme (glucoamylase) on graft copolymers with diverse starch content has a combined character. Affinities of glucoamylase to starch and starch to water and the role of starch as a spacer between polyacrylonitrile chains represent the main factors participating in the combined binding (Slininger *et al.*, 1988).

In utilizing starch as a carrier for enzymes/proteins and drugs much effort was made in the development of polyacryl starch spherical microparticles (or microspheres) with diameter ≤ 1 µm. A multistep procedure was developed for their preparation in several modifications. The basic steps are: (1) etherification of maltodextrin (Artursson *et al.*, 1984; Laakso *et al.*, 1986, 1987*b*; Stjärnkvist *et al.*, 1989) or hydroxy-ethylstarch (Artursson *et al.*, 1984) with glycidyl acrylate or esterification of malto-dextrin with acrylic chloride (Laakso & Sjöholm, 1987*a*), (2) homogenization of phases in a water oil emulsion (Artursson *et al.*, 1984; Laakso *et al.*, 1986; Laakso & Sjöholm, 1987*a*; Laakso *et al.*, 1987b; Stjärnkvist *et al.*, 1989), (3) polymerization (sometimes copolymerization) (Stjärnkvist *et al.*, 1989) and simultaneously intramolecular crosslinking of the water phase droplets to microparticles (Artursson *et al.*, 1984; Laakso *et al.*, 1986, 1987*b*; Laakso & Sjöholm, 1987*a*), (4) chemical activation of polyacryl starch microparticles (Laakso *et al.*, 1987*a*), and finally (5) covalent binding of peptide-drug conjugate onto the microparticles (Laakso *et al.*, 1987*b*). The last two points may be classified to be controversial.

The above papers systematically investigated the degree of acroylation of starch; degree of crosslinking and, consequently, the length of hydrocarbon crosslinks and the length of arm-spacer binding the peptide drug conjugate onto microparticles. The polyacryl starch microparticles might find a broad range of applications, particularly as encapsulation material for biocatalysts.

1.3.3 Other Naturally Occurring Polymers With and Without Chemical Modification

1.3.3.1 Dextran

Dextran was one of the first extracellular microbial polysaccharides to be produced commercially and it has been used in the pharmaceutical industry as a blood plasma volume expander. Introduction of crosslinked dextran gels in 1959 initiated the development of one of the most important purification methods for biological molecules. Dextran and its derivatives represent at present the basic polysaccharide matrices employed in a wide range of separation and purification techniques (Yalpani, 1986). In the 1960s dextran became attractive also as a drug carrier and protein carrier. As implied by its use as blood plasma volume expander, dextran exhibits a well documented low toxicity in human and animal organism (Yalpani, 1986). Early in the 1980s the idea appeared (Hirtenstein & Clark, 1983) and was developed (Nilsson, 1987*a*) of culturing anchorage dependent cells on small spheres (microcarriers) using the beaded ion-exchange gel, DEAE- Sephadex A-50 as a microcarrier (Hirtenstein & Clark, 1983).

As for starch, two modes of utilization were also elaborated for dextran as drug/enzyme carrier. In the first mode dextran or starch form defined and stable compounds and exhibit no significant toxicity or organ and cellular tropism. Dextran chemically coupled to the drug (Molteni, 1985) or the enzyme (Torchilin, 1983; Maksimenko *et al.*, 1987) is able to transfer the active substances in biologically unchanged form to the target site. Many drugs or enzymes conjugated to dextran exhibit improved chemical and biological stability as well as beneficial changes in their pharmacokinetics. Bonds between dextran and the drug (enzyme) may be labile (ester) and in this case the resulting conjugates act as slow-release systems (Molteni, 1985). On the other hand, when drugs, enzymes or other biologically active substances are linked to dextran by nonpolar bonds, their serum lifetimes are prolonged and their pharmaco-biological activity is increased so that the conjugate can reach selectively the system of reticuloendothelial tissues. In all these conjugates (either by esteric or nonpolar linkage), the polysaccharide carrier will improve the chemical and physico-chemical stability of the drug (Molteni, 1985).

In the second mode of immobilization crystallized dextran spheres (Fig. 1.11) are prepared for use as a matrix for entrapment of biologically active substances, which, when released, retain their biological activity. The method of preparation, using a high-pressure homogenization tech-

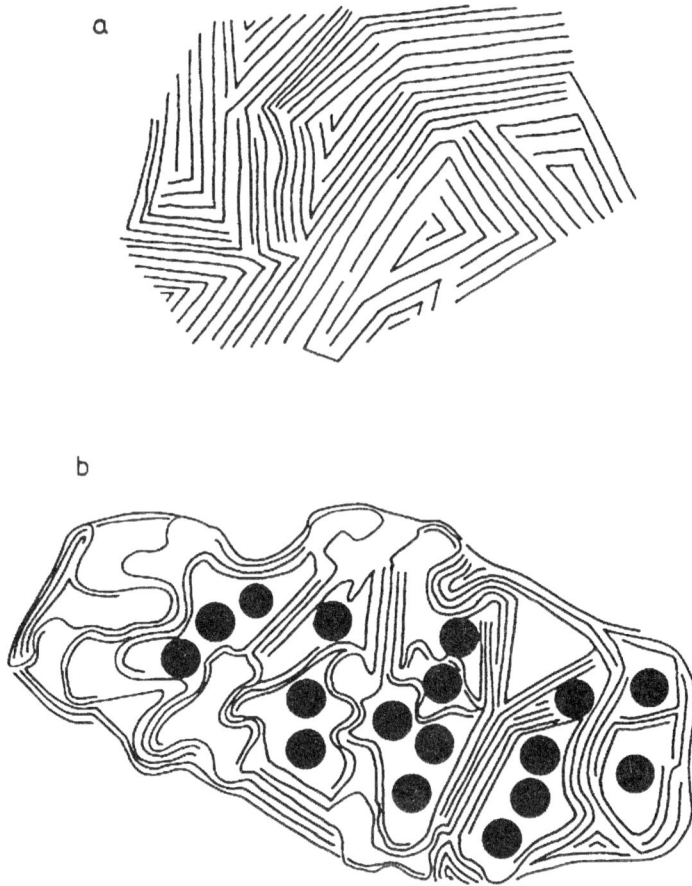

Fig. 1.11 (a) A schematic view of a perfect crystalline polymer matrix, in which the polymers are attached to each other via noncovalent bonds, such as hydrogen bonds, van der Waals forces, or in some cases ionic bonds. (b) A schematic view of a partially crystalline structure, in which drugs, enzymes/proteins, or magnetic particles are entrapped into the matrix (Schröder, 1985).

nique, is easy to scale up. With the use of commercially available devices, one could easily fabricate tons of crystallized dextran (or starch) spheres per day. Unsubstituted as well as substituted (DEAE and sulfate) dextrans with various molecular weights from 1×10^4 up to 50×10^4 are used (Schröder, 1985) for the purpose. A less common technique of noncovalent immobilization of biologically active substances is their encapsulation in biodegradable microspheres of polyacryldextran prepared by emulsion polymerization (Edman *et al.*, 1980).

Dextrans are basically linear water-soluble polysaccharide composed of (1→6)-linked α-D-glucopyranosyl units (Fig. 1.12), although they may frequently be branched.

Insoluble and porous supports based on crosslinked dextran and possessing molecular sieving properties were originally developed as supports for size-exclusion chromatography and are still commercially available as Sephadex®. Sephadex gels are prepared by crosslinking water-soluble dextran with epichlorohydrin. A broad range of well-defined water regain and molecular exclusion limits and consequently molecular sieving properties have been obtained by regulating the content of linear dextran and the degree of crosslinking. Because of these properties Sephadex gels have gained wide acceptance since the very beginning of development of enzyme immobilization (Goldman *et al.*, 1971; Zaborsky, 1973; Porath & Axén, 1976; Chibata, 1978; Kennedy, 1978; Manecke *et al.*, 1979; Sturgeon, 1982). The schematic structure of Sephadex is demonstrated in Fig. 1.13.

Sephadex gels have to be activated for use as supports of immobilized enzymes. One of procedures most commonly and extensively used is the cyanogen bromide activation method which yields a cyclic *trans*-2,3-imido-carbonate derivative of Sephadex (Porath & Axén, 1976). Some other covalent coupling methods of immobilization of enzymes involve the use of cyanuric chloride, benzoquinone, epichlorohydrin, chlorotriazine dyes and periodate for activation of hydroxyl groups (direct coupling), carbodiimides and hydrazines for carboxyl derivatives, and 2-pyridine disulfide for thiol derivatives (indirect coupling) (Goldman *et al.*, 1971; Zaborsky,

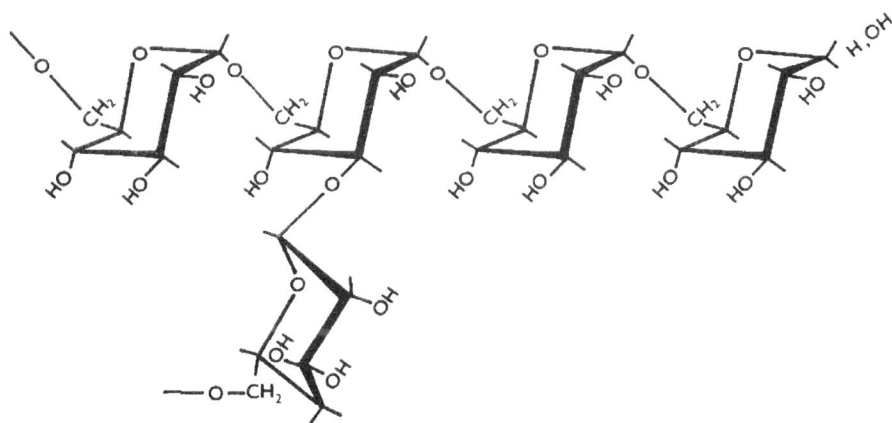

Fig. 1.12 Branched structure of the dextran (from *L. mesenteroides* B512-F) (Yalpani, 1988) based on about 95 per cent of 1,6-linked α-D-glucose in the main chain and side chains. The side chains are linked at the branching point by 1,3-linked α-D-glucose whereby most of the side chains are more than one glucose unit long.

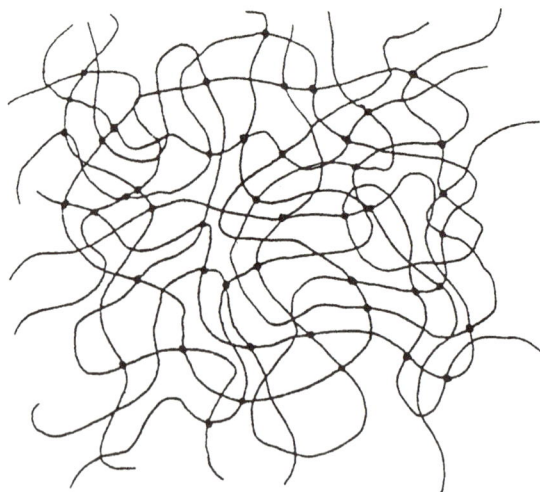

Fig. 1.13 A schematic representation of a Sephadex network formed from random chains (Arnott *et al.*, 1974).

1973; Porath & Axén, 1976; Chibata, 1978; Kennedy, 1978; Manecke *et al.*, 1979; Sturgeon, 1982; Kennedy & Cabral, 1987*a*). Ionic derivatives of Sephadex are also commercially available and they are used in enzyme coupling by ionic interactions (Goldman *et al.*, 1971; Zaborsky, 1973; Chibata, 1978). The scale of positively charged Sephadexes applied as microcarriers for tissue cultures can be further extended (Klein & Kressdorf, 1989). An actualized survey of activation methods applied to Sephadex together with the subsequent use in immobilization of enzymes and other macromolecules or biopolymers was published by Yalpani (1986) and Scouten (1987).

Yalpani (1986) divided the dextran-activating reactions to selective and non-selective ones. Most reactions used for activation of Sephadex are considered to be non-selective. To this group of reactions evidently belong also graft copolymerization reactions utilizing nonmodified Sephadex as matrix (D'Angiurio *et al.*, 1980*b*, 1982). In these cases vinylated derivatives of the enzyme (vinyl-enzyme) act as activation reagent.

Yalpani (1986) classifies among selective reactions oxidations (with aqueous bromine or dimethyl sulfoxide acetic anhydride), reducing end-group modifications (e.g. the treatment of the aldehyde group of the reducing end with sodium cyanide and subsequent hydrolysis of the resulting nitrite to corresponding carboxyl derivative), and other, miscellaneous modifications which rely on the limited presence of protected primary hydroxyl functions, in particular the selective modification of the nonreducing end groups via catalytic oxidation. This type of reaction yields a carboxylic acid function at C-6 or preferential acetylation at C-2 with acetic anhydride-pyridine (Yalpani, 1986).

Other aspects of the chemistry and application of dextran derivatives have already been described (Pitha, 1978; Bamford *et al.*, 1986; Melton *et al.*, 1987) and reviewed (Rogovin *et al.*, 1972; Schmid, 1979).

1.3.3.2   Algal Polysaccharides. Agar, Agarose, Alginates, Carrageenans
Algal (marine) polysaccharides are the most widely used supports for immobilization of biocatalysts. Agar, alginate and carrageenan are algal extracts; among them agar is the most familiar example of this group of gelling substances.

1.3.3.3   Agar
Agar was introduced as a support material for electrophoresis in the early 1960s and more recently it has been employed for the immobilization of bacterial cells and enzymes in the form of crosslinked gels (Porath & Axén, 1976). Later, in the early 1980s, agar gels were again successfully used particularly as thermo-hardening hydrogels. Attention was focused on entrapment of plant cells (Brodelius, 1983; Gisby *et al.*, 1987), chloroplasts and photosynthetic bacteria (Karube *et al.*, 1980; Smith *et al.*, 1982; Ikemoto & Mitsui, 1984) and later also to for co-immobilization of chloroplasts and hydrogenases (Ikemoto & Mitsui, 1984). In spite of these attractive applications agar is, however, better known as a raw material for agarose preparation than as a carrier for immobilization of biocatalysts.

Agar is a complex water-soluble polysaccharide naturally occurring as a mixture of at least two polysaccharides. The gelling agent among them is the agarose (Meer, 1980).

The basic principles of preparation of agar gels in bead form were elaborated by Porath & Axén (1976) and their co-workers. Older procedures based on thermal sol–gel transition (a thawing–cooling program) were enriched by new ones based on crosslinking of the agar gel with the aid of epichlorohydrin, divinyl sulfone, glutaryl dichloride and further bifunctional reagents of suitable chain length. 'Mixed gels' were created by incorporating other kinds of particles (e.g. silica, polyacrylamide) in the agar beads (Porath & Axén, 1976; Kuu & Polack, 1983).

The method for preparation of agar gel by thermal processing was further developed in detail by Brodelius (1983) and others (Kierstan & Coughlan, 1985). These authors proposed to process the gels obtained by cooling of a heated agar in two different ways, i.e. by moulding beads in a mould or by producing a 'gel-block' with subsequent mechanical disintegration into smaller particles (Brodelius, 1983). The latter method is the simpler one but it results in irregular particles of varying sizes. The former method, on the other hand, gives more uniform beads upon moulding of a heated suspension of agar and cells in a mould (Brodelius, 1983). Agar-entrapped biocatalysts (organelles) may also be prepared in the form of films subsequently cut into uniform pieces (Brodelius, 1983; Kierstan & Coughlan, 1985). Recently the resonance nozzle immobilization technique was tested also with agar as a representative of thermohardening hydrogels (Buitelaar *et al.*, 1988).

Crude agar is inexpensive material (Meer, 1980). It is probable that for industrial applications of immobilized biocatalysts extensive purification of agar will not be necessary. However, what will certainly be necessary is a further improvement of its mechanical properties. Extremely dense and hard agar gels were also prepared but this was done at the expense of decreased permeability (Porath & Axén, 1976).

Enzymes could be fixed only to the surface and the outermost layer of the beads of these extreme types of gels. It should also be pointed out, that agar (and agar polysaccharides) are, like synthetic polymers, resistant to microbial attack. The chemical activation of agar is extraordinarily efficient. Matrices with 1–1·5 mmoles of very reactive substituents (imidocarbonate, oxirane, vinylsulfonyl) per gram of dry substance can occasionally be obtained. Moreover, reactions used for derivatization of cellulose, starch, dextran, etc., may also be used for preparing derivatives of agar (Porath & Axén, 1976). Despite all its positive properties including its economic advantage, agar has not reached such a broad application in immobilization techniques as did its extremely purified form, the agarose.

1.3.3.4 *Agarose*

By far the most widely used matrix is bead agarose (Goldman *et al.*, 1971; Zaborsky, 1973; Porath & Axén, 1976; Chibata, 1978; Kennedy, 1978; Sturgeon, 1982; Kennedy & Cabral, 1987*a*; Scouten, 1987) which became predominant in the 1970s mainly because of its good availability. Besides that, a large number of partially prepared matrices, based on agarose, is commercially available (Scouten, 1987). Since the nonderivatized agarose already exhibits hydrophilicity and a minimum of nonspecific adsorptions, many researchers have expected that the same properties will persist if not be improved, also in derivatized agarose species.

Commercial bead agarose and its derivatives have been used exclusively for immobilization of enzymes. At the beginning of the eighties, some authors started to use agarose also for immobilization of cells and organelles (Karube *et al.*, 1980; Smith *et al.*, 1982; Brodelius, 1983; Ikemoto & Mitsui, 1984; Gisby *et al.*, 1987). The process was mostly a physical entrapment performed in the same way as in the case of agar.

Agarose is a linear polysaccharide with repeating agarobiose units which consists of alternating residues of $(1{\to}3)$ linked-β-D-galactose (A) and $(1{\to}4)$-linked 3,6-anhydro-α-L-galactose, i.e. the L-galactose anhydride (B) (Fig. 1.14). Agarose represents the basic gel-forming component of commercial agar. Other species present both in raw agar and in commercial preparations include compounds derived from the β-D-galactose by substitution in position 6, e.g. the 6-methyl ether and 6-sulfate, and the compounds obtained by substitution of anhydride in position 2 (the 2-sulphate derivative). Consequently, chains of agarose residues may deviate from the ideal repeating $(AB)_n$ structure (Fig. 1.14). They are believed to play an important role in the gelation mechanism of agarose by preventing perfect ordering and hence in precipitation of agarose chains during the sol–gel transformation (Clark & Ross-Murphy, 1987).

Agarose is obtained from natural agar by dispergation of a 6–10 per cent hot-water solution of agar in an organic solvent in the presence of a suitable emulsifier. Thermoreversible gelation of agarose occurs when hot agarose solutions are cooled down to 40°C. The resulting hydrocolloid particles have a structure consisting of a regular array of micron-sized, spherical, polymer-rich regions, thought to be formed by a nucleate-free phase separation (Meer, 1980; Kennedy & Cabral, 1983*a*; Clark & Ross-Murphy, 1987). The existence of a wide spectrum of junction zones is considered (Fig. 1.15) (Meer, 1980; Clark & Ross-Murphy, 1987).

Figure 1.16 shows a diagram of the proposed structure of the agarose network. Pentagonal pores are the essential feature of agarose supports.

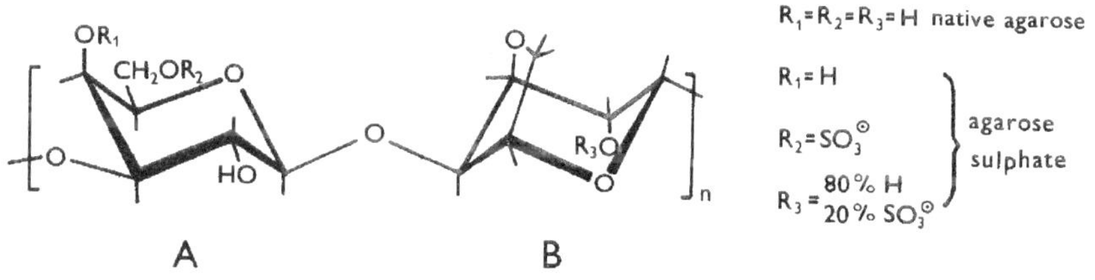

Fig. 1.14 Idealized AB repeat unit of agarose polymer based on 1,3-linked β-D-galactose residue (A) and 1,4-linked anhydro α-L-galactose residue (B). Possible patterns of substitution involving sulphate groups are also indicated (Clark & Ross-Murphy, 1987; Kennedy & Cabral, 1987*a*).

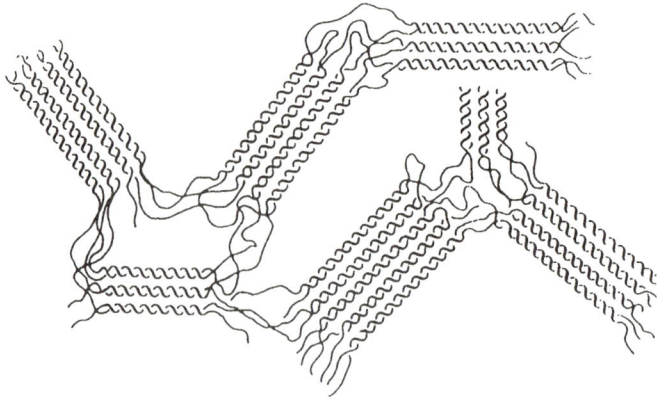

Fig. 1.15 Model for agarose network formation. Crosslinks involve both double helix formation and substantial association of double helices to form microcrystalline junction zones (Clark & Ross-Murphy, 1987).

These pores are large enough to be readily penetrated by a protein with a molecular mass of several millions, yet are strong enough to permit the shaping of agarose into definite spherical particles with good flow characteristics. However, the stability of the pores is dependent on hydrogen bond formation between the three strands of the triple helix of agarose chains (Scouten, 1981). Anything that is capable of disrupting these

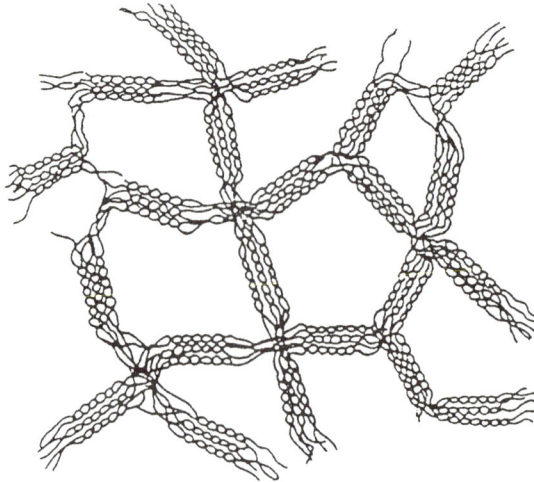

Fig. 1.16 Schematic representation of agarose gel network. The aggregates in agarose gel may actually contain $10–10^4$ helices rather than the smaller number shown here (Arnott *et al.*, 1974).

bonds will disrupt the entire network, and this will result in dissolving the soluble monomeric agarose. Thermal instability of agarose is caused by disruption of these hydrogen bonds. Hydrogen bonds may be disrupted also by urea, guanidine hydrochloride, chaotropic agents, and certain detergents which may be applied as eluants in a washing procedure. The strength of the bonds and, therefore, the porosity and size of the beads are altered by change in ionic strength (Scouten, 1981).

This peculiarity of agarose is normally not too severely limiting in practice. To produce an agarose derivative that is free from such susceptibility to agents disrupting hydrogen bonds, covalent crosslinks between various strands of the agarose network should be formed. Several procedures for crosslinking have been reported and commercial agaroses crosslinked with epichlorohydrin, divinyl sulfone and glutaryl dichloride are available. Despite their restricted porosity, the resulting crosslinked agarose beads are mechanically much stronger, have better flow characteristics, and are chemically more resistant, e.g. they can be employed with a far wider range of buffers and solvents (Porath & Axén, 1976; Scouten, 1981).

Commercial agarose beads contain up to 0·37 per cent sulfur; this indicates that a number of sulfate groups should be considered in this matrix. Removal of ionic groups by sodium borohydride is a reasonable pre-treatment. Chemical reduction of agarose beads may be provided with either $NaBH_4$ in alkaline solution or $LiAlH_4$ in dioxane or diethyl ether (Scouten, 1981).

Crosslinked agarose has to be activated for use as supports of immobilized enzymes. Although activation by cyanogen bromide is one of the most widely used methods and is associated with high coupling yields (Porath & Axén, 1976), its full complexity has been recognized only recently. A variety of inert or only slightly reactive by-products and intermediates may be obtained by cyanogen bromide treatment depending on the reaction conditions.

It has been shown that the activated products of Sepharose® are mainly imidocarbonates and inert carbamates, as well as some cyanate esters. It is noteworthy that the relative proportions of the activated species of Sepharose® differ substantially from those of Sephadex®. Upon coupling of enzyme to the activated polymer, the resulting products may be N-acylurea, imidocarbonates, or carbamate derivatives. Other potential problems of cyanogen bromide-activation include toxicity of the reagent, excess of reactive groups over the required amount, blocking after coupling, cross-linkages, high pH for coupling, introduction of ionic groups. For all the reasons outlined above the method is not ideal for large-scale applications (Yalpani, 1986). It should be noted that alternative cyanylating agents have recently been developed for Sepharose®, and these seem to be more advantageous in many respects (Kohn & Wilchek, 1984; Wilchek *et al.*, 1984; Miron & Wilchek, 1987).

In order to overcome some of the shortcomings of application of cyanylating agents, procedures using selective reagents with leaving-groups properties have been developed. 2,2,2-Tri-fluoroethanesulfonyl chloride (tresyl chloride) (Nilsson & Mosbach, 1987*b*) and its derivatives (Scouten *et al.*, 1986), pentafluorobenzene-sulfonyl chloride and its derivatives (Scouten *et al.*, 1986, 1987), and colored sulfonyl chlorides (Scouten *et al.*, 1987) appear to be the best activating agents among organic sulfonyl chlorides, but they are relatively expensive. Preserving excellent activating properties and leaving-group properties very similar to tresyl chloride, 2-fluoro-1-methyl pyridinium toluene-4-sulfonate (Ngo, 1986) and 1,1'-carbonyldiimidazole (Hearn, 1986, 1987) were proposed as relatively inexpensive reagents for activation of Sepharose. Because of the relatively low price, low level of ligand (enzyme) leakage and the possibility of scaling up these procedures to larger systems, the use of the last activating agents (Hearn, 1987) satisfies many of the criteria set in the basic literature (Goldman *et al.*, 1971; Zaborsky, 1973; Chibata, 1978; Kennedy & Cabral, 1983*a*, 1987*a*; Scouten, 1987).

Numerous monographs (Goldman *et al.*, 1971; Zaborsky, 1973; Porath & Axén, 1976; Chibata, 1978; Kennedy, 1978; Sturgeon, 1982; Kennedy & Cabral, 1983, 1987*a*; Scouten, 1987) deal with chemical activation of crosslinked agarose, particularly Sepharose®. Many of these activation methods are generally used with polysaccharides (Table 1.16). An overview of activation reactions (Manecke *et al.*, 1979) has also been published. Nevertheless, they fail to mention a procedure which utilizes non-modified polysaccharides including commercial crosslinked agarose (Sepharose®) for covalent coupling of enzymes. Last step in this procedure is graft copolymerization of polysaccharide with a vinyl derivative of the enzyme (vinyl-enzyme) (D'Angiuro *et al.*, 1980*a*, 1982).

In conclusion, agarose gels are mechanically stable and have greater pore size than other gels. Unlike other polysaccharides, agar and agarose are resistant to microbial degradation. Agar-degrading enzymes (agarases) have been found only in certain microorganisms living in seaweed (Kennedy & Cabral, 1983*a*). Despite their superior properties as macro-porous hydrophilic and nonadsorbing supports, agarose gels suffer from several disadvantages: they cannot be heat sterilized, and they disintegrate in strong alkaline solutions. Even at neutral pH values there is a possibility that they may be solubilized (Kennedy & Cabral, 1983*a*). It is also impossible to use agarose gels with many organic solvents because

Table 1.16
Selected Reagents to Activate Matrix Hydroxyl Functions

Activation method	Stability of bond	Group that reacts	Reagent toxicity	Activation time (hr)	Type of Bond
Trezyl chloride, sulfonyl chlorides	Excellent	Thiols, amines	Low	0·1–1·0	Secondary amines
Cyanogen bromide	Poor	Amines	High	0·1–0·4	Isourea of imido-carbonate
Bisoxiranes (epoxides)	Excellent	Thiols	Moderate	5–18	Secondary amine
Epichlorohydrin	Excellent	Thiols, amines	Moderate	2–24	Secondary amine
Triazine	Good	Amines, thiols, hydroxyls	High	0·5–2	Triazine-ether
Benzoquinone	Good	Amines	Low	1–2	Anilinyl
Divinylsulfone	Poor in base, fair at pH 7·0	Amines	High	0·5–2	Secondary amine
Glutaraldehyde	Excellent	Amine	Moderate	5–18	Secondary amine (?)
Phosphoryl chloride	Excellent	Amine	High	18	Amide
4-Nitrophenyl chloro-formate	Good	Amine	Moderate		Urethane
N-Hydroxysuccinimidyl chloroformate	Good	Amine	Moderate		Urethane
Carbonyl diimidazole	Good	Amine	Moderate		Urethane
1-Cyano-4-dimethyl-aminopyridinium tetrafluoro borate	Poor	Amines	Moderate	20 min	Isourea
2-Fluoro-3-methylpyridi-nium tosylate	Excellent	Amine/thiol	Moderate		Secondary amine

Taken from Scouten (1987).

their structure changes drastically and irreversibly under such conditions (Scouten, 1981).

1.3.3.5 Alginates
Algin and alginates have been known for 60 years in food and other industries because of their unique colloidal behavior and their ability to thicken, stabilize, emulsify, suspend, form films, and produce gels (Cotrell & Kovacs, 1980). It is supposed that alginate-type systems may be used as safe, non-toxic, readily available media or transporting agents which are suitable even for the administration of drugs (Johnson & Medlen, 1985) in forms securing, e.g. controlled drug release (Stockwell *et al.*, 1986). Along with the expected medical use, calcium alginate gel

systems attract interest as media for the controlled release of herbicides (Connick, 1982) and pesticides (Scher, 1977).

Alginates were probably the first polymers utilized for entrapment of biocatalysts by ionotropic gel formation or polymerization of polyelectrolytes by multivalent ions (Kierstan & Bucke, 1977). Subsequently, hydrocolloidal gels became applied more frequently than any other materials for physical entrapment of biocatalysts by ionic network formation. With increasing use of hydrocolloidal gels, alginate came to be one of the most favored ones (Hulst & Tramper, 1989). With the development of co-immobilization technique (Hahn-Hägerdal, 1983, 1988) this material became more and more frequently used also in enzyme immobilization (Hartmeier, 1985).

Immobilization in alginate gels is a safe, fast, mild, simple, cheap and versatile technique which may be applied in a wide range of biocatalysts (Scher, 1977; Birnbaum *et al.*, 1983; Bucke, 1987). Most satisfactory results may be obtained by using comparatively low concentrations of high-molecular-weight grade sodium alginate. That is what makes alginate gels commercially so advantageous because it minimizes the costs of alginate, which is sold by weight (Bucke, 1987).

However, this mild, simple and rapid technique has some disadvantages. The destruction of calcium alginate gels by high concentrations of K^+ and Mg^{2+} ions, phosphate and chelating agents as well as the tendency of beads to swell in the presence of monovalent cations are problems that may be overcome very simply but in doing so the cost of the immobilization process increases. Limitation of oxygen transfer within the gels may be a problem or an advantage depending on the type of biocatalyst entrapped. Entrapment using calcium alginate gel is inapplicable to most enzymes since the pore size of the gel is so large that enzymes leak out the gel network (Scher, 1977; Bucke, 1987).

Molecular properties. Alginates are salts of alginic acid and occur as intracellular material in brown algae. In molecular terms, alginic acid is composed of two distinct types of monosaccharides, i.e. 1,4-linked β-D-mannuronic acid (M) and 1,4-linked α-L-guluronic acid (G). These residues are present in varying proportions depending on the source of the alginic acid. It turned out that alginic acid (and its salts) is a block co-polymer, containing both MM... and GG... homopolymer blocks and mixed blocks containing irregular sequences of M and G units. The M and G residues and their modes of linkage are shown in Fig. 1.17.

The M/G ratios in alginic acids vary from source to source and even within a single alga. However, the majority of laboratory chemical sup-

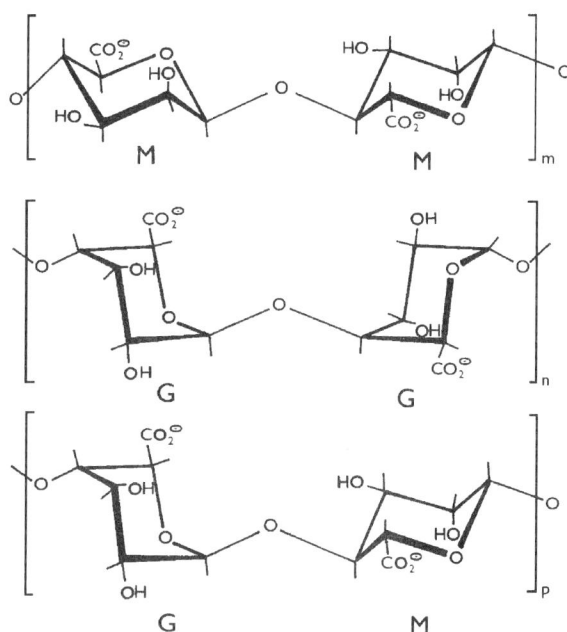

Fig. 1.17 Structures of mannuronate, guluronate and mixed blocks constituting alginate polymers. Mannuronate blocks are based on 1,4-linked β-D-mannuronic acid (M); guluronate blocks on 1,4-linked α-L-guluronic acid (G). Mixed blocks are linked as shown (Clark & Ross-Murphy, 1987; Kennedy & Cabral, 1987*a*).

pliers offer alginic acid and sodium alginate without mentioning the source or specifying their product in more detail. Hence, most reports on the use of alginate as immobilization material have been based upon experiments with uncharacterized commercial samples.

Structural description of alginates in purely sequential terms is clearly difficult to specify statistically. Besides specification of the M/G ratio in a particular case, there have to be considered also details concerning block length and block arrangement of the material (Clark & Ross-Murphy, 1987). Nevertheless, ^{13}C and ^{1}H NMR spectroscopy provides a possibility for sequential analysis particularly via measurement of monad, diad, triad and higher order frequencies (Grasdalen *et al.*, 1981; Skjåk-Braek *et al.*, 1986). This method was used, e.g., in investigation of tailoring of alginates by enzymatic modification *in vitro*. Alginates with a broad variety of initial composition and sequential structure were modified with a mannuronan-C-5 epimerase. Polymers with high content of guluronic acid and, accordingly, with enhanced ability to form gels with calcium ions, were obtained in this way (Skjåk-Braek *et al.*, 1986).

Gelling of alginates. Alginates do not form thermoreversible gels. Network formation is induced in solutions of a sodium salt when divalent cations such as Ca^{2+} are introduced. Replacement of Na^+ by Ca^{2+} ions leads to the formation of stiff, brittle (often turbid) gels whose rheological character and turbidity depend on the type of the alginate (e.g., the M/G ratio) and the mode of introduction of the divalent ions. Ca^{2+} ions may be introduced by diffusion, using dialysis (Skjåk-Braek *et al.,* 1989) or *in situ* by adding inorganic calcium salts of low solubility and simultaneously controlling carefully the pH value (Clark & Ross-Murphy, 1987; Flink & Johansen, 1985). Smidsrød *et al.* (1972, 1974) found that the type of divalent cation had an influence on gel strength. It was established that, according to their capability to produce strong gels, Group II cations may be ordered in the following sequence: $Ba^{2+} > Sr^{2+} > Ca^{2+} \gg Mg^{2+}$.

Final gel properties are very much dependent on the mode of calcium addition to an alginate solution. If calcium is added too rapidly, the result is spot gelation and a discontinuous gel structure. The rate of calcium addition can be controlled by using a slow-dissolving calcium salt, e.g. calcium citrate (Flink & Johansen, 1985) or dicalcium phosphate and/or by the addition of a sequestrant, such as tetrasodium pyrophosphate (Cotrell & Kovacs, 1980) or an acidifier such as glucono-δ-lactone (Flink & Johansen, 1985).

Calcium alginate gels formed by diffusion of calcium ions into solutions of sodium alginate exhibit various degrees of inhomogeneity, in the sense that the polymer concentration is much higher at the surface than in the center of the gels (Skjåk-Braek *et al.*, 1989). Parameters controlling the non-uniform distribution of the polymer in alginate gel include: concentration and molecular size of the polymer and the concentration in the outer reservoir of ion-inducing gel formation. L-Guluronate residues, if present in a large amount, slightly increase the inhomogeneity, whereas the presence of non-gelling cations strongly decreases it (Skjåk-Braek *et al.*, 1989).

Calcium alginate gels are prepared usually in the form of beads. There are many experimental details the modification of which may optimize both the bead structure and the bead formation process (Cotrell & Kovacs, 1980; Birnbaum *et al.*, 1983; Bucke, 1987). Less frequent are fibrous and sheet forms which can be achieved by vet spinning of alginate directly in aqueous solutions (Kobayashi *et al.*, 1986).

Divalent cation-induced gelling of alginates in solution reflects their specific ion binding accompanied by a conformational change. It appears that calcium ions bind preferentially to guluronate (G) blocks (Fig. 1.18). This process has been shown to be highly cooperative if the number of guluronate residues exceeds 20. No evidence for a similar high specificity

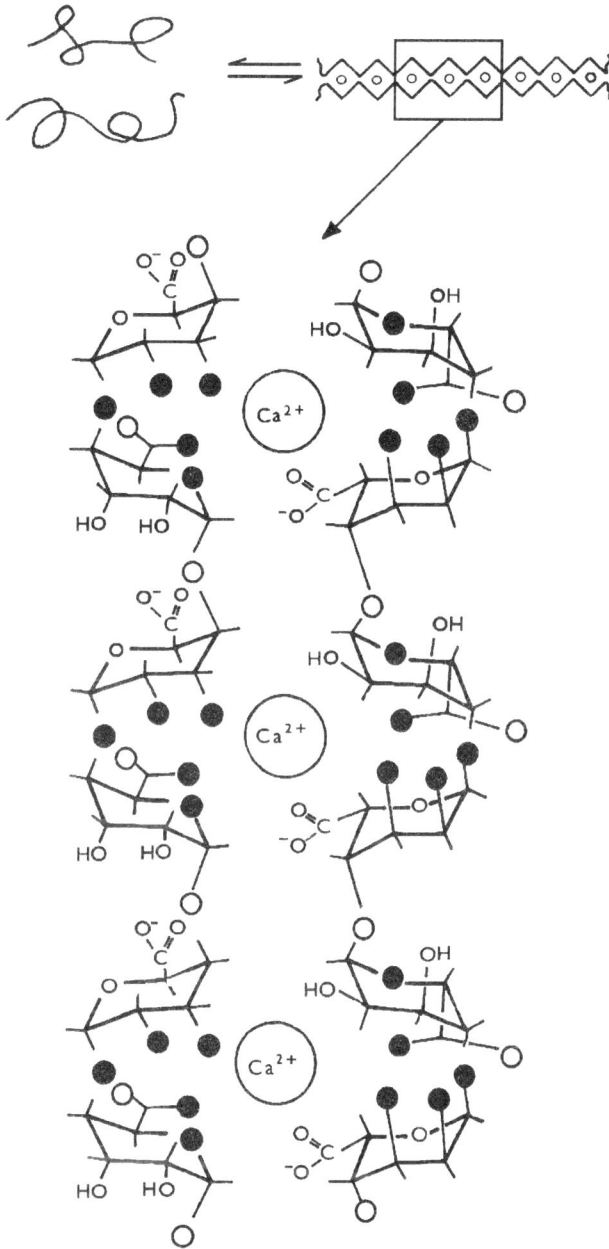

Fig. 1.18 'Egg-box' model for Ca^{2+} induced dimerization of poly-L-guluronate. Oxygen atoms involved in chelation are shown as filled circles (Morris, 1986).

or cooperativity has been found in the case of mannuronate, which apparently exhibits standard polyelectrolyte behaviour (Clark & Ross-Murphy, 1987).

Thus, guluronate sequences in alginates associate into matched crystalline aggregates in a two-fold ribbon-like form. Ions cooperatively bound during the process are placed inside the electronegative cavities, like eggs in an egg-box. It is assumed that the association proceeds as a side-by-side association of two chains only (Fig. 1.19). Higher aggregates, it appears, are likely to occur only under forcing conditions of high Ca^{2+} concentration (Clark & Ross-Murphy, 1987). Thus, calcium alginate gel appears as an inert three-dimensional polymer network with relatively large interconnected interstitial spaces. The dimensions of these spaces will vary with the type of alginate used.

Many grades of sodium alginates selected for different tasks are commercially available. High L-guluronate content is required for gel

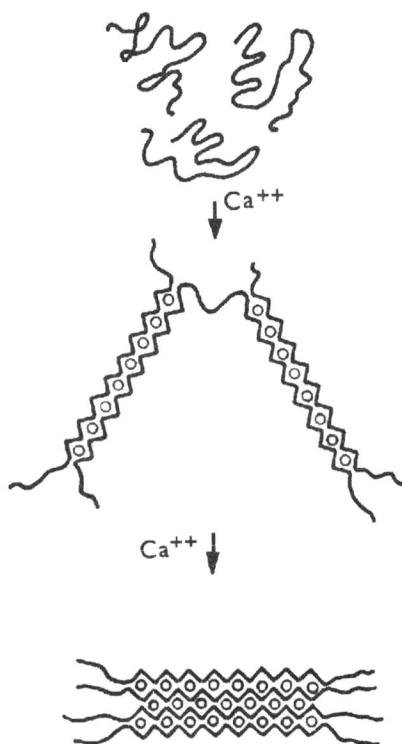

Fig. 1.19 Model for alginate network formation proposed as a two-stage process involving (i) initial dimerization (corresponding to the 'strong association') and (ii) subsequent Ca^{2+} induced aggregation of these preformed dimers (Clark & Ross-Murphy, 1987).

strength. The maximum content is 70 per cent (Bucke, 1987). That is why the rheological consequences of network formation have been explored. Based upon special measurements Smidsrød (Smidsrød & Haug, 1972; Smidsrød, 1974) reported that the rigidity of alginate gels may be considered as a function of polymer concentration, molecular weight and M/G ratio. Systems rich in mannuronate gave very weak gels of high turbidity, while gels rich in guluronate were rigid and much less opaque. The guluronate gels were found to be very brittle, while gels containing an increased mannuronate content showed greater extensibility. Increased deformability in this case could be ascribed to the presence of increasing amounts of MM and MG sequences between junction zones (Clark & Ross-Murphy, 1987; Smidsrød & Haug, 1972; Smidsrød, 1974).

Since the gelling properties of alginate, vital to its use as immobilization material, depend strongly upon its monomeric composition, sequential arrangement, and the lengths of the G-blocks, a careful selection of starting material is necessary to secure the best reproducible results.

Martinsen *et al.* (1989) used a wide range of chemically well characterized alginates to study how the mechanical and swelling properties of calcium alginate beads depend upon the intrinsic properties of alginic polymers, such as monomeric composition, block-structure and molecular size. They found that the beads with the highest mechanical strength, lowest shrinkage, best stability towards monovalent cations, and highest porosity were made from alginate with a content of L-guluronic acid higher than 70 per cent and an average length of the G-blocks exceeding 15. For 'high G' alginates with molecular weights over 2.4×10^5 it follows that their gel strength will be independent of the molecular weight (Martinsen *et al.*, 1989). Smidsrød (Smidsrød & Haug, 1972; Smidsrød, 1974) found also that with other characteristics being equal, the gel rigidity modulus rises sharply with increasing molecular weight until DP reaches roughly 500, then it becomes almost constant.

Physical properties of calcium alginate gels. Typical range of size of calcium alginate beads represents the diameter of 0·5–3·5 mm. Klein *et al.* (1983) proposed that the size of the particles is primarily determined by surface tension of the alginate solution and not by inner diameter of the needle. They introduced various ways to prepare smaller particles. The most effective method appeared to be that with a controlled stream of compressed air concentric to the extrusion die. Simultaneously, an unspecified surfactant was used to reduce surface tension. This technique produced particles with diameters as small as 10 μm. Controlled drying is another way to achieve a decrease in particle size (Klein *et al.*, 1979a, 1983).

It may be concluded that the above discussed techniques for immobilization of a biocatalyst into alginate gel beads make possible the size control of particles. A more perfect method for preparation of monodispersed biocatalyst beads is the resonance nozzle immobilization technique (Hulst *et al.*, 1985; Woodward, 1988). This technique consists of breaking up a jet of the cell/presupport mixture in uniform droplets by means of mechanical vibration (Hulst *et al.*, 1985). The production capacity of the resonance nozzle technique is about two orders of magnitude larger than that of the dripping technique (Hulst *et al.*, 1985). A perfection of the latter technique permits the preparation of gel beads varying but little (<10%) in their diameter (Woodward, 1988). Also the original dripping technique was improved (Rehg *et al.*, 1986) and adapted (Matulovic *et al.*, 1986) to reach bead size control (e.g. by application of an air stream along the needles) and to promote the scale-up of the immobilization technique with alginate.

Examination of mechanical properties of hydrogels such as calcium alginate gel (CAG) beads should establish definitely the resistances of gels to both compression and fracture. The compression behavior of single catalyst particles (Klein & Wagner, 1978; Klein & Waushausen, 1979*b*) and columns packed with beads of CAG (Cheetham, 1979) was studied. Experiments showed that critical compression force(s) (CCF) for bead rupture may be optimized and brought up to levels of typical industrial polymers such as poly(styrene)/divinylbenzene ion-exchange resins (Table 1.17). The CCF value of CAG beads may be increased by a value exceeding 10 N can readily be expected (Klein & Waushausen, 1979*b*). Dry deformation of CAG beads is plastic (D<0·1) and the complete deformation due to drying is small (Klein & Waushausen, 1979b).

Columns packed with ductile beads of CAG are compressed comparatively easily, and display creep effects, but they are relatively resistant to

Table 1.17
Critical force (in Relative Units) for the Rupture of a Single Polymeric Bead in a Compression Experiment

Polymer	Force
Polyacrylamide	10
PMAAm	20
Carrageenan	10
Alginate	100–800
Epoxide	600–1000
Polystyrene/divinylbenzene resin	≤900

All the data were taken from Klein & Wagner (1978).

fragmentation. In contrast, more rigid materials such as glass or ceramic beads or many ion exchange resins are more difficult to compress, creep effects are very small, but their fragmentation is comparatively easy (Cheetham, 1979). Columns of CAG pellets containing entrapped microbial cells are resistant to compression over long periods. Compressibility rises with decrease in the alginate concentration used to form the pellets and with increases in temperature and the concentration of cells or other particulate materials. Nevertheless, the compressibility appears to be unaffected by relative dimensions of the column. Pellets were not fractured unless very high pressures were used. Deformations were only partially reversible (Cheetham, 1979). When the pressure applied to the column increases, the creep ate will increase too but it can be again decreased by pumping fluid up the column.

In the form of small beads, CAG was reported to have a relatively low resistance to substrate diffusion (Tanaka *et al.*, 1984*b*; Scott *et al.*, 1989). Despite this, the diffusion of reactants, especially large molecules, into the inner sphere of the gels was generally assumed to be a rate-limiting step. Therefore, gel matrices having either a small size or particular shapes have been designed. Among the forms suitable in this respect, the fine fibrous structure seems to be of particular importance. This form may be easily achieved by wet spinning of an aqueous admixture of sodium alginate and the biocatalyst with the aid of a divalent metallic ion solution used as coagulation bath (Kobayashi *et al.*, 1987). The resulting short-cut fibers may be transformed into paper sheets in an additional procedure (Kobayashi *et al.*, 1987).

Gel hardening. When a risk of disrupting immobilized cells appears on supplying media or substrate solutions which contain materials capable of chelating calcium, the beads may be made resistant to disruption. This procedure has become known as 'hardening' although in consequence of its application the beads do not necessarily become physically harder (Bucke, 1987). Physical treatment or chemical modifications involving graft copolymerization and procedures combining both approaches are applied as hardening technique. Materials for hardening include besides alginates also the derivatives of alginic acid (e.g. propylene glycol ester), as well as further polymeric polyelectrolytes, e.g. poly(ethyleneimine), (PEI), or composite materials (colloidal silica).

The mechanical stability of CAG beads is not too high, but it can be considerably increased by promoting the process of partial drying. An essential feature of this shrinking process is that no reswelling of the small beads occurs (Klein *et al.*, 1979*a*).

Substitution of strontium or barium for calcium results in restricted application (because the resulting gels are probably not acceptable for use in production of food products). Both strontium and barium alginate are more resistant to chelation than calcium alginate.

Fukushima *et al.* (1988) succeeded in preparing a hard and flexible gel by using a mixture of alginate and colloidal silica (AS gel). Gel matrix of the alginate is formed on the surface of the droplet and the colloidal silica inside the droplet cannot disperse outside. This results in the formation of a gel matrix of silica along with that of alginate. The physical strength of AS gel increases with the amount of colloidal silica added. No difference could be recognized in the diffusion coefficient of glucose between AS and Al (alginate) gels. AS gels had the ability to retain proteins such as bovine serum albumin and γ-globulin (Fukushima *et al.*, 1988).

The CAG beads may be hardened using polymers possessing a branched structure. CAG beads coated with polymers such as PEI or poly(propylene) have, at least, higher resistance to phosphate ions (Veliky & Williams, 1981; Joung *et al.*, 1987) and even to mechanical compression. It is advisable to use polymeric polyelectrolyte, e.g. PEI, as a polymer for hardening. Polymers of this kind not only crosslink intramolecularly the negatively charged carbohydrate groups of the alginate but they also crosslink the latter groups with that on cell surface (Bucke, 1987). Further hardening with glutaraldehyde liquidates, however, the viability of the cells (Birnbaum *et al.*, 1982).

Superior results were obtained with a procedure in which periodate-oxidized sodium alginate was mixed with cells and extruded into calcium chloride solution and, subsequently, crosslinked with PEI (Birnbaum *et al.*, 1982). The same authors described also a stabilization with carbodiimide and PEI and obtained a preparation very stable in sodium phosphate and highly active but expensive for use on a large scale.

Immobilization using CAG is inapplicable to most enzymes since the pore size of the gel is so large that enzymes leak out from the support. With CAG beads coated with PEI and partially quaternized PEI (QPEI) Tanaka *et al.* (1984a) found that QPEI-coated beads have excellent features for immobilization of glucoamylase ($M_W = 9.7 \times 10^4$). However, in a short period (~ 10 h) of repeated incubations of the entrapped enzyme, its initial activity decreased considerably, mainly because of shrinkage of the gel. To stabilize a loosely networked structure of the gel Kokufuta *et al.* (1988) applied both an additional polyanion (potassium poly(vinylalcohol sulfate) in the mixture with sodium alginate and a polycation (trimethylammonium glycol chitosan iodide) in the precipitation bath containing calcium chloride. The resulting CAG was stabilized with a polyelectrolyte complex (PEC) consisting of both the synthetic polymeric polyanion and the polycation. It was found (Kokufuta *et al.*, 1988) that the system consisting of β-amylase entrapped in PEC-stabilized calcium alginate gel remained stable without serious loss in activity or a large leakage of enzyme from the support.

Gels of alginate have also been prepared by using a polycation instead of calcium. Gels produced by mixing propylene glycol ester of alginic acid with PEI were found to be strong and thermostable but brittle (Mohamed & Salleh, 1982). They were superior to beads produced by substituting gelatin for PEI (Salleh & Mohamed, 1982). Propyleneglycol alginate (PGA) applied in the above experiments is the only organic derivative of alginic acid currently available on the market. PGA has improved acid stability and resists precipitation by calcium and other polyvalent metal ions (Cotrell & Kovacs, 1980).

Application of PGA with low DS between 5 and 8·5 per cent of alginate (up to a final concentration of 10 per cent) does not change the mechanical stability of chitosan-alginate capsules (Daly & Knorr, 1988). Increase of esterification above 30 per cent resulted in capsules that were very fragile and could not be handled. No capsules were formed when more than 70 per cent of the carboxyl groups were esterified.

The effectiveness of immobilization of biocatalysts, particularly that of enzymes, could be increased considerably by using alginate as polymer for graft copolymerization. Thus copolymerization of alginate with acrylamide yielded a copolymer with high density of both the free carboxyl groups originating from alginate and the amide groups originating from poly(acrylamide). Conversion of these two functional groups via hydrazide into acyl azide provided a carrier with high binding capacity towards enzymes. The carrier activated in this way was capable of binding enzymes via amide bonds (Kumaraswamy *et al.*, 1981). Graft copolymerization of alginate with hydroxyethyl methacrylate yields a hybrid material with interesting capsule-forming properties (Stevenson & Sefton, 1987).

Graft copolymerization is one of the possibilities for modifiying the hydrophobic nature of alginate gels. There are doubts, however, whether a hydrophobic matrix is really an advantage, when water is the only solvent (Klein, 1988*b*).

Other aspects of design of matrix alginate gel type have also been systematically discussed (Tampion & Tampion, 1987). One of these aspects is, e.g. the structural contribution of alginate (primary structure and size of the soluble precursors) in the construction and operation of a biotechnological system (Klein, 1988*a,b*).

1.3.3.6   Carrageenans

Carrageenans are used commercially as thickening, suspending, and gelling agents. Typical applications are: thickener or 'binder' in toothpaste; suspending agent for cocoa in chocolate milk; gelling agent for milk puddings, water-gel desserts, air-freshener gels (Guiseley *et al.*, 1980), etc. At the end of the 1970s, carrageenans became attractive as carriers for immobilized cells (Takata *et al.*, 1977) and they remain attractive up to now (Brodelius, 1983; Chibata *et al.*, 1987; Kennedy & Cabral, 1987*a*; Linsefors & Brodelius, 1987; Scott, 1987; Tampion & Tampion, 1987; Woodward, 1988). The application of carrageenans for immobilization of enzymes (Tosa *et al.*, 1979) is, however, rare. Carrageenans as carriers for biocatalysts are not used so frequently as alginates.

Carrageenans have properties very similar to agarose and alginate, which render them a suitable matrix for immobilization of biocatalysts. Of the three known types of carrageenans, lambda, kappa and iota, only the last two are suitable as a support for biocatalyst immobilization; the κ-form is considered to be most suitable for immobilization of cells. This concerns particularly κ-carrageenan-immobilized microbial cells where remarkable catalytic activities of multienzyme systems were achieved (Chibata *et al.*, 1987). This capability of κ-carrageenan-immobilized microbial cells remains also outstanding in comparison to several polymers that form gel lattice suitable for entrapping microbial cells.

Gel formation. Like agar, κ-carrageenan may be brought to gel consistency by cooling. Gelation may also occur by contact with a solution containing gel-inducing reagents such as K^+, NH_4^+, Ca^{2+}, Cu^{2+}, Mg^{2+}, Fe^{3+}, amines, and water-miscible organic solvents. Both procedures, cooling and/or contact with an aqueous solution containing K^+ or NH_4^+ are very easy to carry out. Further, these procedures are very mild and immobilized preparations have a high catalytic activity (Chibata *et al.*, 1987).

Carrageenans which form hydrocolloidal gels are doing so because of double-helix formation (Fig. 1.20). At temperatures over the melting point of the gel thermal agitation overcomes the tendency to form helices, and the polymer exists in solution as a random coil. On cooling, a three dimensional polymer network builds up with double helices forming junction zones between the polymer chains (gel I). Further cooling leads to aggregation of these junction zones (gel II) (Rees, 1972; Guiseley *et al.*, 1980). The effect of sulfation on gelling properties can be explained sterically also on the basis of the double-helix secondary structure (Rees, 1972).

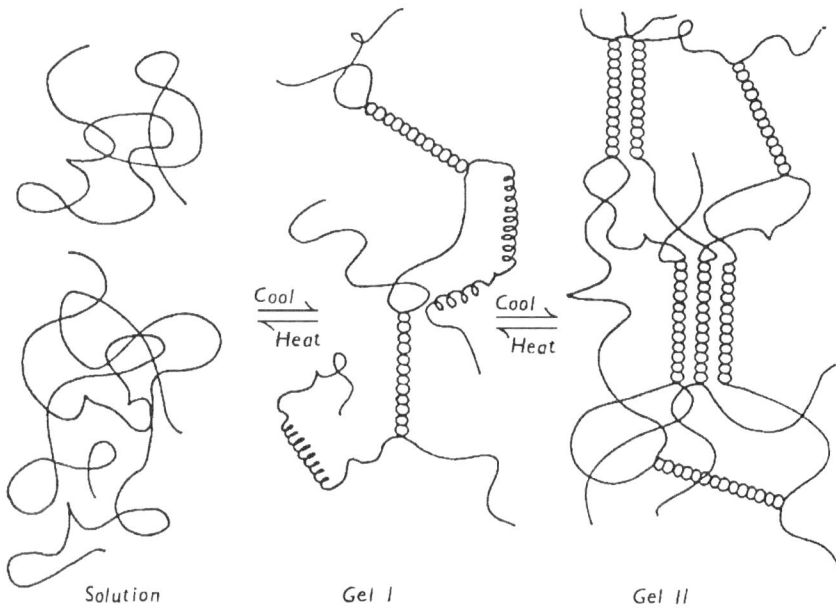

Fig. 1.20 Sol-gel transformation of kappa- and iota-carrageenans (Guiseley *et al.*, 1980).

In practice, the above mechanism of interchain association seems to cease after the formation of small soluble clusters (or domains) typically consisting of about 10 chains, with exchange of partners at 'kinking' residues (Morris, 1986). Generally, the higher the 3,6-anhydro-D-galactose (3,6-AG) content and the lower the ester sulfate level (Fig. 1.21), the less readily will the carrageenan hydrate (Tye, 1989).

Further crosslinking of these domains into a cohesive gel structure involves side-by-side association of the double helices. In the case of highly charged carrageenans, such as λ- or κ-, helix-helix aggregation and consequent gelation may occur only in the presence of cations (typically K^+ or Ca^{2+}). Cations are capable of suppressing electrostatic repulsions between the participating chains by being packed within the aggregate structure (Morris, 1986) (Fig. 1.22).

Physical properties of carrageenan gels. Several methods for entrapment of a biocatalyst in the carrageenan gel were outlined. These methods are mostly valid also for other gelling polysaccharides, such as alginate or agar, and provide entrapped biocatalysts in gel bead form consisting of beads 2 to 5 mm in diameter (Brodelius, 1983).

R = OH   : κ – Carrageenan
R = OSO₃⁻ : ι – Carrageenan

λ – Carrageenan

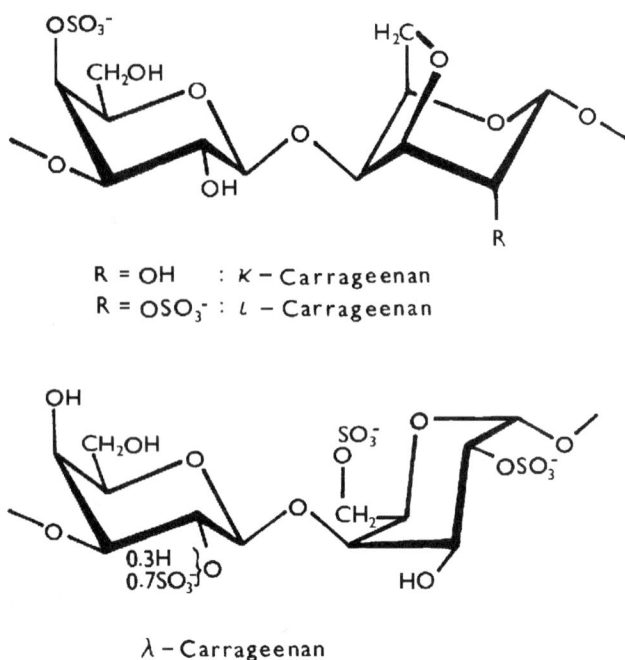

Fig. 1.21 Idealized AB repeat unit of kappa- and iota-polymers based on 1,3-linked β-galactose residue (A) and 1,4-linked 3,6-anhydro-α-D-galactose residue (B) (Clark & Ross-Murphy, 1987). The sequence is broken occasionally by residues of general type B' (Clark & Ross-Murphy, 1987). Repeat unit of lambda-polymer is represented below (Kennedy & Cabral, 1987a).

Concerning the mechanical properties, the control of geometry, beads uniformity and the diameter particles is highly recommended. The data in Table 1.17 represent relative values of the mechanical stability in compression experiment designed by Klein and Wagner (1978). The relative values indicated in Table 1.17 represent the compression forces required for bead rupture. According to the values presented in Table 1.17 carrageenan gel beads together with polyacrylamide beads are the most fragile among all beads under study. Of all carriers which are used for entrapment of cells, only the epoxy beads and the CAG beads (Klein & Wagner, 1978) reach mechanical strength typical for industrial polymers such as poly(styrene)/divinylbenzene ion-exchange resins, etc.

Further data concerning the mechanical stability of gels may be obtained from potassium carrageenan gel in the form of 2-cm cubes. Compression and shear at a constant rate of force application as well as at varying deformations were measured. In comparison to calcium alginate

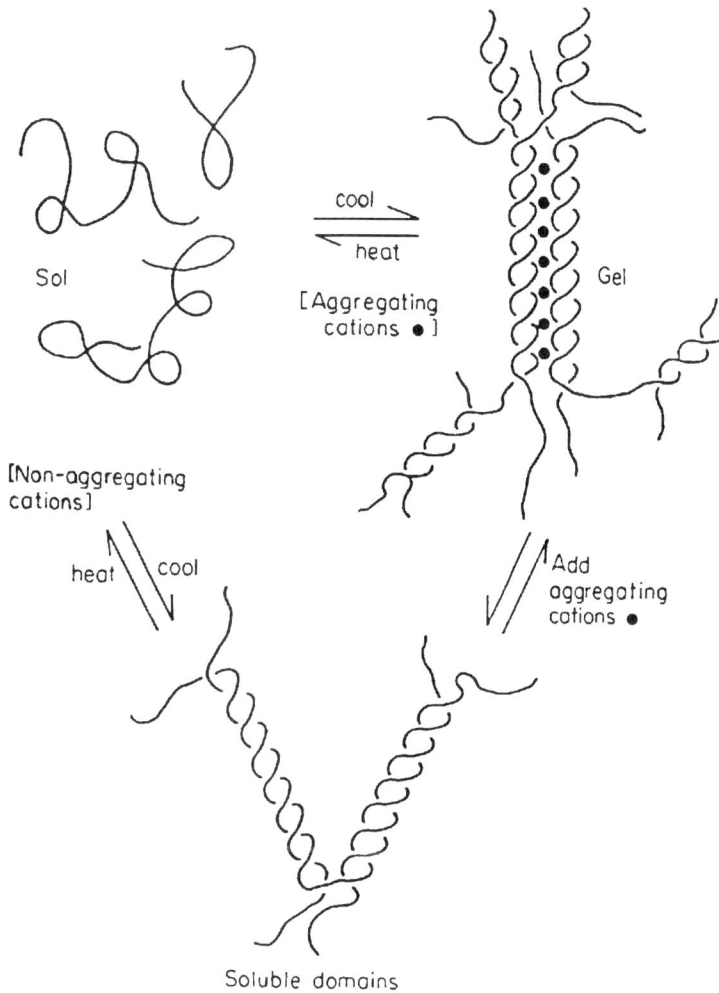

Fig. 1.22 'Domain model' of carrageenan gelation (Morris, 1986).

and gelatin, κ-carrageenan provided the least firm and elastic gel which decomposed relatively soon into a large number of small pieces (Munoz *et al.*, 1986). Fiszman *et al.* (1985) reported about a method for improvement of two parameters (maximum rupture strength F_{max} and apparent Young's modulus E_{app}) of compression behaviour of κ-carrageenan-based gels. Addition of sucrose (up to 30 per cent) increased both, F_{max} and E_{app}, while the effect of cellulose applied as additive, proved to be dependent on concentration of other components.

The mechanical strength of spherical beads of κ-carrageenan containing entrapped cells was also improved by treatment with amines (Chao *et al.*, 1986). The two-step procedure applied in preparation of cured beaded biocatalyst resembled the suspension-polymerization. Results of a critical compression test indicate that both linear and branched PEI are suitable curing agents (Table 1.18). In addition, PEI-treated carrageenan gel beads exhibited superior resistance also to heat and abrasion.

As concerns characteristics of various solutes to (Nguyen & Luong, 1986; Scott *et al.*, 1989) and from (Nguyen & Luong, 1986) spherical κ-carrageenan gel beads, tests have been performed with biocatalyst beads over a large range of bead diameters, gel concentrations, concentration of biocatalyst and other additives, as well as molecular weights of the solutes. At low gel concentration with entrapped biocatalyst a low-molecular-weight solute will have a diffusion coefficient

Table 1.18
Physical Property Tests of Amine-Cured κ-Carrageenan Gel[a]

Gel-curing agent[b]	Critical compression force, lb	Thermostability[c] (60°C/121°C)
KCl (control)	4·9	−
1,6-Hexanediamine	6·1	−
Hexamethylenetetramine	1·5	−
Triethylenetetramine	19·2	−
Trifunctional aziridines;		
XAMA-2	8·2	+
XAMA-7	8·9	+ (at pH 4 only)
Linear polyethylenimine (PEI)		
(MW 150·000)	10·7	+ (at pH 4, 7, 9)
Branched PEI		
CORCAT P-12 (MW 1200)	22·9	+
CORCAT P-150 (MW 10·000)	23·1	+
CORCAT P-600 (MW 60·000)	18·6	+
Aldrich (MW 50·000–100·000)	16·6	+ (at pH 4, 7, 9)

[a]Testing samples, containing 3·7 per cent of κ-carrageenan gel and 2 per cent of yeast cells, were prepared in both disk form (4·5 × 11 mm) and bead form (1–2 mm), and cured by soaking in the amine solution at 25°C for 1 h. After washing, the gel disks were measured for the critical compression force and the gel beads were tested for thermal stability of 60°C (16 h) and 121°C (20 min) in 0·1 M buffer solution of pH 4, 7, and 9, respectively.
[b]5 per cent aqueous solution of neutralized with 6N HCl.
[c] −, turbid supernatant; +, clear supernatant.
All the data were taken from Chao *et al.* (1976).

approaching that measured in water (Nguyen & Luong, 1986). With increasing gel concentration and especially with high loading of micro-organisms and other additives the diffusion coefficients are significantly decreased (Nguyen & Luong, 1986; Scott *et al.*, 1989). A series of tests was made with both carrageenan and alginate gels over a concentration range of 1 to 4 per cent of carrageenan and 1 to 3 per cent of alginate (Table 1.19). In contrast to carrageenan gels, alginate gels exhibited significantly lower glucose diffusion coefficients, particularly at higher gel concentration (Nguyen & Luong, 1986). This effect may be ascribed to the significant difference between pore structures of alginate and carrageenan gels.

Carrageenan gels allow relatively rapid diffusion of low-molecular weight solutes (Nguyen & Luong, 1986; Scott *et al.*, 1989). As expected, diffusion coefficients will decrease with increasing molecular weights of the applied solute (Table 1.20). The trend will be valid up to the molecular weight of α-lactalbumin (14 200) (Scott *et al.*, 1989). With albumin (67 000) no diffusion into the gel beads could be observed (Nguyen & Luong, 1986). This is completely in agreement with the suggestion made by Tosa *et al.* (1979), i.e. that the pore size of carrageenan gel matrix is small enough to prevent high-molecular-weight compounds, such as enzymes/proteins, from leaking out from the gel lattice.

Some biocatalysts, e.g. microbial cells, entrapped within κ-carrageenan gel beads, require higher operational stability. For this reason the beads were treated with hardening agents such as glutaraldehyde and hexamethylenediamine (Tosa *et al.*, 1979).

Mechanical properties of κ-carrageenan gels could be modified significantly by enrichment with galactomannans (D-galacto-D-mannoglucans) (Guiseley, 1989). Tosa *et al.* (1979) reported that the concentration of

Table 1.19

Effect of Gel Material on Glucose Diffusion in Nominal 2-mm-Diameter Gel Beads at 30°C.

Gel material	*Average diffusion coefficient* $\times 10^6$ (cm^2 s^{-1})^a
1% κ-Carrageenan	6·7
2% κ-Carrageenan	6·6
4% κ-Carrageenan	6·7
1% Alginate	6·6
2% Alginate	5·5
3% Alginate	5·0

^{a}Each value is the average of the diffusion coefficients of glucose into and out of the same gel beads with a standard deviation of $0·3 \times 10^{-6}$ cm^2 s^{-1}.
All the data were taken from Scott *et al.* (1989).

P. Gemeiner, L. Rexová-Benková, F. Švec & O. Norrlöw

Table 1.20
Diffusion Coefficients of Several Different Solutes in 4% κ-Carrageenan Gel Beads at 30°C
as Compared with Water

Solute	Molecular weight	Diffusion coefficient $\times 10^6$ (cm^2 s^{-1})	
		in carrageenan gel[a]	in water
Ethanol	46	10·1	11·0[b]
Glucose	180	6·7	6·8
L-Tryptophan	204	5·8	6·6
α-Lactalbumin	14 200	1·2	1·0

[a]Each value is the average of the diffusion coefficients of the solute into and out of the same beads with a standard deviation of $0·5 \times 10^{-6}$ cm^2 s^{-1} or less.
[b]Minimum value adjusted for temperature from Scott *et al.* (1989).
All the data were taken from Tanaka *et al.* (1984*b*); Scott *et al.* (1989).

κ-carrageenan used for microbial cell immobilization may also be low-ered by addition of locust bean gum (LBG) without any reduction in gel strength. Audet *et al.* (1988) described a resistance of κ-carrageenan LBG gel against lactic acid during fermentation of entrapped lactic acid bac-teria. A ratio of κ-carrageenan to LBG close to 2:1 is generally recom-mended as the most favourable for synergistic effects on mechanical properties (Miles *et al.*, 1984; Fiszman *et al.*, 1985; Cairns *et al.*, 1986). Tara gum was also studied as a possible replacement for carob gum (LBG) in the formation of κ-carrageenan-galactomannan mixed gels (Cairns *et al.*, 1986). Nevertheless, with tara gum higher total polymer concentrations are needed to match either the modulus or the fracture stress of the LBG-carrageenan system.

Generally, immobilization in κ-carrageenan may be a good alternative to immobilization in alginate, when the latter cannot be applied in prac-tice. Like alginate κ-carrageenan has the particular advantage of easy tailoring of diverse shapes (bead, cube, membrane) of the biocatalyst for particular application purposes. When it is necessary to investigate the characteristics of entrapped cells after immobilization, κ-carrageenan gel may be rapidly dissolved in saline (Scott *et al.*, 1989). In contrast to alginate, κ-carrageenan particles produced by the dripping technique are less spherical and uniform (Hulst & Tramper, 1989). Application of the resonance nozzle immobilization technique not only eliminates the latter shortcomings but permits an upscaling of preparation of κ-carrageenan particles (Buitelaar *et al.*, 1988; Woodward, 1988).

1.3.3.7   *Pectins, Pectinates, Pectates*

Pectins are used mainly as gelling agents in acidic foods. Jams, jellies, and marmalades still absorb the major part of the world's production. New uses are being developed for pectin as a gelling agent, a thickener and a protective colloid in foods. The fact that pectin is a natural product with a long record of safety, favors the use of pectin in applications particularly in food, where other hydrocolloids might be used at a lower cost (Pedersen, 1980).

Like starch, pectins have as yet found only minor applications as carriers. In contrast to starch, however, the true reasons for the low interest of investigators and users in pectins is not known. This is even more strange since pectins, similar to starch, are industrially isolated and widely used. Concerning world production and unit price pectins are comparable to alginates (Yalpani & Sanford, 1987).

Pectins are cell-wall polysaccharides which have a structural role in plants. Although pectins are branched in their native form, when extracted they are predominantly linear polymers based on a 1,4-linked α-D-galacturonate backbone, interrupted randomly by 1,2-linked L-rhamnose (Clark & Ross-Murphy, 1987).

Pectin is an acidic polysaccharide and is in many respects a landbased counterpart of alginate. Like alginate it is based on a diaxially linked backbone, in this case composed of α-D-galacturonate residues, and again forms gels with Ca^{2+}. Gel formation studies indicate a very strong binding of these counterions, even in very dilute solutions (Thibault & Rinaudo, 1985).

Unlike alginates, however, pectins form gels in several ways. This may be ascribed to variability of chemical structure of marketed pectins. Polygalacturonic acid as the principal constituent is partly esterified with methoxyl groups, and the free acid groups may be partly or fully neutralized with monovalent ions (Na^+, K^+, NH_4^+). The degree of methoxylation (DM) has an essential influence on the properties of pectin, especially on its solubility and its requirements for gelation which are directly derived from the solubility (Pedersen, 1980).

The DM of 50 per cent divides commercial pectins into high-methoxyl (HM) pectins and low-methoxyl (LM) pectins. These two groups of pectin are gelled by different mechanisms (Fig. 1.23). HM pectins require a minimum amount of soluble solids (glucose) and a pH within a moderate range around 3·0 in order to form gels. LM pectins require the presence of a controlled amount of calcium ions for gelation, and need neither sugar nor acids (Pedersen, 1980).

DM values of HM pectins control their relative speed of gelation as

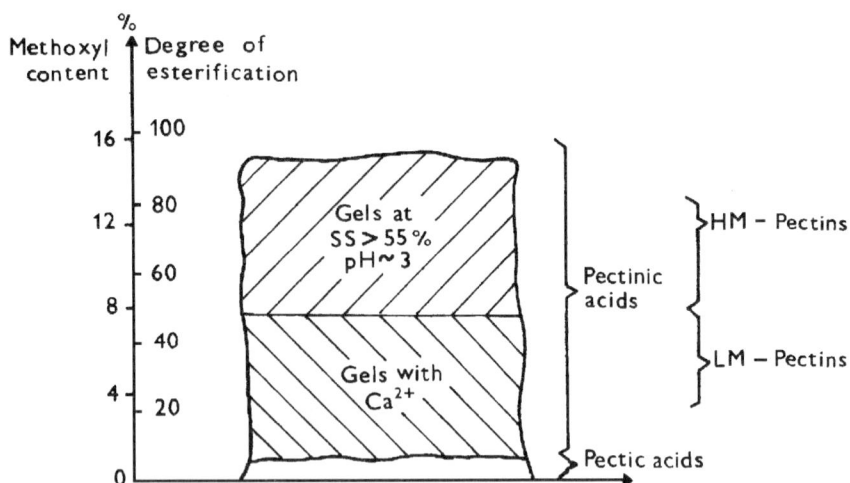

Fig. 1.23 Nomenclature of pectic substances as related to degree of esterification (methoxyl content) (Pedersen, 1980).

reflected by the designations slow-set and rapid-set HM pectin (Fig. 1.23). Moreover, the degree of methoxylation of LM pectins also controls their calcium sensitivity (Pedersen, 1980).

The highest DM that can be achieved by extraction of the natural raw material is about 75 per cent (Fig. 1.24) and provides the so-called rapid-set HM pectin. HM pectin with a DM from 75 to 90 per cent is produced by the methoxylation of HM pectin with methanol. Pectins with a DM from 0 to 70 per cent are produced by controlled demethoxylation the manufacturing process (Pedersen, 1980).

Entrapment of biocatalysts within both ionic pectinate (Navarro *et al.*, 1983, 1984; Berger & Rühlemann, 1988) and pectate (Berger & Rühlemann, 1988; Gemeiner *et al.*, 1989c; Tóth *et al.*, 1989) beads was described using calcium (Navarro *et al.*, 1983, 1984; Berger & Rühlemann, 1988; Gemeiner *et al.*, 1989c; Tóth *et al.*, 1989) and aluminium (Berger & Rühlemann, 1988) as counter cations.

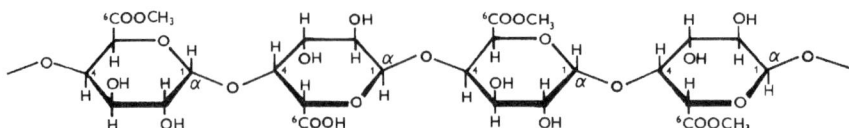

Fig. 1.24 Segment of high-methoxyl (HM) pectin, degree of methoxylation (DM) of 75 per cent (Pedersen, 1980).

LM-pectin gelation in the presence of calcium ions, and at higher pH values, resembles the behavior of alginate. Galacturonic acid blocks in pectins are almost mirror images of the guluronic acid blocks of alginates (Fig. 1.25). Hence it was anticipated that, as for the alginates, the crosslinking of pectate (pectinate) molecules occurs according to the 'egg-box' model. However, in the fiber of polygalacturonic acid the sequences do not take the form of the buckled two-fold ribbon necessary for 'egg-box' formation but they form three-fold helices instead (Morris *et al.*, 1982; Clark & Ross-Murphy, 1987). Unlike alginate, pectin contains only one type of uronic acid and the inter-chain association occurs in a different way.

Serious limitation on junction zone formation is caused by occasional residues of 1,2,-linked L-rhamnose in the polymer backbone. These are sterically incompatible with incorporation in the 'egg-box' structure, and therefore introduce rather a 'kink' which terminates the inter-chain association. While polyuronates are very resistant to acid hydrolysis, the rhamnosyl linkage is comparatively labile. On mild hydrolysis the chain is therefore cleft preferentially at the above mentioned 'kinking' residues. The resulting fragments have a very narrow distribution of molecular eight, corresponding to an interrupted sequence length of approximately 25 galacturonate residues between single 'kinks'. This is comfortably above the minimum sequence length of approximately 15 residues required for co-operative binding of calcium. Calcium pectate gel network (Fig. 1.26) therefore incorporates two solubilizing features: rhamnosyl 'kinks' which sharply delimit potential junction zones, and

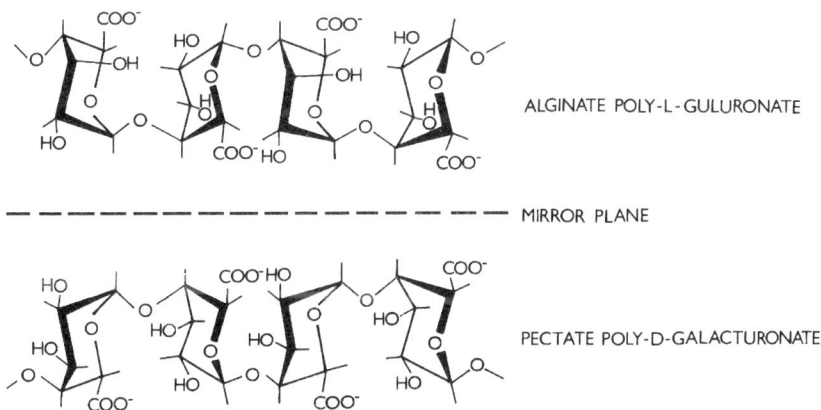

Fig. 1.25 Structural analogy between two polysaccharides with ribbon-like ordered structures: alginate poly-L-guluronate and pectate poly-D-galacturonate (Morris *et al.*, 1982).

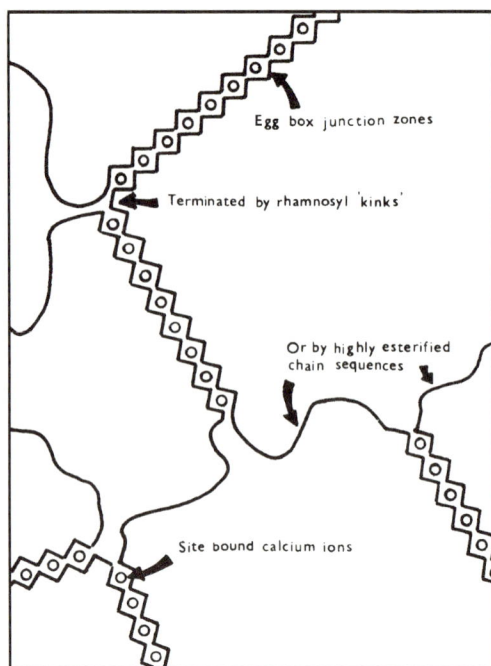

Fig. 1.26 Schematic representation of calcium pectate gel (Morris, 1986).

esterification, which provides a more subtle mechanism for control of physical properties (Morris, 1986; Clark & Ross-Murphy, 1987).

Gelation of LM-pectins with calcium ions is temperature dependent. LM-pectins must therefore be combined with the soluble calcium source at a temperature above the gelling temperature of the system (Pedersen, 1980).

Gelation of pectate by diffusion of calcium ions into solution of pectate exhibits inhomogeneity similar to those observed with alginate, i.e. the polymer concentration is much higher at the surface than in the center of the gel (Skjåk-Braek *et al.*, 1989).

Similarly to alginate, ionotropic gelation of pectate is very simple and inexpensive, and also provides extremely mild and physiological immobilization conditions. Beads can be easily produced by dropping polymer/cell suspension into a crosslinking solution. Crosslinking will occur with Ca^{2+} or Al^{3+}. The matrix has to be formed around the cells *in situ* (Pedersen, 1980; Tóth *et al.*, 1989).

Calcium pectate as well as aluminum pectate gel beads are much more insensitive to small mono- and multivalent anions (citrate, phosphate, lactate, gluconate, chloride) as well as calcium (aluminium) complexing agents, which diminish the stability of calcium and aluminium alginate gel beads (Berger & Rühlemann, 1988).

Formation of a three-dimensional network of calcium pectate gel (CPG) beads might be discussed also with respect to molecular properties of the soluble precursor. Table 1.21 summarizes some of those potassium pectates prepared from apple pectin (I) and citrus pectin (II) (Gemeiner *et al.*, 1989*c*). Both pectins were deesterified by either (i) sodium hydroxide (Gemeiner *et al.*, 1989*c*, 1991) or (ii) enzymatically using a pectinesterase (Gemeiner *et al.*, 1991).

Table 1.22 compares deformation behavior and other physical properties of CPG and calcium alginate gel (CAG) beads prepared from soluble precursors presented in Table 1.21 (Gemeiner *et al.*, 1991). From data given in both tables it follows that none of the single molecular properties of soluble precursors details clearly the deformation behavior of CPG beads. However, as far as other molecular properties of precursors are identical, the influence of the uronan content dominates. Moreover, the enzymatic deesterification of pectins seems to be more advantageous for these purposes (Powell *et al.*, 1982).

Potassium pectate prepared by enzymatic deesterification of apple pectin forms, after being crosslinked with calcium ions, a gel with many properties remarkable for cell entrapment (Gemeiner *et al.*, 1991). However, CPG beads should be stabilized and hardened when used continuously; thus, preformed beads might be treated with polyethyleneimine (PEI) followed by glutaraldehyde (Kurillová *et al.*, 1992). In this way, a stabilized layer was formed which substantially increased the stability of

Table 1.21
Molecular Properties of Soluble Pectates and Alginates

Polysaccharide	*Content of D-galacturonic acid* (%)	*Ion exchange capacity* (mmol g^{-1})	*Intrinsic viscosity* $[\eta]/(cm^3 g^{-1})$	*Constant of stability* log K	*Elution volume* (ml)
Pectate Ia	78·8	3·68	211	4·18	82·5
Pectate IIa	94·7	4·12	122	4·18	87·0
Pectate Ib	68·0/79·2	3·70	480	4·16	84·0; 93·0
Pectate IIb	88·5	4·13	390	4·30	84·0; 78·0
Alginate I	—	4·37	905	3·87	95·4
Alginate II	50·0	4·61	474	3·54	78·0; 69·0

Pectates I and II were prepared from marketed apple and citrus pectins. Starting sample was deesterified either by sodium hydroxide (a) or enzymatically (b) by a tomato pectinesterase Gemeiner *et al.* (1991).
Alginates I and II were from Janssen Chimica and Fluka, respectively.

Table 1.22

Deformation Behavior and Other Physical Parameters/Properties of Calcium Pectate and Calcium Alginate Gel Beads

Precursors for ionotropic gelation	Polymer concentration (%)		Diameter of the bead (mm)	Wet weight of the bead (mg)	Deformation behavior load time (g.s)	Penetration modulus A_{120} (MPa)
	initial	final				
Pectate Ia	5	5·82	—	10·28	3 972	—
Pectate IIa	5	6·15	—	11·50	2 232	—
Pectate Ib	5	5·73	2·92	10·63	11 550	0·22
Pectate IIb	5	6·01	3·02	11·65	10 950	0·33
Alginate	5	7·57	2·78	9·99	2 265	0·10

Pectates I and II were prepared from marketed apple and citrus pectins. Starting sample was deesterified either by sodium hydroxide (a) or enzymatically (b) by a tomato pectinesterase (Gemeiner *et al.*, 1991).
Alginate was from Fluka.

the CPG beads towards Ca^{2+}-complexing reagents (phosphate, citrate), pH, as well as to mechanical stress. Stabilization and hardening of the CPG beads did not substantially change the morphologic properties of the beads including the porosity, pore size distribution, size-exclusion limit (cutoff), effective diffusion coefficient of the low-molecular-weight solute or the original plasticity.

Standard CPG beads have relatively wide pore size distribution with the smallest pores less than 1 nm and the largest over 100 nm (Fig. 1.27). On the contrary, hardened gel has a narrower pore size distribution where the substantial part of pore sizes present is in the range of 5–50 nm (Fig. 1.27). Values of pore sizes of CPG beads presented by Kurillová *et al.* (1992) are close to results of Andresen *et al.* (1977) who reported pore size of CAG beads in the range 5–200 nm. Radial anisotropy of inner structure was proved in the case of CAG beads (Bittner *et al.*, 1987) and cylinders (Skjåk-Braek *et al.*, 1989), where the smallest pores with several nanometers in magnitude were located in a thin layer at the outer surface of gel (Klein *et al.*, 1983; Heinze *et al.*, 1990). It can be supposed that also in the case of standard CPG the smallest pores are located at the outer surface of the gel and after crosslinking they disappear. The absence of pores smaller than 5 nm for the hardened gel is a result of the crosslinking being preferably performed in the surface layer of the gel bead.

Entrapped bacterial cells reduced the network density of CPG beads determined from mechanical measurements and negatively affected the

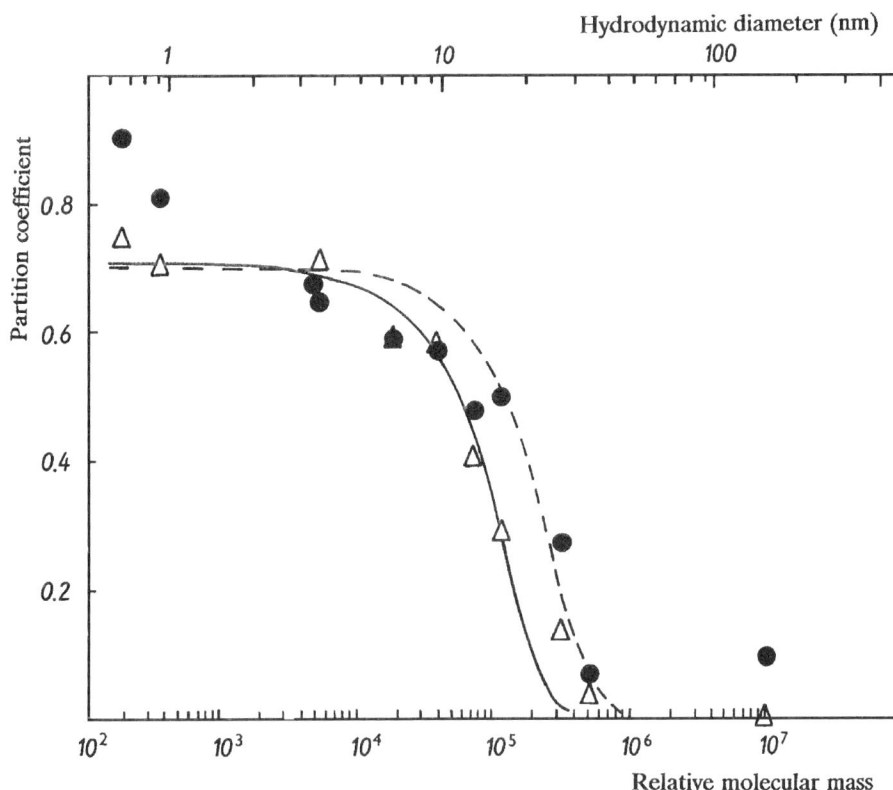

Fig. 1.27 Pore size measurement. Dependence of partition coefficient K of standard (●) and hardened (△) CPG beads on relative molecular mass M_r or hydrodynamic diameter d_h, respectively. Solid curve for hardened gel and dashed curve for standard gel represent dependences obtained by fitting experimental data using logistic model (Kurillová *et al.*, 1992).

resistance of CPG material to deformation, even when stabilized with PEI and glutaraldehyde (Kurillová *et al.*, 1992).

The action of glutaraldehyde was chosen for such time (1 min), as to be long enough for stabilization and hardening of CPG beads, but at the same time short enough for not affecting the viability/metabolic activity of bacterial cells. The storage stability of the stabilized and hardened CPG beads with entrapped cells was exceptional. After storage at 4°C with repeated intermittent operation at 30°C in phosphate buffer for one month, entrapped *Trigonopsis variabilis* maintained 98 per cent of their transformation (D-amino acid oxidase) activity against cephalosporin C. During the continuous operation of CPG beads with entrapped *S. cerevisiae* in a packed-bed reactor in very hard conditions (60°C and 2 M sucrose) the physical and biocatalytic properties were impaired whereas the stabilized and hardened CPG beads with entrapped cells were sufficiently stable in the same conditions (Kurillová et al., 1992).

The discovery of new synergistic gelling systems may have a potential commercial value. Mixtures of alginate and HM-pectins form thermoreversible gels below pH 3·8 without any addition of sugar. D-Glucono-δ-lactone is recommended for use as a slow acidifier in the cold. With this substance gelling occurs under conditions where neither alginate nor the HM-pectin gels alone (Morris & Chilvers, 1984; Toft et al., 1986). The strength of the synergistic gels increases with decreasing pH and increasing content of L-guluronic acid residues in the alginates (Toft et al., 1986).

Gelling effect was correlated with the sequential distribution in the alginate chain of two monomers, D-mannuronic acid (M) and L-guluronic acid (G) residues. The results indicate that 'blocks' of at least four contiguous G-units are necessary for gelling to occur (Toft et al., 1986). Mixed alginate-pectin systems may have pharmaceutical or medical applications in areas where a solution which will become more viscous or gel-like when brought into an acidic environment, such as in the stomach, is required (Morris & Chilvers, 1984).

Pectin, polygalacturonic acid and calcium polygalacturonate were used in graft copolymerization to entrap and/or entrap and covalently couple enzymes. When the maleic anhydride moiety, e.g. acrylonitrile-comaleic anhydride was used for copolymerization, calcium polygalacturonate had to be applied with the aim to reduce the solubility of the graft copolymer system. This material exhibited suitable coupling properties against enzymes and suitable retention of catalytic activity (Beddows et al., 1986). High binding capacity towards enzymes exhibited graft copolymers consisting of poly(hydroxymethyl acrylate)-co-pectin and activated with p-benzoquinone or with p-toluenesulfonyl chloride (Beddows et al., 1984). Attachment of a spacer (1,6-diaminohexane) may induce an extremely high efficacy of enzyme binding that reaches 400 mg of enzyme/g of carrier (Beddows et al., 1986).

Pectate was applied as a carrier also in metal-chelate immobilization of biocatalysts. For the support matrix, a crosslinked pectate was used in the role of the polysaccharide backbone, with iron as ligand (Vijayalakshmi et al., 1979, 1980), and iminodiacetate as polydentate spacer (Vijayalakshmi et al., 1980). The immobilization method was based on orienting metal hydroxides on the surface of an inert matrix. It was known that polygalacturonic acids and their salts function as highly selective cation exchangers and the stability constants of transition-metals are particularly high. Moreover, the metal-binding capacity of an insoluble polygalacturonic acid can further be increased by coupling polydentate ligands (Vijayalakshmi *et al.*, 1980).

1.3.3.8 Other Water-Soluble Gums, Multicomponent Mixtures, Mixed Gels.
Among other water-soluble gums gellan exhibits the highest potential industrial applicability particularly as a gelling agent alternating to agar and carrageenans (Tako *et al.*, 1989). The sensitivity of gelation to acetyl content and salt composition of the gelation media permit the use of gellan as a thermo-setting, cold-setting, thermo-reversible or thermo-irreversible gelling agent (Chilvers & Morris, 1987). Synergistic interactions have been reported between gellan gum and gelatin resulting in high strength gels (Shim, 1985). Multicomponent mixtures of gellan gum, xanthan gum and galactomannans provide a basis for varying the hardness, elasticity and cohesiveness of gels (Pettit, 1986).

In its native form gellan is acetylated, containing on average one O-acetyl substituent per chemical repeat unit. The position of the acetyl substituent is not known with certainty but it is believed to be attached at C(6) on one or both (1→3) and (1→4) glucose residues (Jansson *et al.*, 1983). Thus, the polymer represents a linear anionic heteropolysaccharide composed of tetrasaccharide repeat units (Fig. 1.28).

Gellan polysaccharide contains 6 per cent of O-acetyl groups. Both the acetylated and the deacetylated polysaccharide form thermo-reversible gels with diverse mechanical properties beginning from soft and elastic ones for acetylated form up to hard and brittle ones for the fully deacetylated polysaccharide (Chilvers & Morris, 1987). Gelation itself is sensitive to the type, valency and concentration of cations present in salts added to the dispersion prior to gelation (Grasdalen & Smidsrød, 1987).

Chilvers and Morris published (1987) a detailed study dealing with coacervation in gelatin–gellan gum mixtures. The procedure may be controlled by the pH value and this offers a chance for distinguishing between simple and complex coacervation. Data originating from the above experiments have been used to investigate the potential uses for encapsulation of oils or solid core materials.

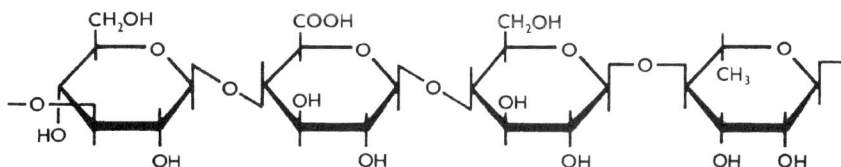

Fig. 1.28 Repeat unit of gellan gum (Chilvers & Morris, 1987).

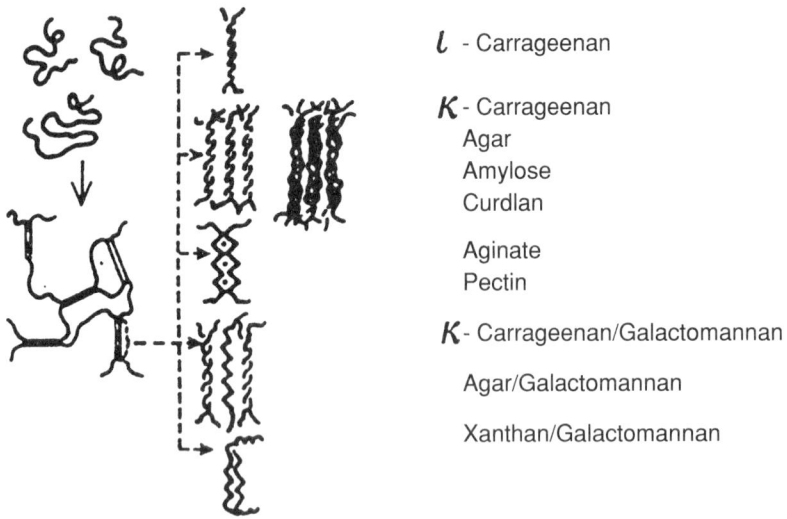

L - Carrageenan

K - Carrageenan
    Agar
    Amylose
    Curdlan

Aginate
Pectin

K - Carrageenan/Galactomannan

Agar/Galactomannan

Xanthan/Galactomannan

Fig. 1.29 Different mechanisms of crosslinking in a polysaccharide gel (Dea, 1987).

Broad perspectives open in the utilization of mixed polysaccharide reactions, especially the specific and permanent chain–chain polymer interactions' rigid gel structures. Current studies of mixed polysaccharide interactions are concerned with either aggregation of different types of ordered ribbons, or aggregation of multiple helices with ordered ribbons. Galactomannans and a number of structurally related 1,4-linked β-D-glycans are centrally involved in these interactions. The lower part of Fig. 1.29 depicts the interaction of mixed polysaccharides in gel formation.

Generally, the extent of mixed polysaccharide interaction decreases with increasing level of substitution along the individual polysaccharide. This concerns particularly galactose substitution in galactomannans, O-acetylation in xanthans, O-sulfation in carrageenans, O-methylation in agars, etc. (Dea, 1987).

1.3.3.9 Chitin, Chitosan

Chitin, partially deacetylated chitin and chitosan contain amino groups which have been used as reactive groups in an attempt to develop inexpensive carriers for biocatalyst immobilization (Muzzarelli, 1977, 1980; Yamaguchi *et al.*, 1982; Muzzarelli, 1984), and microcarriers for cell cultivation (Klein & Kressdorf, 1989). Chitin is an abundant by-product of the fishing (crab, shrimp, and prawn) and fermentation industries. Chitin and chitosan are readily available commercial products.

Chitin, somewhat like cellulose, is insoluble in water. However, many of its derivatives are soluble in water and produce solutions (Muzzarelli, 1977) sometimes capable of gelation (Moore & Roberts, 1980). Chitosan and its derivatives may be used to form membranes (Blair *et al.*, 1987). Chitosan and trimethylammonium glycol chitosan iodide represent components employed for the formation of coacervate capsules (Daly & Knorr, 1988; Kokufuta *et al.*, 1988) and polyelectrolyte complex systems (Knorr & Daly, 1988) suitable for encapsulation/entrapment of biocatalysts. Silica gel particles coated with chitin and chitosan represent a second generation of carriers combining the advantages of an inorganic support (mechanical stability, regular porosity, etc.) with the high coupling efficiency of the coating polymer (Muzzarelli, 1984).

The term chitin is used for designation of fibrous $(1 \rightarrow 4)$-linked 2-acetamido-2-deoxy-β-D-glucan (Fig. 1.30) the acetylation of which is usually incomplete to various extents. Chitin exists in three polymorphic forms with various degrees of crystallinity. The term chitosan is preferred when the nitrogen content is higher than 7 per cent w/w, and it is generally used for artificially deacetylated chitins (Fig. 1.30) (Muzzarelli, 1977, 1984). Water-soluble chitosan can be obtained from chitin by deacetylation in concentrated sodium hydroxide solutions.

Chitosan is soluble also in several diluted organic acids, such as acetic and formic acid, but its solubility in mineral acids is limited. Using chitosan solution (pH < 6), gel formation will occur with a number of different multivalent anion counterions. By adding a chitosan solution dropwise into a crosslinking solution having a pH < 6 a real ionotropic gel is formed. The NH_2 groups of chitosan are protonated and an ionic crosslinking occurs. Bead formation in a crosslinking solution of a pH >7·5 can also take place but an ionotropic gel forms initially. Then, if the pH of the chitosan spheres is also > 7·5, a precipitate forms. At pH values greater than 7·5, chitosan is totally deprotonated and becomes water-insoluble (Vorlop & Klein, 1987).

Crosslinking of chitosan with high-molecular-weight counter-ions (Table 1.23) results in capsules while crosslinking with low-molecular

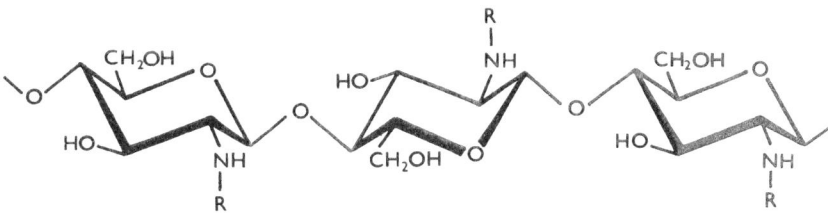

Fig. 1.30   Repeat unit of chitin (R = $COCH_3$) and chitosan (R = H) (Muzzarelli, 1984).

Table 1.23
Possible Counter-ions for the Ionotropic Gelation of Chitosan

Polycation	Counter-ions
Chitosan	Low molecular weight
	Pyrophosphate
	Tripolyphosphate
NH_3	Tetrapolyphosphate
	Octapolyphosphate
	Hexametaphosphate
	$[Fe(CN)_6]^{4-}$, $[Fe(CN)_6]^{3-}$
	High molecular weight
	Poly(1-hydroxy-1-sulfonate-2-propene
	Poly(aldehydocarbonic acid)
	Alginate

Taken from Vorlop & Klein (1987).

weight counter-ions (Table 1.23) results in globules in which the cells are entrapped in a real network (Vorlop & Klein, 1987; Klein & Kressdorf, 1989). Using more hydrophobic counterions (Table 1.24) it is possible to prepare hydrophobic carriers (Vorlop & Klein, 1987).

Enzymes may be bound to chitin by adsorption but this is usually followed by crosslinking with glutaraldehyde (Muzzarelli, 1977, 1984; Stanley *et al.*, 1988). Covalent linkage onto a carbonyl derivative, obtained by a previous treatment with glutaraldehyde, is also possible (Muzzarelli, 1977, 1984; Iyengar & Rao, 1979).

Chitosan, in a soluble form, may be mixed with an enzyme solution and a gel may be formed either by adding a multifunctional crosslinking agent, usually glutaraldehyde, or by ionotropic gelation with several multivalent anionic counter-ions (Kennedy & Cabral, 1987a; Vorlop & Klein, 1987; Klein & Kressdorf, 1989) (Table 1.22). Chitosan biocatalysts prepared by ionotropic gelation are, unlike calcium alginate and potassium carrageenan ones, stable in phosphate buffered media. Mechanical

Table 1.24
Hydrophobic Counter-ions for the Ionotropic Gelation of Chitosan

Polycation	Hydrophobic counter-ions
Chitosan	Octyl sulfate
	Lauryl sulfate
	Hexadecyl sulfate
NH_3	Cetylstearyl sulfate

Taken from Vorlop & Klein (1987).

stability of chitosan beads is comparable to that of calcium alginate beads (Vorlop & Klein, 1987; Klein & Kressdorf, 1989).

In other techniques (Leuba & Widmer, 1977) repecipitated chitosan obtained from chitosan acetate or epichlorohydrin-crosslinked chitosan is first treated with glutaraldehyde to obtain a carbonyl derivative which is capable of linking the enzyme covalently.

1.4   MINERAL CARRIERS

Since the early 1950s when pioneering papers on adsorption of enzymes on inorganic supports by Zittle (1953) and McLaren & Packer (1970) appeared, the use of inorganics, natural and processed ones became a common practice in the immobilization of enzymes and other biologically active substances such as antibodies, antigens, hormones or cells. The distinct advantages of some inorganic over organic supports, in particular mechanical strength, resistance to solvents and microbial attack, availability on a large scale and mostly a low price were factors that made inorganic materials attractive for both laboratory and industrial applications. Many different kinds of minerals and other inorganic materials such as alumina, bentonite, calcium phosphate, caolinite, activated carbon, ceramics, glass, hydroxylapatite, silicas, metal salts and oxides have been used as carriers for immobilization of biologically active molecules. Cells were mostly immobilized using controlled pore glass, ceramics, metal hydroxides (Weetall *et al.*, 1974; Kennedy *et al.*, 1976; Navarro *et al.*, 1976; Messing *et al.*, 1979*b*; Marcipar *et al.*, 1980) and such bulk materials as bricks, clay and sand particles. Some examples of enzyme and microbial cell immobilization are summarized in Tables 1.25 and 1.26.

1.4.1   General Characteristics of Inorganic Carriers

The carrier is the main contributing component to the performance of immobilized enzyme or cell. The most essential requirements in the selection of a carrier for the immobilized system application are those of its stability under the conditions of attachment and application, easy handling, economics of its preparation and application, and regeneration. Inorganic carriers mostly meet these requirements.

Inorganic structures are generally rigid. They are often built of solid microparticles forming strong agglomerates resisting pressure and mechanical degradation. Due to structural rigidity they show no compaction and large pressure increase in flow columns even at extremely

Table 1.25
Inorganic Supports Used for Enzyme Immobilization

Support 1	Enzyme 2	References 3
Alumina	Amylase	Allen *et al.*, 1979
	Polygalacturonase	Pifferi & Preziuso, 1987*a*
		Pifferi *et al.*, 1987*b*
	Cellulase	Shimizu & Ishihara, 1987
	Glucoamylase	Rokugawa *et al.*, 1980
	Glucose oxidase	Weetall, 1970*a*
	Urease	Grunwald *et al.*, 1979
Bentonite	Beta-amylase	Monsan & Durand, 1971
	Cellulase	Monsan & Durand, 1971
	Invertase	Monsan & Durand, 1971
	Ribonuclease A	Monsan & Durand, 1971
	Trypsin	Monsan & Durand, 1971
	Urease	Monsan & Durand, 1971
Ceramics	Cellulase	Shimizu & Ishihara, 1987
	Polygalacturonase	Rexová-Benková *et al.*, 1986
Glass		
beads	Acetylcholinesterase	Baum *et al.*, 1972
	Trypsin	Konečný & Sieber, 1980
	Acetate-kinase	Manners *et al.*, 1987
	Phosphotransacetylase	Manners *et al.*, 1987
fiber	Papain	Kennedy & Pike, 1980*b*
porous	Alkaline phosphatase	Weetall, 1969; Weetall & Messing, 1972*b*; Zingaro & Uziel, 1970; Milton & Koji, 1989
	L-Aminoacid oxidase	Weetall & Baum, 1970*b*
	Aminopeptidase	Royer & Andrews, 1973
	Arylsulphatase	Weetall, 1971
	Aldehyde dehydrogenase	Lee, 1978
	Carbamylphospholinase	Marshall, 1973
	Cellulase	Shimizu & Ishihara, 1987
	Chymotrypsin	Tamizawa & Bender, 1974; Robinson *et al.*, 1971; Zemanová *et al.*, 1981
	Ficin	Weetall, 1970*a*
	Beta-galactosidase	Robinson *et al.*, 1971; Wierzbicky *et al.*, 1974
	Glucoamylase	Marsh *et al.*, 1973; Emery & Cardoso, 1978; Cabral *et al.*, 1981
	Glucose isomerase	Strandberg & Smiley, 1972
	Glucose oxidase	Weetall, 1970*a*; Weibel *et al.*, 1973; Rovito & Kittrell, 1973
	Hexokinase	Shapira *et al.*, 1972
	Invertase	Mason & Weetall, 1972
	Lactate dehydrogenase	Cho & Swaisgood, 1974; Dixon *et al.*, 1973; Widmer *et al.*, 1973
	Leucine aminopeptidase	Royer & Andrews, 1973

(continued)

Table 1.25—*contd.*

Support 1	Enzyme 2	References 3
Glass		
porous	Papain	Weetall, 1969, 1970*a*; Weetall & Mason, 1973; Zemanová *et al.*, 1981
	Pectinesterase	Borrego *et al.*, 1989; Romero *et al.*, 1989
	Pectin lyase	Bourdillon *et al.*, 1977
	Pepsin	Line *et al.*, 1971; Ivanov *et al.*, 1990
	Peroxidase	Weetall & Messing, 1972*b*
	Phospholipase	Adamich *et al.*, 1978
	Polygalacturonase	Romero *et al.*, 1989
	Pronase	Royer & Green, 1971
	Steroid esterase	Grove *et al.*, 1971
	Trypsin	Messing & Weetall, 1970*c*; Weetall, 1969, 1970*a*; Zemanová *et al.*, 1981
	Urease	Messing & Weetall, 1970*c*; Weetall & Hersh, 1970*c*;
zirconia coated	Beta-galactosidase	Weetall *et al.*, 1974
	Glucoamylase	Weetall & Havewala, 1972*a*
	Glutamate dehydrogenase	Srivastava *et al.*, 1987
	Papain	Weetall & Mason, 1973
Hydroxylapatite	Glucose oxidase	Messing & Weetall, 1970; Weetall, 1970*a*
Iron oxide	Chymotrypsin	Robinson *et al.*, 1971
	Beta-galactosidase	Robinson *et al.*, 1971
Nickel oxide	Glucose oxidase	Messing & Weetall, 1970; Weetall & Hersh, 1970*c*
	Peroxidase	Messing & Weetall, 1970
Quartz	Papain	Ivanov *et al.*, 1990
Sand	Trypsin	Puvanakrishnan & Bose, 1980
Silica	Alkaline phosphatase	Messing & Weetall, 1970
	Beta-galactosidase	Gonzales *et al.*, 1980
	Glucose oxidase	Messing & Weetall, 1970
	Glucoamylase	Emery & Cardoso, 1978; Wojcik *et al.*, 1987
	Chymotrypsin	Zemanová *et al.*, 1981
	Hydrogenase	Hatchikian & Monsan, 1980
	Papain	Weetall, 1970*a*; Zemanová *et al.*, 1981
	Polygalacturonase	Stratilová *et al.*, 1987, 1989
	Pepsin	Voivodov *et al.*, 1979
	Trypsin	Weetall, 1970*a*; Zemanová *et al.*, 1981
	Nuclease	Rokugawa *et al.*, 1980
Silicon dioxide		
(nonporous)	Chymotrypsin	Fusek *et al.*, 1987
	Polygalacturonase	Rexová-Benková *et al.*, 1990 Stratilová *et al.*, 1989
Stainless steel	Xylosidase	Carvajal *et al.*, 1978
Titania	Cellulase	Shimizu & Ishihara, 1987
	Papain	Kennedy *et al.*, 1980*a*
	Ribonuclease	Dale & White, 1979

Table 1.26
Inorganic Supports Used for Microbial Cell Immobilization

Support 1	*Microorganism* 2	*References* 3
Alumina	*Sacchar. cerevisiae*	Kana *et al.*, 1989
Anthracite	Mixed culture	Linko & Linko, 1983
Brick (porous)	*Saccharomyces carlbergensis*	Ghose & Bandyopadhydy, 1980
Ceramics	*Candida tropicalis*	Marcipar *et al.*, 1980
	Rhodotorula sp.	Marcipar *et al.*, 1980
	Trichosporon sp.	Marcipar *et al.*, 1980
	Sacchar. cerevisiae	Sitton & Gaddy, 1989
	Phanerochaete chrysosporium	Cornwell *et al.*, 1990
Clay	*Candida tropicalis*	Marcipar *et al.*, 1980
	Rhodotula sp.	Marcipar *et al.*, 1980
	Sacchar. cerevisiae	Marcipar *et al.*, 1980
	Trichosporon sp.	Marcipar *et al.*, 1980
Cordierite	*A. niger*	Messing & Oppermann, 1979*a*
	Sacchar. amuraceae	Messing *et al.*, 1979*b*
Diatomaceous earth	*Sacchar. carlbergensis*	Grinbergs *et al.*, 1977
	Penic. chrysogenum	Gbewonyo *et al.*, 1987
Diatomite	*Sacchar. cerevisiae*	Moo-Young *et al.*, 1980
Glass	*Anabaena cylindrica*	Lambert *et al.*, 1979
	Denitrifying mixed bacteria	Spier & Whiteside, 1976
	Fusarium moniliforme	Heinrich & Rehm, 1981
	Sacchar. cerevisiae	Rouxhet *et al.*, 1981
	Sacchar. cerevisiae	Hecker *et al.*, 1990
	Aspergillus niger	Messing *et al.*, 1979*b*
	Bacillus subtilis	Messing & Oppermann, 1979*a*
	E. coli	Messing & Oppermann, 1979*a*
	Penic. chrysogenum	Messing & Oppermann, 1979*a*
	Sacchar. cerevisiae	Messing & Oppermann, 1979*a*
	Sacchar. amuraceae	Messing *et al.*, 1979*b*
	Animal cells	Looby & Griffiths, 1990
	Pichia farinosa	Baumann *et al.*, 1990
Silica gel on glass silica	*Sacchar. cerevisiae*	Carturan *et al.*, 1989
	Sacchar. carlbergensis	Navarro & Duran, 1977*d*
Silicon	*E. coli*	Oriel, 1988
Hydrous titanium (IV)	*Acetobacter* sp.	Kennedy, 1979
oxide	*Acetobacter* sp.	Kennedy *et al.*, 1980*a*
	E. coli	Kennedy *et al.*, 1976
	Sacchar. cerevisiae	Kennedy *et al.*, 1976
	Serratia marcenses	Kennedy *et al.*, 1976
Hydrous zirconium	*E. coli*	Kennedy *et al.*, 1976
(IV) oxide	*Serratia marcenses*	Kennedy *et al.*, 1976
Sand	Denitrifying mixed bacteria	Marcipar *et al.*, 1980
Stainless steel	*Bac. subtilis*	Atkinson *et al.*, 1979
	Sacchar. cerevisiae	Atkinson *et al.*, 1979
	Streptomyces griseus	Atkinson *et al.*, 1979

high flow-rates. Due to a lack of deformation of the support structure, the tertiary structure of enzymes immobilized on the surface of these carriers is more protected.

Inorganic materials can be applied in aqueous as well as organic systems without changing the pore size and volume. No preswelling before use is therefore required.

Unlike gel structured organic carriers, inorganic materials are more stable toward heat. Dimensional stability reflected in low coefficients of expansion makes possible the regeneration of most inorganic materials by simple pyrolysis process. Because of these characteristics, inorganic supports are suitable for use, mainly in industry.

An evaluation of inorganic supports compared to organic ones with respect to immobilization of cells is, however, restricted to those methods where the cells are bound by adsorption or by a chemical binding. Binding capacity of inorganic carriers for cells is mostly low in comparison with that of organic supports. For instance, for microbial cells values of 248 and 2 mg g^{-1} respectively were reported for wood chips and porous silica (Durand & Navarro, 1978). The adsorption capacity of bulk materials like sand, bricks, ceramics, etc. is still substantially lower.

1.4.2 Factors Affecting the Properties of Immobilized Species

An important influence on loading and activity of immobilized enzymes, especially those acting on high-molecular substrates is ascribed to the texture of a support (Weetall, 1976). Pore size and particle size play a decisive role in the choice of the proper carrier for the enzyme or cells to be immobilized. Pore size determines the access of the enzyme and diffusion of the substrate into the support structure. The pore should be of such a size that both the enzyme or cell and the substrate and the product can enter the pores. The correlation of immobilized enzyme activity with pore size gives a sharp maximum (Fig. 1.31) (Messing, 1974; Stratilová *et al.*, 1989) corresponding to pores large enough for entrance of both enzyme and substrate.

Inorganic supports including porous glass and porous ceramics differ extremely in porosity. Therefore the optimum carrier has to be investigated in each case. In connection with the use of inorganic supports the mean pore diameter is one of the most frequently mentioned factors. For materials with nonuniform pores this characteristic is however, unreliable (Stratilová *et al.*, 1989). As shown with silica supports (Table 1.27), mean pore diameter determined by mercury porosimetry, may differ by two orders of magnitude depending on whether its estimation was based on

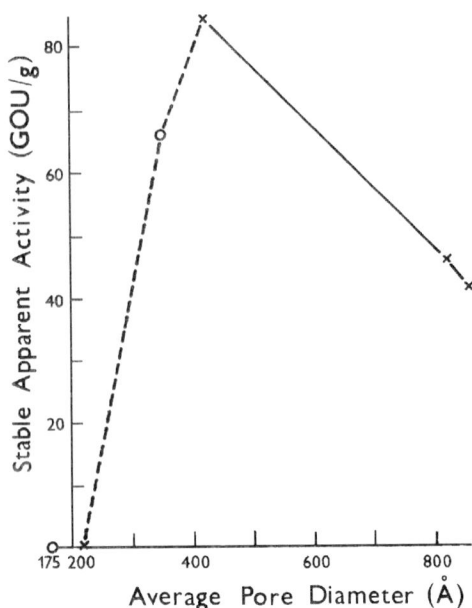

Fig. 1.31   Glucose oxidase activity versus carrier pore diameter.

area or volume. Producers usually characterize their supports by mean pore diameter according to volume. In the case of materials with nonuniform pores this determination gives higher values. A better insight into how the texture of a support affects activity and loading of an enzyme can be obtained by correlating these characteristics with surface area and

Table 1.27
Mean Pore Diameter (nm) of Silica Supports

Support	Claimed[a]	Determined[b]		Calculated[c]
		A	V	
Kieselgel 100	10	7·0	8·0	9·2
Fractosil 200	19	9·8	12·4	11·6
Fractosil 500	43	16·2	32·8	20·0
Fractosil 1000	100	6·4	96·0	37·6
Fractosil 2500	300	5·0	238·4	54·6

[a]by the producer
[b]by mercury porosimetry
[c]as 2 × total volume: total area
A based on the area
V based on the volume

Table 1.28
Properties of the Immobilized Endopolygalacturonase

	Total surface $(m^2 g^{-1})$	*Surface area* $(m^2 g^{-1})$ *and volume* $(ml g^{-1})$ *in pores of the diameter broader than[a]* (nm)					*Enzyme bound* $(mg g^{-1})$	*Relative activity* (%)
	Total volume $(ml g^{-1})$	10	20	30	45	100		
Kieselgel 100	565 (1·30)	95 (0·55)	26 (0·31)	10 (0·22)	3·3 (0·16)	1·0 (0·12)	1·38	4
Fractosil 200	281 (0·82)	147 (0·57)	6 (0·17)	3 (0·15)	1 (0·14)	0·2 (0·12)	2·55	1·4
Fractosil 500	133 (0·72)	83 (0·66)	57 (0·56)	31 (0·40)	4 (0·15)	0·3 (0·10)	2·88	10·2
Fractosil 1000	121 (1·14)	50 (1·05)	43 (1·03)	40 (1·01)	34 (0·95)	11·1 (0·52)	2·56	10·0
Fractosil 2500	58 (0·80)	18 (0·75)	14 (0·74)	13 (0·73)	12 (0·79)	10·5 (0·69)	2·57	8·4

[a]based on the area.
The numbers in parentheses represent volume in the pores.

distribution of pores. The data for endopolygalacturonase bound to silica supports of different porosity (Table 1.28) demonstrate that the amount of the enzyme bound is a function not only of pore diameter but also of the surface area. For the enzyme access into pores with diameter less than 20 nm is rather difficult. The highest amount of the enzyme was bound by the support Fractosil 500 having the largest percentage area in pores with diameter greater than 20 nm. With further increase in pore size, the loading of the enzyme decreases, due to smaller surface area of large pored particles (Filbert, 1975).

A drawback of the application of porous materials in immobilization procedures is the kinetically limited access of the substrates or reactants to immobilized enzyme and removal of the product formed due to internal diffusion which thus becomes a rate limiting step of operation. The role of the latter increases with the size of the reactants and with decreasing pores of the support.

The relative size of the support pore to the enzyme and substrate molecule may greatly affect activity and the mode of action of enzymes acting on high-molecular substrates. In dependence on pore volume, size exclusion principle, transport phenomena, and diffusion limitations govern the behavior of the immobilized enzymes (Borrego *et al.*, 1989; Stratilová *et al.*, 1989; Borrego *et al.*, 1989). For instance in the case of endopolygalacturonase bound to silica supports of different porosity, size-exclusion principle was the main factor that determined the loading and mode of action of the enzyme bound in pores with diameter less than 45 nm. Due to reduced access of substrate molecule to the enzyme in the pores the catalytic action was restricted to external parts of macro-molecular substrate, the length of the reacting segment being propor-tional to pore size. In pores with diameter between 45 and 100 nm ex-ternal diffusion represented the main factor controlling the action of the enzyme. This effect could be, however, reduced by increasing the rate of substrate transfer.

Optimum operation of an enzyme requires unlimited access of the sub-strate to enzyme molecule. This requirement can be met using macro-porous supports. Similar effect can be achieved using non-porous supports where all enzyme is bound on the surface (Ivanov *et al.*, 1990; Rexová-Benková *et al.*, 1990). A drawback of the application of nonporous materials in immobilization procedures is their small surface area, and consequently, low binding capacity. This drawback can be overcome using a microparticular material, such as pyrogeneous, highly dispersed silicon dioxide Cabosil LM-50 (Cabot Comp. USA) which because of its small particle size (10–100 nm) has a large surface area (approx. 200 m^2 g^{-1}). Its binding capacity is comparable with that of porous supports (Fusek *et al.*, 1987). It is, however, suitable only for batch operation because the small particles cause problems in column operations. Under certain con-ditions, the aggregation of submicron particles can occur in solution, making the immobilized enzyme macromolecules inaccessible for substrate. The laser high scattering method revealed that aggregates are 2–3 μm in diameter (Ivanov *et al.*, 1990). The particle size of a support should be chosen depending on application.

To achieve an optimal use of immobilized cells, the geometry of matrix is an important factor. In most cases a spherical shape will be the first choice. The relationship between the pore morphology and accumulation of stable and viable cells is determined by the mode of reproduction of the microorganism. As shown by Messing *et al.* (1979*b*), with microbes that reproduce by fission (*E. coli, Bacillus subtilis*) maximum accumula-tion of biomass occurs when the pore diameter is 1–5 times the major dimension of the microorganism. The microorganisms reproducing by budding (*Saccharomyces cerevisiae*) require 4-fold, and those exhibiting mycelial growth and producing spores (*Aspergillus niger*) require 16-fold largest dimensions.

1.4.3 Inorganics Most Commonly Used as the Supports of Enzymes and Cells

1.4.3.1 Silica

Silica is a polymer built of $(SiO_2)_x$ units (Iler, 1979). This also includes porous glass (quartz, crystalline) which has many similarities with silica. SiO_2 is a common part of the earth's crust in the form of a pure mineral (quartz) or a part of a mineral. The naturally occurring SiO_2 is in the crystalline form, while the synthetic form is amorphous. There are three major ways of manufacturing silica: in gas phase, in solution (precipitated) and the gelling of silicic acid (colloid). The colloid way represents an important way of manufacturing biotechnical carriers, therefore it will be described closer.

The main ingredients for manufacturing silica are waterglass and diluted sulphuric acid. The concentration and composition are critical for the properties of the silica. When the two ingredients are rapidly mixed a hydrosol is formed: $(Na_2O.SiO_2) + H_2SO_4 + H_2O = Si(OH)_4 + Na_2SO_4$.

The formed sols condense into a raw gel. By treating the raw gel with water of different pH and temperature, pore volume and surface area can be specified. The washing of the gel results in a hydrogel containing 65–70 per cent water. This water is adsorbed to the surface of the gel and can be removed by heating at above 150°C. During the drying, the structure will shrink forming a highly densed gel (Fig. 1.32).

The amorphous silica is in equilibrium with a water solution. Solubility of the silica in water is in the range of 70–150 ppm at 25°C. The solubility depends on pH, its minimum is at pH 7–8. At higher pH silica is quickly dissolved. This gives a practical upper limit for silica use at pH 8. The particle size has also an effect on the solubility. Smaller particles dissolve more easily than larger ones.

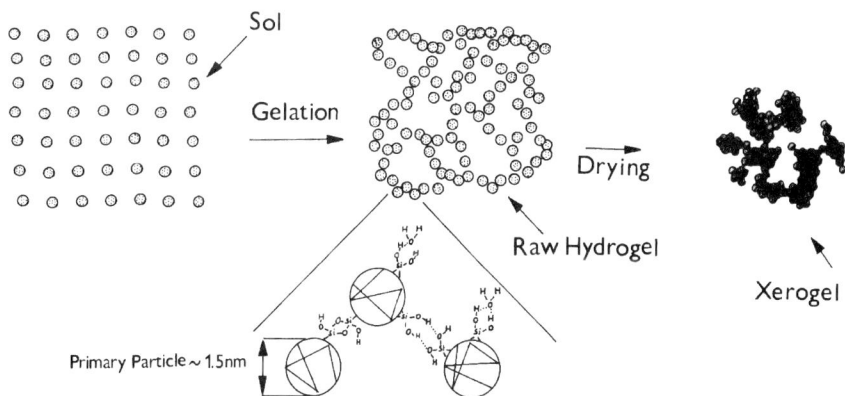

Fig. 1.32 Silica gel formation.

Table 1.29
IR Adsorption Bands of Silica's Hydroxyl Groups

Type of groups	Adsorption band IR (cm^{-1})
Single –OH on the surface	3750
Paired –OH on the surface	3540
Internal –OH hydrogen bonds	3650
Molecular H$_2$O	3400, 1627

The silica surface contains hydroxyl groups – 4–5 –OH groups nm^{-2}. For the wide porous silicas the concentration of –OH groups is higher. The different kinds of hydroxyl groups can be identified by IR absorption measurements (Table 1.29).

Heating the silica causes a dehydratation and siloxane bonds are formed. If silica is heated at above 400°C, siloxane bonds are formed at a higher rate resulting in a slightly rehydratable surface. The dehydration gives a more hydrophobic surface. The surface is normally weakly negatively charged due to impurities. The negative charge can be strengthened by introducing co-ions like aluminum (Fig. 1.33).

The surface can be altered by organic molecules in order to change its properties. This can be done by:

1. Adsorption of molecules on the surface
2. Covalent binding of molecules
3. Coating of the surface by a polymer, e.g. organic silanes (Arkles, 1977).

Silanes with different functional groups can be chosen giving the surface almost any property. Different functional silanes can also be mixed forming a more complex surface. The silanization is normally carried out

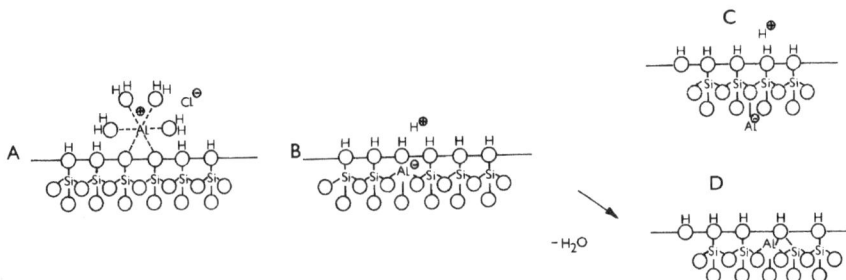

Fig. 1.33 The alumina silica surface in different transition states A–D. (A) adsorbed Al^{3+} coordination number 6; (B) aluminosilicate site Brønsted acid (C) permanent Brønsted acid site, and (D) dehydrated Lewis acid site.

in dry toluene (Unger, 1979) or in gas phase (Wikström *et al.*, 1988) forming a monolayer, or in water phase (Regnier & Noel, 1976) forming a multilayer. The silanization in water phase gives a polymer coating structure where the silica is used only as carrier without affecting the surface properties. Another example represents the coating of silica with polyethyleneimine (Alpert & Regnier, 1979) which is crosslinked to form a shell giving an ion-exchange gel. Interactions between silica and proteins are well known from chromatographic data; generally the strongest interaction with silica shows the proteins with higher isoelectric point (Messing, 1975*b*; Erlandsson *et al.*, 1986). It is not always an advantage to have a strong interaction between enzyme and carrier as this can create deformation of the enzyme, followed by a loss of activity. The interaction can be balanced either by changing the surface or by adding molecules to the solution to weaken the interactions.

Silica can be chosen as the support for the following reasons: defined pores, high surface area, low pressure drop, high mechanical stability, low adsorption power, low price. It is solvent resistant, chemically and microbially inert, easily modified, and non-toxic.

1.4.3.2 Glass

Glass represents a model system which may be employed as a standard, as much of the recent literature has been devoted to this material for studying immobilized enzyme performance. Many studies of enzyme immobilization were performed with controlled-pore glass by adsorption (Messing, 1969, 1970*a,b*) and various covalent techniques (Weetall, 1969; Baum *et al.*, 1972; Mason & Weetall, 1972). Porous glass can be prepared at any desired pore diameter over the range of 30–2000 Å. Careful control of various physical and chemical treatments enables glasses to be produced which exhibit extremely narrow pore size distribution. One of the problems in the application of silica-based supports is their solubility in aqueous solutions above pH 7·5. Therefore, controlled-pore glass is often coated with zirconia to create alkali stable surfaces.

Controlled-pore glass is produced and marketed in underivatized form as well as bearing functional groups such as aminopropyl, carboxypropyl, glycidoxy, phenylhydrazine, hydroxysuccinimide, thiol, amino aryl. Glycophase coated glass (Glycophase G^T controlled pore glass, Pierce, USA) is a support covered by a hydrophilic non-ionic carbohydrate (glyceryl groups) monolayer. The support enables uniform specific binding of proteins via amino groups (Manjón *et al.*, 1985). Enzyme electrodes with enzymes coupled to glass have been developed in many laboratories and have been commercialized.

1.4.3.3   Ceramics

Ceramics can be produced in a variety of forms, from membranes to microspheres, and have been used for immobilization of enzymes as well as cells. Ceramic hollow fibres and membranes were used for immobilization of cells to produce lactic acid (Bunch *et al.*, 1990). The comparative study has shown that the latter material performs better but the former is easier to work with. Salter *et al.* (1990) elaborated a novel method of cell immobilization, using highly porous ceramic microbeads containing a single, large mesopore, where the shape of the particle results in hydrodynamic forces depositing cells within the mesopore of the rigid carrier. A column of such particles with immobilized *S. cerevisiae* was used to produce ethanol in high yield.

1.4.3.4   Other Oxides

Hydrous oxides of the transition metals represent another type of material for immobilization of proteins or cells (Kennedy *et al.*, 1976, 1977, 1981; Kennedy & Pike, 1980*b*). Hydrous Ti(IV), Zr(IV), Al(III), Fe(III), V(III) and Sn(II) oxides forms insoluble complexes with enzymes. The immobilization is based on replacement of hydroxyl groups on the surface of the metal hydrous oxide by hydroxyl groups of serine or threonine, or carboxyl groups of glutamic or aspartic acid or amino group of lysine residues from the enzyme or cell to be immobilized, this resulting in the formation of covalent bonds. The choice of the support is governed by the pH range required for the application. For instance, Ti(IV) hydrous oxide is more effective in acidic conditions, Zr(IV) hydrous oxide is more effective at neutral or higher pH.

Metal ions play an important role in enzymatic reactions as coenzymes. A carrier often adsorbs the metal ions of the enzyme. This results in a loss of activity for the enzymes containing a metal ion as coenzyme. Titania exhibits a very powerful reducing surface which can be used to keep enzymes that are likely to oxidate in the reduced state. Examples of these enzymes are urease (sulfhydryl groups) and glucose oxidase (redox prosthetic groups).

1.4.4   Immobilization

The described inorganic carriers contain surface hydroxyl groups that can be used to couple or adsorb enzymes. With supports not containing reactive groups only mechanical adhesion can occur. Adsorption of the enzyme to a carrier is the easiest way of immobilization. Just by mixing the enzyme and the carrier a link is formed. By adjusting pH, solvent, temperature and ion concentration of the solvent the strength of the link can be adjusted. The method has two advantages, it is cheap and non-toxic.

Immobilization by covalent attachment to inorganic supports involves reactions that are similar to the covalent attachment of enzymes to organic supports. The covalent bonding procedure involves the formation of an activated support followed by the reaction with the enzyme to be immobilized. Materials such as glass, silica, ceramic materials and others are most frequently activated with organosilicon compounds of general formula R-Si-X where R is the organofunctional group, X designates hydrolyzable groups (Messing, 1975*b*). The organofunctional group is separated from the silicon atom by a propyl chain and X is an alkoxy group (methoxy, ethoxy). Most commonly used silane coupling agents are commercially available e.g. γ-aminopropyltriethoxysilane, N-(β-aminoethyl)-γ-aminopropyltrimethoxysilane, γ-chloropropyltriethoxysilane, γ-glycidoxypropyltrimethoxysilane, γ-mercapto-trimethoxysilane, 3-cyanopropyltrimethoxysilane. The silanized supports can either react directly with the enzyme to be immobilized, or the supports are further modified with organic intermediates before coupling reaction.

The reaction with γ-aminopropyltriethoxysilane gives an amino group functionalized matrix to which enzyme or other substances are mostly attached by use of the condensation reagents. Silanization with the γ-glycidoxypropyltrimethoxysilane leads to the introduction of oxirane groups. This group reacts with nucleophiles and thus allows the coupling of compounds containing amino, hydroxyl, or thiol groups.

Epoxy-silica may, however, be further derivatized. Oxirane groups may be hydrolyzed to diols by treatment with acid. Diol silica obtained has also been used for immobilization of biologically active compounds (Walters, 1982; Crowly *et al.* 1986, Hearn, 1986). Activation of the hydroxyl groups was accomplished with carbodiimidazole.

The aldehyde silica obtained after periodate oxidation of the diol is suitable derivative for the attachment of proteins in good yield. However, the Schiff base formed during the reaction has to be reduced in order to stabilize the bond. Treatment of diol silica with tresyl chloride gives tresylate silica which is a suitable support for binding compounds via amino groups.

Reduction of aldehyde silica with sodium borohydride results in primary hydroxyl-containing silica (Ernst-Cabrera & Wilchek, 1986) which is then amenable to activation with any of the conventional reagents available for activating polysaccharides. Ernst-Cabrera and Wilchek recommended the activation of the supports with chloroformates (p-nitrophenyl chloroformate, N-hydroxysuccinimide chloroformate) or 1,1-carbodiimidazole.

Amino functionalization of silica supports may also be achieved by activation of the support with gaseous BCl_3 to replace silanol groups by strongly reactive B-Cl groups. In the second step, the chlorine atoms are substituted by amino groups of an aliphatic diamine, e.g. ethylene diamine or hexamethylene diamine (Wojcik *et al.*, 1987).

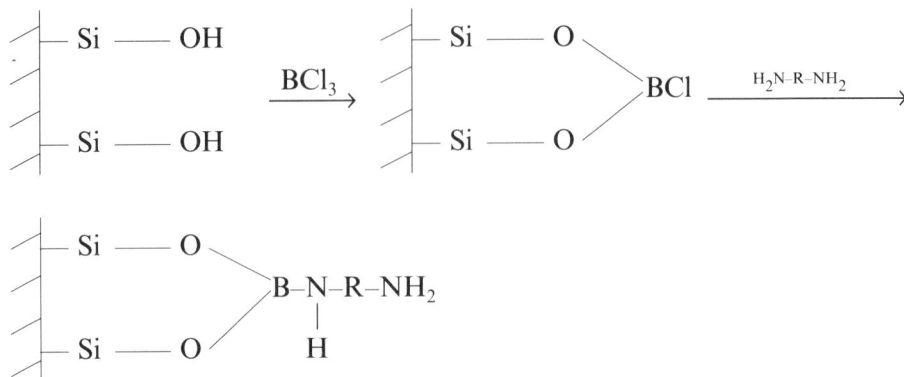

$$
\begin{array}{c}
\text{—Si —— OH} \\[4pt]
\text{—Si —— OH}
\end{array}
\quad \xrightarrow{BCl_3} \quad
\begin{array}{c}
\text{—Si —— O} \\
\qquad\qquad\quad \searrow \\
\qquad\qquad\qquad BCl \\
\qquad\qquad\quad \nearrow \\
\text{—Si —— O}
\end{array}
\quad \xrightarrow{H_2N-R-NH_2}
$$

$$
\begin{array}{c}
\text{—Si —— O} \\
\qquad\qquad\quad \searrow \\
\qquad\qquad\qquad \text{B–N–R–NH}_2 \\
\qquad\qquad\qquad\quad | \\
\qquad\qquad\quad \nearrow\quad\ \text{H} \\
\text{—Si —— O}
\end{array}
$$

Inorganic surfaces can be activated with transition metal salts (Emery *et al.*, 1972; Kennedy & Cabral, 1987*b*). This procedure was used for immobilization of glucoamylase on porous glass (Cabral *et al.*, 1981) and ground borosilicate glass (Emery *et al.*, 1972), nuclease P_1 on porous glass and silica gel (Rokugawa *et al.*, 1980), papain (Kennedy & Pike, 1980*b*) etc. Cabral *et al.* (1981) have shown that the interaction of the transition metal salt, namely titanium(IV) chloride with the silanol groups of the inorganic support is not very strong, and chelation occurs only to a small extent depending on the hydrophobicity of the support, i.e. on the free silanol groups available on the surface of the inorganic support.

Immobilization of cells to inorganic supports can be achieved by physical adsorption or by any method that involves the formation of covalent bond. Although a number of immobilization techniques have been applied to whole cells it is recognized that no one particular process can be classified as an ideal general method for cell immobilization.

Adsorption to a solid support offers a simple and mild method to immobilize cells that has as a major advantage the fact that the cells remain alive and their enzymatic activities are not affected (Hattori & Furusaka, 1960; Navarro & Durand, 1977; Moo-Young *et al.*, 1980; etc). However, due to the relatively weak binding forces involved a release of cells from the support occurs. Gerson & Zajic (1979) published a detailed analysis of the thermodynamic requirements with regard to the surface energies of cells and carriers. It was further shown that ionic strength, pH and inorganic support composition are the factors affecting the interactions (Navarro & Durand, 1977; Vijayalakshmi *et al.*, 1979; Marcipar *et al.*, 1980; Zemanová *et al.*, 1981). Beyond van der Waals forces, hydrogen bonds, electrostatic interactions with negatively charged carriers such as glass or ceramics occur (Arcuri *et al.*, 1980). Immobilization of microorganisms (*Saccharomyces cerevisiae, Bacillus subtilis, E. coli*) on ceramic supports was described in which modification of the ceramics with N-(2-aminoethyl)3-aminopropyltrimethoxysilane and related silane coupling agents caused a shift of the zeta potential to positive, and the coupling could be effected with coulometric interactions (Kawase *et al.*, 1989). Diatomaceous earth, known as celite or kieselguhr was considered as a suitable carrier for immobilization of microbial cells (Gbewonyo *et al.*, 1987). The pore size distribution enables use of the material for microbial cell entrapment. The cells *Saccharomyces cerevisiae* were immobilized successfully by inclusion in silica hydrogel without changing biological activity (Rouxhet *et al.*, 1981). The immobilization was not connected with the formation of chemical bonds between the cell and the silica hydrogel matrix.

Covalent binding method of immobilization is not widely used for cells because of the toxicity of the coupling agents used, which can cause the loss of both the viability of the cell and/or of enzyme activity. In the case of inorganic carriers, mainly silica and ceramics activated by silanization have been used.

1.5   REFERENCES

Adamich, M., Voss, H. F. & Dennis, E. A. (1978). *Arch. Biochem. Biophys.*, **189**, 417–423.

Alfrey, T., Bohrer, J. J. & Mark, H. (1952). Copolymerization. High Polymers, Vol. VIII. Interscience, New York.

Allen, B. R., Charles, M. & Coughlin, R. W. (1979). *Biotechnol. Bioeng.*, **21**, 1639–1648.

Alpert, A. J. & Regnier, F. (1979). *J. Chromatogr.*, **185**, 375–392.

Andresen, I. L., Skipnes, O., Smidsrød, O., Ostgaard, K. & Hemmer, P. C. (1977). *ACS Symp. Ser.*, **48**, 361–381.

Arcuri, E. J., Worden, R. M. & Shumate, S. E. (1980). *Biotechnol. Lett.*, **2**, 499–504.

Arkles, B. (1977). *Chemtech*, **Dec**. 766.

Arnott, S., Fulmer, A., Scott, W. E., Dea, I. C. M., Moorhouse, R. & Rees, D. A. (1974). *J. Mol. Biol.*, **90**, 269–284.

Artursson, P., Edman, P., Laakso, T. & Sjöholm, I. (1984). *J. Pharm. Sci.*, **73**, 1507–1513.

Atkinson, B., Black, G. M., Lewis, P. J. S. & Pinches, A. (1979). *Biotechnol. Bioeng.*, **21**, 193–200.

Audet, P., Paquin, C. & Lacroix, C. (1988). *Appl. Microbiol. Biotechnol.*, **29**, 11–18.

Bahadur, A., Bahadur, P. & Riess, G. (1985). *Makromol. Chem.*, **186**, 1387–1394.

Baldrian, J., Peška, J., Štamberg, J., Ludwig, J. & Paul, D. (1978). In *Chromatography of Synthetic and Biological Polymers*, Vol. 1: *Column Packings, CPG, GF and Gradient Elution*, ed. R. Epton. Ellis Horwood, Chichester, pp. 109–121.

Bamford, C. H., Middleton, I. P. & Al Lamee, K. G. (1986). *Polymer*, **27**, 1981–1985.

Bang, W.G., Behrendt, U., Lang, S. & Wagner, F. (1983). *Biotechnol. Bioeng.*, **25**, 1013–1025.

Battista, O. A. (1975). In *Cellulose Technology Research*, ed. A. F. Turbak. American Chemical Society, Washington, pp. 1–8.

Baum, G., Ward, F. B. & Weetall, H. H. (1972). *Biochim. Biophys. Acta*, **268**, 411–414.

Bauman, E. K., Goodson, L. H., Guilbault, G. & Kramer, D. N. (1965). *Anal. Chem.*, **37**, 1378–1381.

Bauman, E. K., Goodson, L. H. & Thompson, J. R. (1967). *Anal. Biochem.*, **19**, 587–592.

Baumann, U., Bisping, B. & Rehm, H.-J. (1990). *Food Biotechnol.*, **4**, 99.

Beddows, C. G., Gil, M. H. & Guthrie, J. T. (1984). *Polym. Bull.*, **11**, 1–6.

Beddows, C. G., Gil, M. H. & Guthrie, J. T. (1986). *Biotechnol. Bioeng.*, **28**, 51–57.

Berger, R. & Rühlemann, I. (1988). *Acta Biotechnol.*, **8**, 401–405.

Bernfeld, P. & Wan, J. (1963). *Science*, **142**, 678–679.

Birnbaum, S., Pendleton, R., Larsson, P.-O. & Mosbach, K. (1982). *Biotechnol. Lett.*, **3**, 393–400.

Birnbaum, S., Larsson, P.-O. & Mosbach, K. (1983). In *Solid–Phase Biochemistry*, ed. W. H. Scouten, J. Wiley & Sons. New York, pp. 679–762.

Bittner, H. R., Schmid, W. & Sernetz, M. (1987). *Enzyme Engineering*, Vol. 8 (Laskin, A. I., Mosbach, K., Thomas, D. and Wingard, Jr., L. B., eds.). *Ann, N.Y. Acad. Sci.*, **501**, New York, pp. 302–305.

Black, G. M., Webb, C., Metthews, T. M. & Atkinson, B. (1984). *Biotechnol. Bioeng.*, **26**, 134–144.

Blair, H. S., Guthrie, J., Law, T. K. & Turkington, P. (1987). *J. Appl. Polym. Sci.*, **33**, 641–656.

Borrego, F., Tari, M., Manjon, A. & Iborra, J. L. (1989). *Appl. Biochem. Biotechnol.*, **22**, 129–140.

Bourdillon, C., Bourgeois, J. P. & Thomas, D. (1977). *Biotechnol. Bioeng.*, **21**, 1877–1879.

Bovara, R., Carrera, G., Cremonesi, P. & Mazzola, G. (1981). *Anal. Biochem.*, **112**, 239–243.

Breier, A., Gemeiner, P., Ziegelhöffer, A., Turi Nagy, L. & Štofaníková, V. (1987). *Colloid Polymer Sci.*, **265**, 933–937.

Brodelius, P. (1983) In *Immobilized Enzymes and Cells*, Vol. I, ed. B. Mattiasson. CRC Press, Cleveland, pp. 27–55.

Brown, E. & Joyeau, R. (1973). *Tetrahedron Lett.*, **8**, 601–602.

Brown, E. & Joyeau, R. (1974). *Polymer*, **15**, 546–552.

Buchholz, K., ed. (1979). *Dechema Monographs*, Vol. 84: '*Characterization of Immobilized Biocatalysts*'. Verlag Chemie, Weinheim.

Buchholz, K. & Klein, J. (1987). *Methods in Enzymology*, Vol. 135: '*Immobilized Enzymes and Cells, Pt. B*', ed. K. Mosbach. Academic Press, Orlando, pp. 3–30.

Bucke, C. (1987). *Methods in Enzymology*, Vol. 135: '*Immobilized Enzymes and Cells, Pt. B*', eds. K. Mosbach. Academic Press, Orlando, pp. 175–189.

Buitelaar, R. M., Hulst, A. C. & Tramper, J. (1988). *Biotechnol. Tech.*, **2**, 109–114.

Bunch, A. W., Higton, G., Ball, J. & Knowles, C. J. (1990). In *Physiology of Immobilized Cells*, ed. J. A. M. de Bont, J. Visser, B. Mattiasson & J. Tramper. Elsevier, Amsterdam, pp. 127–135.

Butler, L. G. (1975). *Arch. Biochem. Biophys.*, **171**, 645–650.

Büttner, W., Becker, M., Rupprich, Ch., Boeden, H.-F., Henklein, P., Loth, F. & Dautzenberg, H. (1989). *Biotechnol. Bioeng.*, **33**, 26–31.

Cabral, J. M. S., Cardosso, J. P. & Novais, J. M. (1981). *Enzyme Microb. Technol.*, **3**, 41–45.

Cairns, P., Morris, V. J., Miles, M. J. & Brownsey, G. J. (1986). *Food Hydrocolloids*, **1**, 89–93.

Cantarella, M., Cantarella, L. & Alfani, F. (1988). *Brit. Polym. J.*, **20**, 477–485.

Carturan, G., Camposhini, R., Dire, S., Scardi, V. & De Alteriis, E. (1989). *J. Mol. Catal.*, **57**, L13–L16.

Carvajal, N., Martinez, J., de Oca, F. M., Rodriguez, J. & Fernandez, M. (1978). *Biochim. Biophys. Acta*, **527**, 1–7.

Cerny, L. C., Stasiw, D. M., Cerny, E. L., Baldwin, J. E. & Gill, B. (1982). *Clin. Homerheol.*, **2**, 355–365.

Chambers, R. P., Ford, J. R., Allender, J. H., Baricos, W. H. & Cohen, W. (1974). In *Enzyme Engineering*, Vol. 2, eds E.K. Pye & L.B. Wingard, Jr. Plenum Press, New York, pp. 195–202.

Chambers, R. P., Cohen. W. & Baricos, W. H. (1976). In *Methods in Enzymology*, Vol. 44: '*Immobilized Enzymes*', ed. K. Mosbach. Academic Press, New York, pp. 291–317.

Chandler, B. V. & Johnson, R. L. (1979). *J. Sci. Agric. Food*, **30**, 825–832.

Chang, T. M. S. (1976). In *Methods in Enzymology*, Vol. 44: '*Immobilized Enzymes,* ed. K. Mosbach. Academic Press, New York, pp. 201–218.

Chang, T. M. S. (1985). In *Methods in Enzymology*, Vol. 112: '*Drug and Enzyme Targeting*', eds K. J. Widden & R. Green. Academic Press, Orlando, 1985, pp. 195–203.

Chang, T. M. S. (1987). In *Methods in Enzymology*, Vol. 136: '*Immobilized Enzymes and Cells, Pt. C*', ed. K. Mosbach. Academic Press, Orlando, pp. 67–82.

Chao, K. C., Haugen, M. M. & Royer, G. P. (1986). *Biotechnol. Bioeng.*, **28**, 1289–1293.

Charles, M., Coughlin, R. W., Allen, B. R., Paruchuri, E. K. & Hasselberger, F. X. (1974). *Adv. Exp. Med. Biol.*, **42**, 213–234.

Cheetham, P. S. J. (1979). *Enzyme Microb. Technol.*, **1**, 183–188.

Chen, L. F. & Tsao, G. T. (1976). *Biotechnol. Bioeng.*, **18**, 1507–1516.

Chen, L. F. & Tsao, G. T. (1977). *Biotechnol. Bioeng.*, **19**, 1463–1473.

Chen, L. F., Gong, C. S. & Tsao, G. T. (1981). *Starch/Stärke*, **33**, 58–63.

Chibata, I., ed. (1978). *Immobilized Enzymes. Research and Development*. J. Wiley & Sons, New York.

Chibata, I., Tosa, T., Sato, T. & Takata, I. (1987). *Methods In Enzymology*, Vol. 135: '*Immobilized Enzymes and Cells, Pt. B*', ed. K. Mosbach. Academic Press, Orlando, pp. 189–198.

Chilvers, G. R. & Morris, V. J. (1987). *Carbohyd. Polym.*, **7**, 111–120.

Cho, I. C. & Swaisgood, H. (1974). *Biochim. Biophys. Acta*, **334**, 243–256.

Chun, M., Dickopp, G. & Sernetz, M. (1980). *J. Solid-Phase Biochem.*, **5**, 211–221.

Chun, M. & Sernetz, M. (1980). In *Enzyme Engineering*, Vol. 5, eds. H. H. Weetall & G. P. Royer. Plenum Press, New York, pp. 457–460.

Clark, A. H. & Ross-Murphy, S. B. (1987). In *Advances In Polymer Sciences*, Vol. 83. Springer–Verlag, Berlin, pp. 57–192.

Comfort, A., Albert, E. C. & Langer, R. (1989). *Biotechnol. Bioeng.*, **34**, 1366–1373.

Connick, Jr., W.J. (1982). *J. Appl. Polym. Sci.*, **27**, 3341–3348.

Cornwell, K. L., Tinalnd–Butez, M. F., Tardonne, P. J., Cabasso, I. & Hammel, K. (1990). *Enzyme Microb. Technol.*, **12**, 916–920.

Cotrell, I. W. & Kovacs, P. (1980). *Handbook of Water-Soluble Gums and Resins*, ed. R. L. Davidson. McGraw-Hill, New York, Chpt. 2.

Couderc, R. & Baratti, J. (1980). *Biotechnol. Bioeng.*, **22**, 1155–1173.

Crook, E. M., Brocklehurst, K. & Wharton, C. W. (1970). In *Methods in Enzymology*, Vol. 19: '*Proteolytic Enzymes*', eds G. E. Perlman & L. Lorand. Academic Press, New York, pp. 963–978.

Crowly, S. C., Chan, K. C. & Walters, R. R. (1986). *J. Chromatogr.*, **359**, 359–368

Dale, B. E. & White, D. H. (1979). *Biotechnol. Bioeng.*, **21**, 1639–1648.

Daly, M. M. & Knorr, D. (1988). *Biotechnol. Progr.*, **4**, 76–81.

D'Angiuro, L., Cremonesi, P., Mazzola, G., Focher B. & Cantafi, R. (1978). *Cell. Chem. Technol.*, **12**, 685–692.

D'Angiuro, L., Cremonesi, P., Mazzola, G., Focher, B. & Vecchio, G. (1980*a*). *Biotechnol. Bioeng.*, **22**, 2251–2272.

D'Angiuro, L., Focher, B., Cremonesi, P., Mazzola, G. & Vecchio, G. (1980*b*). *J. Polym. Sci.: Polym. Symp.*, **68**, 69–78.

D'Angiuro, L., Mazzola, G., Vecchio, G., Focher, B. & Cremonesi, P. (1980*c*). *J. Appl. Biochem.*, **2**, 208–217.

D'Angiuro, Vecchio, G., Mazzola, G., Cantafi, R. & Cremonesi, P. (1982). *Angew. Makromol. Chem.*, **104**, 129–143.

Da Silva, M. A. & Gil, M. H. (1988). In *Analytical Uses of Immobilized Biological Compounds for Detection, Medical and Industrial Uses*, eds. G. G. Guilbault & M. Mascini. Reidel Publishing Company, New York, pp. 177–185.

Dautzenberg, H., Loth, F., Fechner, K., Mehlis, B. & Pommerening, K. (1985*a*). *Makromol. Chem. Suppl.*, **9**, 203–210.

Dautzenberg, H., Loth, W., Wagenknecht, W., Phillip, B. (1985*b*). *Papier*, **39**, 85, 601–607.

Dea, I. C. M. (1987) In *Industrial Polysaccharides. The Impact of Biotechnology and Advanced Methodologies*, eds S. S. Stivala, V. Crescenzi & I. C. M. Dea. Gordon & Breach Science Publishers, New York, pp. 367–385.

Dea, I. C. M. (1989). *Pure Appl. Chem.*, **61**, 1315–1322.

Dean, P. D. G., Johnson, W. S. & Middle, F. A., eds (1985). In *Affinity Chromatography. A Practical Approach*. IRL Press Ltd, Oxford, pp. 1–31.

Dickensheets, P. A., Chen., L. F. & Tsao, G. T. (1977). *Biotechnol. Bioeng.*, **19**, 365–375.

Dinelli, D., Marconi, W. & Morisi, F. (1976). In *Methods in Enzymology*, Vol. 44: 'Immobilized Enzymes', ed. K. Mosbach. Academic Press, New York, pp. 227–243.

Dixon, J., Andrews, P. & Butler, L. G. (1979). *Biotechnol. Bioeng.*, **21**, 2113–2123.

Dixon, J. E., Stolzenbach, F. E., Berenson, J. A. & Kaplan, N. O. (1973). *Biochem. Biophys. Res. Commun.*, **52**, 905–912.

Dočolomanský, P., Gemeiner, P., Mislovičová, D., Štefuca, V. & Danielsson, B. (1993). *Biotechnol. Bioeng.*, **43**, in press.

Drobník, J., Labský, J., Kudlvasrová, H., Saudek, V. & Švec, F. (1982). *Biotechnol. Bioeng.*, **24**, 487–493.

Dunkle, R. L. & Shasha, B. S. (1988). *Environ. Entomol.*, **17**, 120–126.

Durand, G. & Navarro, J. M. (1978). *Process Biochem.*, **13**, 14–23.

Dušek, K. (1969). *Collect. Czech. Chem. Commun.*, **34**, 1891–1903.

Dušek, K. (1971). In *Polymer Networks: Structural and Mechanical Properties*, ed. A. J. Chompf. Plenum Press, New York, pp. 245–260.

Edman, P., Ekman, B. & Sjöholm, I. (1980). *J. Pharm. Sci.*, **69**, 838–842.

Emery, A. N. & Cardoso, J. P. (1978). *Biotechnol. Bioeng.*, **20**, 1903–1929.

Emery, A. N., Hough, J. S., Novais, J. M. & Lyons, P. T. (1972). *Chem. Eng. (London)*, **259**, 71–76.

Engelkirchen, K. (1987) In *Methoden der Organischen Chemie (Houben-Weyl)* Band E 20/Teil 3: 'Makromolekulare Stoffe', eds. H. Bartl & J. Falbe. Thieme Verlag, Stuttgart–New York, pp. 2043–2134.

Erlandsson, P., Hansson, L. & Isacsson, R. (1986). *J. Chromatogr.*, **185**, 375–383.

Ernst-Cabrera, K. & Wilchek, M. (1986). *Anal. Biochem.*, **159**, 267–272.

Filbert, A. M. (1975). In 'Immobilized Enzymes for Industrial Reactors', ed. R. A. Messing. Academic Press, New York, pp. 39–61.

Fiszman, S. M., Costell, E. & Duran, L. (1985). *Rev. Agroquim. Technol. Aliment.*, **25**, 591–598.

Flink, J. M. & Johansen, A. (1985). *Biotechnol. Lett.*, **7**, 765–768.

Focher, B., Cremonesi, P., D'Angiuro, L., Vecchio, G. & Mazzola, G. (1975). *Cell. Chem. Technol.*, **9**, 135–143.

Freeman, A. & Aharonowitz, Y. (1981). *Biotechnol. Bioeng.*, **23**, 2747–2759.

Fukushima, Y., Okamura, K., Imai, K. & Motai, H. (1988). *Biotechnol. Bioeng.*, **32**, 584–594.

Furusaki, S. & Asai, N. (1983a). *Biotechnol. Bioeng.*, **25**, 2209–2219.

Furusaki, S., Seki, M. & Fukumura, K. (1983b). *Biotechnol. Bioeng.*, **25**, 2921–2928.

Fusek, M., Čapka, M. & Turková, J. (1987). *Biotechnol. Lett.*, **9**, 561–566.

Gbewonyo, K., Meier, J. & Wang, D. I. C. (1987). In 'Methods in Enzymology',

Vol. 135: '*Immobilized Enzymes and Cells, Pt. B*', ed. K. Mosbach. Academic Press, Nev York, pp. 318–332.

Gemeiner, P., Kuniak, L'. & Zemek, J. (1980). *Collect. Czech. Chem. Commun.*, **45**, 2847–2854.

Gemeiner, P. & Viskupič, E. (1981*a*). *J. Biochem. Biophys. Methods*, **4**, 309–319.

Gemeiner, P. & Zemek, J. (1981*b*). *Collect. Czech. Chem. Commun.*, **46**, 1693–1700.

Gemeiner, P. & Breier, A. (1982). *Biotechnol. Bioeng.*, **24**, 2573–2582.

Gemeiner, P. & Pašteka, M. (1983). *Appl. Biochem. Biotechnol.*, **8**, 381–393.

Gemeiner, P., Polák, C., Breier, A., Petruš, L. & Beneš, M. J. (1986). *Enzyme Microb. Technol.*, **8**, 109–114.

Gemeiner, P., Barteltová, L., Šoltés, L. & Breier, A. (1987). *Enzyme Microb. Technol.*, **9**, 44–46.

Gemeiner, P. (1988*a*). In *Enzýmové inžinierstvo*, ed. P. Gemeiner. Alfa Publishers, Bratislava, pp. 126–183 (in Slovak).

Gemeiner, P., Breier, A. & Beneš, M. J. (1988*b*). In *Progress in Biotechnology*, Vol. 4: '*Interbiotech '87. Enzyme Technologies*', eds A. Blažej & J. Zemek. Elsevier, Amsterdam, pp. 261–274.

Gemeiner, P. & Zemek, J. (1988*c*). In '*Interbiotech '88: 'Industrial BioChemicals*'. Incheba, Bratislava, pp. 46–47.

Gemeiner, P., Beneš, M. J. & Štamberg, J. (1989*a*). *Chem. Papers*, **43**, 805–848.

Gemeiner, P., Hrabárová, E., Zacharová, M., Breier, A. & Beneš, M. J. (1989*b*). *Collect. Czech. Chem. Commun.*, **54**, 2375–2385.

Gemeiner, P., Kurillová, L'., Malovíková, A., Tóth, D. & Tomašovičová, D. (1989*c*). *Folia Microbiol.*, **34**, 214–227.

Gemeiner, P., Breier, A. & Mlýnek, J. (1990). *Cell. Chem. Technol.*, **24**, 55–64.

Gemeiner, P., Kurillová, L', Markovič, O., Malovíková, A., Ilavský, M., Štefuca, V., Polakovič, M. & Báleš, V. (1991). *Biotechnol. Appl. Biochem.*, **13**, 334–345.

Gerig, J. T. & Loehr, D. T. (1980). *Int. J. Biol. Macromol.*, **2**, 17–24.

Gerson, D. F. & Zajic, J. E. (1979). *ACS Symp. Ser.*, **106**, 29–57.

Ghommidh, C., Navarro, J. M. & Durand, G. (1981). *Biotechnol. Lett.*, **3**, 93–98.

Ghommidh, C. & Navarro, J. M. (1982). *Biotechnol. Bioeng.*, **24**, 1991–1999.

Ghose, T. K. & Bandyopadhyay, K. K. (1980). *Biotechnol. Bioeng.*, **22**, 1489–1496.

Giordano, C., Esposito, R., Mazzola, G., Vecchio, G., Pluvio, M., Cirillo, D., Capasso, G. B. & Buonanno, G. (1981). *Int. J. Artif. Organs*, **4**, 244–248; cf. Mazzola, G. & Vecchio, G. (1980). *ibid.*, **3**, 120–123.

Giovenco, S., Marconi, W. & Pansolli, P. (1987). *Methods in Enzymology*, Vol. 135: '*Immobilized Enzymes and Cells, Part B*', ed. K. Mosbach. Academic Press, Orlando, 282–293.

Gisby, P. E., Rao, K. K. & Hall, D. O. (1987). In *Methods in Enzymology*, Vol. 135: '*Immobilized Enzymes and Cells, Pt. B*', ed. K. Mosbach. Academic Press, Orlando, pp. 440–454.

Goldman, R., Goldstein, L. & Katchalski, E. (1971). In *Biochemical Aspects of Reactions on Solid Supports*, ed. G. R. Stark. Academic Press, New York, pp. 1–78; cf. Cuatrecasas, P. *ibid.*, pp. 79–109.

Goldstein, L., Pecht, M., Blumberg, S., Atlas, D. & Levin, Y. (1970). *Biochemistry*, **19**, 2322–2334.

Gonzales, R., Monsan, P. & Ros, O. (1980). *Interferon Biotech.*, **5**, 229–236.

Grasdalen, H., Larsen, B. & Smidsrød, O. (1981). *Carbohyd. Res.*, **89**, 179–191.

Grasdalen, H. & Smidsrød, O. (1987). *Carbohyd. Polym.*, **7**, 371–394.

Grinbergs, M., Hildebrand, R. B. & Clarke, B. J. (1977). *J. Inst. Brew.*, **83**, 25–29.

Grove, M. J., Strandberg, G. W. & Smiley, K. L. (1971). *Biotechnol. Bioeng.*, **13**, 709–711.

Grubke, E. A. (1989). In *Encyclopedia of Polymer Science and Engineering*, **Vol. 16**, ed. H. Mark. J. Wiley, New York, pp. 443–473.

Grunwald, P., Gunsser, W., Heiker, F. R. & Roy, W. (1979). *Anal. Biochem.*, **100**, 54–57.

Guilbault, G. & Das, J. (1970). *Anal. Biochem.*, **33**, 341–355.

Guiseley, K. B., Stanley, N. F. & Whitehouse, P. A. (1980). *Handbook of Water-Soluble Gums and Resins*, ed. R. L. Davidson. McGraw-Hill, New York, Chpt. 5.

Guiseley, K. B. (1989). *Enzyme Microb. Technol.*, **11**, 706–714.

Guyot, A. (1988). In *Syntheses and Separations Using Functional Polymers*, ed. D. C. Sherrington & P. Hodge. J. Wiley, New York, pp. 1–42.

Hahn-Hägerdal, B. (1983). In *Immobilized Enzymes and Cells*, Vol. I, ed. B. Mattiasson. CRC Press, Cleveland, pp. 79–94.

Hahn-Hägerdal, B. (1988) In *Methods in Enzymology*, Vol. 137: 'Immobilized Enzymes and Cells, Pt. D', ed. K. Mosbach. Academic Press, Orlando, pp. 641–647.

Handa, T., Hirose, A., Akino, T., Watanabe, K. & Tsuchiya, H. (1983). *Biotechnol. Bioeng.*, **13**, 2957–2967.

Hanus, J., Kučera, J., Kodet J. & Masílková, A. (1974). *J. Chromatogr.*, **97**, 270–272.

Hartmeier, W. (1985). *Trends in Biotechnol.*, **3**, 149–153.

Hatchikian, E. C. & Monsan, P. (1980). *Biochem. Biophys. Res. Commun.*, **92**, 1091–1096.

Hattori, T. & Furusaka, C. (1960). *J. Biochem. (Tokyo)*, **48**, 831–839.

Hearn, M. T. W. (1986). *J. Chromatogr.*, **376**, 245–257.

Hearn, M. T. W. (1987). In *Methods in Enzymology*, Vol. 135: 'Immobilized Enzymes and Cells, Pt. B', ed. K. Mosbach. Academic Press, Orlando, pp. 102–117.

Hebeish, A. & Guthrie, J. T. (1981). In *The Chemistry and Technology of Cellulosic Copolymers*. Springer-Verlag, Berlin, pp. 8–9.

Hecker, D., Bisping, B. & Rehm, H.-J. (1990). *Appl. Microbiol. Biotechnol.*, **32**, 627–632.

Heinrich, M. & Rehm, H.-J. (1981). *Eur. J. Appl. Microbiol. Biotechnol.*, **11**, 139–145.

Heinze, T., Klemm, D., Loth, F. & Philipp, B. (1990). *Acta Polym.*, **41**, 259–269.

Heitz, W. & Platt, K. L. (1969). *Makromol. Chem.*, **127**, 113–140.

Heitz, W. (1970). *J. Chromatogr.*, **53**, 37–49.

Hirtenstein, M. & Clark, J. (1983). In *Immobilized Enzymes and Cells*, Vol. I, ed. B. Mattiasson. CRC Press, Cleveland, pp. 57–88.

Hon, D. N.-S. (1985). In *Cellulose and its Derivatives. Chemistry, Biochemistry and Applications*, eds J. F. Kennedy, G. O. Phillips, D. J. Wedlock & P. A. Williams. Ellis Horwood, Chichester, pp. 71–86.

Hon, D. N.-S. (1988). *Polym. News*, **13**, 134–140.

Hong, Y., Kwon, S., Chun, M. & Sernetz, M. (1982). In *Enzyme Engineering*, Vol. 6, eds. I. Chibata, S. Fukui & L. B. Wingard, Jr. Plenum Press, New York, pp. 217–218.

Horák, D., Pelzbauer, Z., Bleha, M., Ilavský, M., Švec, F. & Kálal, J. (1981a). *J. Appl. Polym. Sci.*, **26**, 411–421.

Horák, D., Pelzbauer, Z., Švec, F. & Kálal, J. (1981b). *J. Appl. Polym. Sci.*, **26**, 3205–3211.

Hradil, J. & Švec, F. (1982a). *Enzyme Microb. Technol.*, **3**, 331–335.

Hradil, J. & Švec, F. (1982b). *Enzyme Microb. Technol.*, **3**, 336–340.

Hulst, A. C., Tramper, J., Van't Riet, K. & Westerbeek, J. M. M. (1985). *Biotechnol. Bioeng.*, **27**, 870–876.

Hulst, A. C. & Tramper, J. (1989). *Enzyme Microb. Technol.*, **11**, 546–558.

Hupkes, J. V. & Tilburg, R. (1976). *Starch*, **28**, 356–360.

Ikada, Y. & Tabata, Y. (1988). *J. Bioactive Compatible Polym.*, **1**, 32–46.

Ikemoto, H. & Mitsui, A. (1984). *Adv. Photosynth. Res.*, **2**, 789–792.

Iler, R. K. (1979). *The Chemistry of Silica*. John Wiley & Sons, New York.

Inloes, D. S., Taylor, D. P., Cohen, S. N., Michaels, A. S. & Robertson, C. R. (1983). *Appl. Envir. Microbiol.*, **46**, 264–278.

Iqbal, J. & Saleemuddin, M. (1983). *Ind. J. Biochem. Biophys.*, **20**, 33–38.

Ivanov, A. E., Turková, J., Čapka, M. & Zubov, V. P. (1990). *Biocatalysis*, **3**, 235–241.

Iyengar, L. & Rao, A. V. S. P. (1979). *Biotechnol. Bioeng.*, **21**, 1333–1343.

Jacobelli, H., Bartolin, A. & Guyot, A. (1979). *J. Appl. Polym. Sci.*, **23**, 927–939.

Jansson, P., Lindberg, B. & Sandford, P. A. (1983). *Carbohyd. Res.*, **124**, 135–139.

Johnson, M. & Medlen, J. (1985). *Eur. Polym. J.*, **21**, 147–150.

Joung, J. J., Akin, C. & Royer, G. P. (1987). *Appl. Biochem. Biotechnol.*, **14**, 259–275.

Kamamura, M. & Kaetsu, I. (1983). *J. Appl. Polym. Sci.*, **28**, 3759–3765.

Kamamura, M. & Kaetsu, I. (1984). *Polym. J.*, **16**, 113–117.

Kamide, K. (1987). In *Wood Celluloses*, eds J. F. Kennedy, G. O. Phillips & P. A. Williams. Ellis Horwood, Chichester, pp. 467–473.

Kana, K., Kanellaki, M., Papadimitriou, A., Psarianos, C. & Kontinas, A. A. (1989). *J. Ferment. Bioeng.*, **8**, 213–215.

Karube, I., Otsuka, T., Kayano, H., Matsunaga, T. & Suzuki, S. (1980). *Biotechnol. Bioeng.*, **22**, 2655–2665.

Kato, T., Unno, K. & Goto, A. (1985). In *Methods in Enzymology*, Vol. 112: '*Drug and Enzyme Targeting*', eds K. J. Widder & R. Green. Academic Press, Orlando, pp. 139–150.

Kawaguchi, H., Koiwai, N. & Ohtsuka, Y. (1988). *J. Appl. Polym. Sci.*, **35**, 743–753.

Kawase, M., Matsubara, Y. & Majima, T. (1989). *Zdiryo Gijutsu*, 14–19.

Kennedy, J. F., Barker, S. A. & Rosevear, A. (1973). *J. Chem. Soc., Perkin Trans. I*, 2293–2299.

Kennedy, J. F. (1974a). In *Advances in Carbohydrate Chemistry and Biochemistry*, Vol. 29, eds R. S. Tipson & D. Horton. Academic Press, New York, pp. 306–405.

Kennedy, J. F. (1974*b*). In *Insolubilized Enzymes*, eds. M. Salmona, C. Saronio & S. Garattini. Raven Press, New York, pp. 29–49.

Kennedy, J. F. & Rosevear, A. (1974*c*). *J. Chem. Soc., Perkin Trans.*, 757–62.

Kennedy, J. F., Barker, S. A. & Humphrey, S. (1976). *Nature*, **262**, 242–244.

Kennedy, J. F., Barker, S. A. & White, S. A. (1977). *Carbohyd. Res.*, **54**, 1–12.

Kennedy, J. F. (1978). In *Carbohydrate Chemistry*, Vol. 10, ed J. S. Brimacombe. The Chemical Society, London, pp. 426–497; cf. (1979). *ibid.*, pp. 445–513.

Kennedy, J. F. (1979). *ACS Symp. Ser.*, **106**, 119–131.

Kennedy, J. F., Humphreys, J. D., Barker, S. A. & Greenshilds, R. N. (1980*a*). *Enzyme Microb. Technol.*, **2**, 209–216.

Kennedy, J. F. & Pike, V. W. (1980*b*). *Enzyme Microb. Technol.*, **2**, 288–294.

Kennedy, J. F., Humpreys, J. D. & Barker, S. A. (1981). *Enzyme Microb. Technol.*, **3**, 129–136.

Kennedy, J. F. & Cabral, J. M. S. (1983*a*). In *Solid-Phase Biochemistry*, ed. W. H. Scouten. J. Wiley and Sons, New York, pp. 253–391.

Kennedy, J. F. & Cabral, J. M. S. (1983*b*). In *Immobilized Microbial Cells*, ed. I. Chibata & L. B. Wingard, Jr. Academic Press, New York, pp. 189–280.

Kennedy, J. F. & Cabral, J. M. S. (1987*a*). In *Biotechnology*, eds H.-J. Rehm & G. Reed, Vol. 7a: '*Enzyme Technology*', ed. J. F. Kennedy. VCH Verlagsgesselschaft mbH, Weinheim, pp. 347–404.

Kennedy, J. F. & Cabral, J. M. S. (1987*b*). In *Methods in Enzymology*, Vol. 135: '*Immobilized Enzymes and Cells, Part B*', ed. K. Mosbach. Academic Press, Orlando, pp. 117–130.

Kéry, V., Višváderová, J. & Šimúth, J. (1988). In '*Interbiotech '88: Industrial Biochemicals*'. Czechoslovak Scientific and Technica Society, Bratislava, p. 46.

Kierstan, M. & Bucke, C. (1977). *Biotechnol. Bioeng.*, **19**, 387–397.

Kierstan, M. P. J. & Coughlan, M. P. (1985). In *Immobilized Cells and Enzymes. A Practical Approach*, ed. J. Woodward. IRL Press, Oxford, pp. 39–48.

Klein, J. & Wagner, F. (1978). In *DECHEMA Monogr.* Band. 82: *Proc. 1st. Europ. Congr. Biotechnology, Interlaken.* Verlag Chemie, Weinheim, pp. 142–164.

Klein, J., Vorlop, K. D., Eng, H., Kluge, M. & Washausen, P. (1979*a*). In *DECHEMA Monogr.* Band 84: '*Characterization of Immobilized Biocatalysts*', ed. K. Buchholz. Verlag Chemie, Weinheim, pp. 274–276.

Klein, J. & Waushausen, P. (1979*b*). In *DECHEMA Monogr.* Band 84: '*Characterization of Immobilized Biocatalysts*', ed. K. Buchholz. Verlag Chemie, Weinheim, pp. 277–283.

Klein, J., Stock, J. & Vorlop, K.-D. (1983). *Eur. J. Appl. Microbiol. Biotechnol.*, **18**, 86–91.

Klein, J. (1988*a*). In *Bioreactor Immobilized Enzymes and Cells. Fundamentals and Applications*, ed. M. Moo-Young. Elsevier Applied Science, London, pp. 1–8.

Klein, J. (1988*b*). *Makromol. Chem., Macromol. Symp.*, **19**, 93–105.

Klein, J. & Kressdorf, B. (1989). *Angew. Makromol. Chem.*, **166/167**, 293–309.

Knorr, D. & Daly, M. M. (1988). *Process Biochem.*, **23**, 48–50.

Kobayashi, Y., Matsuo, R. & Kawakatsu, H. (1986). *J. Appl. Polym. Sci.*, **31**, 1735–1747.

Kobayashi, Y., Matsuo, R., Ohya, T. & Yokoi, N. (1987). *Biotechnol. Bioeng.*, **30**, 451–457.

Koch, H. & Roper, H. (1988). *Starch/Stärke*, **40**, 121–131.

Kohn, J. & Wilchek, M. (1984). *Appl. Biochem. Biotechnol.*, **9**, 285–305.

Kokufuta, E., Shimizu, N., Tanaka, H. & Nakamura, I. (1988). *Biotechnol. Bioeng.*, **32**, 756–759.

Kolarz, B. N., Lobarzewski, J., Trochimczuk, A. & Wojaczynska, M. (1989). *Angew. Makromol. Chem.*, **171**, 201–211.

Kumaraswamy, M. D. R., Rao, K. P., Joseph, K. T. & Santappa, M. (1989). *Biotechnol. Bioeng.*, **23**, 1889–1892.

Konečný, J. & Sieber, M. (1980). *Biotechnol. Bioeng.*, **22**, 2013–2029.

Kostner, A. & Mandel, M. (1976). In *Methods in Enzymology*. Vol. 44: '*Immobilized Enzymes*', ed. K. Mosbach. Academic Press, New York, pp. 191–195.

Kramer, D., Lehmann, K., Pennewiss, H. & Plainer, H. (1975). In *Proceedings of 23rd Colloquium 'Protides of Biological Fluids'*. Pergamon Press, Oxford, pp. 505–512.

Kučera, J. (1983). *Biotechnol. Bioeng.*, **25**, 599–602.

Kuga, S. (1980a). *J. Chromatogr.*, **195**, 221–230.

Kuga, S. (1980b). *J. Colloid Interface Sci.*, **77**, 413–417.

Kun, K. A. & Kunin, R. (1968). *J. Polym. Sci. A-1*, **6**, 2689–2701.

Kurillová, L'., Gemeiner, P., Ilavský, M., Štefuca, V., Polakovič, M., Welwardová, A. & Tóth, D. (1992). *Biotechnol. Appl. Biochem.*, **16**, 236–251.

Kuu, E. Y. & Polack, J. A. (1983). *Biotechnol. Bioeng.*, **25**, 1995–2006.

Laakso, T., Artursson, P. & Sjöholm, I. (1986). *J. Pharm. Sci.*, **75**, 962–967.

Laakso, T. & Sjöholm, I. (1987a). *J. Pharm. Sci.*, **76**, 935–939.

Laakso, T., Stjärnkvist, P. & Sjöholm, I. (1987b). *J. Pharm. Sci.*, **76**, 134–140.

Lambert, G. R., Dadey, A. & Smith, G. D. (1979). *FEBS Lett.*, **101**, 125–128.

Lee, C. Y. (1978). *J. Solid-Phase Biochem.*, **3**, 71–83.

Lee, S. B., Shin, H. S., Ryu, D. D. Y. & Mandels, M. (1982). *Biotechnol. Bioeng.*, **24**, 2137–2153.

Lee, S. B., Kim, I. H., Ryu, D. D. Y. & Taguchi, H. (1983). *Biotechnol. Bioeng.*, **25**, 33–52.

Leuba, J. L. & Widmer, F. (1977). *J. Solid-Phase Biochem.*, **2**, 257.

Lilly, M. D. (1976). In *Methods in Enzymology*, Vol. 44: '*Immobilized Enzymes*', ed. K. Mosbach. Academic Press, New York, pp. 46–52.

Line, W. F., Kwong, A. & Weetall, H. H. (1971). *Biochim. Biophys. Acta*, **242**, 194–202.

Linko, Y.-Y., Viskari, R., Pohjola, L. & Linko, P. (1977). *J. Solid-Phase Biochem.*, **2**, 203–212.

Linko, Y.-Y., Poutanen, K., Weckström, L. & Linko, P. (1979). *Enzyme Microb. Technol.*, **1**, 26–30.

Linko, P., Poutanen, K., Weckström, L. & Linko, Y.-Y. (1980). *Biochimie*, **62**, 387–394.

Linko, P. (1982). *Appl. Biochem. Biotechnol.*, **7**, 5-9.

Linko, P. & Linko, Y.-Y. (1983). In *Applied Biochemistry and Biotechnology*, Vol. 4: '*Immobilized Microbial Cells*', ed. I. Chibata & L. B. Wingard, Jr. Academic Press, New York, pp. 54–136.

Linko, Y.-Y. & Linko, P. (1987). In *Methods in Enzymology*, Vol. 135: '*Immobilized Enzymes and Cells, Pt. B*', ed. K. Mosbach. Academic Press, Orlando, pp. 268–282.

Linsefors, L. & Brodelius, P. (1987). In *Methods in Enzymology*, Vol. 135: '*Immobilized Enzymes and Cells, Pt. B*', ed. K. Mosbach. Academic Press, Orlando, pp. 434–439.

Looby, D. & Griffiths, B. (1990). *Trends Biotechnol.*, **8**, 204–209.

Luby, P., Kuniak, L'. & Berek, D. (1971). *J. Chromatogr.*, **59**, 79–85.

Maeda, H., Chen, L. F. & Tsao, G. T. (1979). *J. Ferment. Technol.*, **57**, 238–243.

Maksimenko, A. V., Arkhipova, O. G., Yaglov, V. V. & Torchilin, V. P. (1987). In *Progress in Biotechnology*, Vol. 4: '*Interbiotech '87: Enzyme Technologies*', eds. A. Blažej & J. Zemek. Elsevier, Amsterdam, pp. 509–522.

Manecke, G. & Pohl, R. (1978a). *Makromol. Chem.*, **179**, 2361–2377.

Manecke, G. & Vogt, H.-G. (1978b). *Pure Appl. Chem.*, **50**, 655–666.

Manecke, G., Ehrenthal, E. & Schlunsen, J. (1979). In *DECHEMA Monogr.*, Band 84: '*Characterization of Immobilized Biocatalysts*, ed. K. Buchholz. Verlag Chemie, Weinheim, pp. 49–72.

Manecke, G. & Beier, W. (1982). *Angew. Makromol. Chem.*, **104**, 39–58.

Manecke, G. & Beier, W. (1983). *Angew. Makromol. Chem.*, **113**, 179–202.

Manjøn, A., Llorca, F. I., Bonete, M. J., Bastida, J. & Iborra, J. L. (1985). *Process Biochem.*, 17–22.

Manners, G., Stegers, G., Lambrecht, K., De Laughe, H., Puttemans, K. & Block, C. (1987). *Enzyme Microb. Technol.*, **9**, 285–290.

Månsson, M.-O. & Mosbach, K. (1987). In *Pyridine Nucleotide Coenzymes: Chemical, Biochemical, and Medical Aspects*, Vol. 2B, eds D. Dolphin, R. Poulson & O. Avramovic. J. Wiley & Sons, pp. 217–273.

Marcipar, A., Cochet, N., Brackenridge, L. & Lebeault, J. M. (1980). *Biotechnol. Lett.*, **1**, 65–70.

Marconi, W. & Morisi, F. (1979). In *Applied Biochemistry and Bioengineering*, Vol. 2: '*Enzyme Technology*', eds L. B. Wingard, Jr., E. Katchalski-Katzir & L. Goldstein. Academic Press, New York, pp. 219–258.

Marconi, W., Pansolli, P. & Giovenco, S. (1987). *J. Mol. Catal.*, **40**, 261–265.

Marconi, W. (1989). *React. Polym.*, **11**, 1–19.

Marek, M., Valentová, O. & Káš, J. (1984). *Biotechnol. Bioeng.*, **26**, 1223–1226.

Marsh, D. R., Lee, Y. Y. & Tsao, G. T. (1973). *Biotechnol. Bioeng.*, **15**, 483–492.

Marshall, D. L. (1973). *Biotechnol. Bioeng.*, **15**, 447–453.

Martinsen, A., Skjåk-Braek, G. & Smidsrød, O. (1989). *Biotechnol. Bioeng.*, **33**, 79–89.

Mason, R. D. & Weetall, H. H. (1972). *Biotechnol. Bioeng.*, **14**, 637–645.

Matsumoto, K., Hirayama, Ch. & Motozato, Y. (1982). *Kohunshi Ronbunshu*, **39**, 605–611.

Matulovic, U., Rasch, D. & Wagner, F. (1986). *Biotechnol. Lett.*, **8**, 485–490.

McLaren, A. D. & Esterman, E. F. (1956). *Arch. Biochem. Biophys.*, **61**, 158–173.

McLaren, A. D. & Packer, L. (1970). In *Advances in Enzymology*, Vol. 33, ed. F. F. Nord. J. Wiley & Sons, New York, pp. 245–308.

Means, G. & Feeney, R. E. (1971). *Chemical Modification of Proteins*. Holden-Day, San Francisco, pp. 12–14.

Meer, W. (1980). *Handbook of Water-Soluble Gums and Resins*, ed. R. L. David-son. McGraw-Hill Book Co., New York, Chpt. 7.

Mehltretter, C. L. & Weakly, F. B. (1972). *Biotechnol. Bioeng.*, **14**, 281–282.

Melton, R. G., Wiblin, C. N., Foster, R. L. & Sherwood, R. F. (1987). *Biochem. Pharmacol.*, **36**, 105–112; c.f. *ibid.*, **36**, 113–121.

Messing, R. A. (1969). *J. Amer. Chem. Soc.*, **91**, 2370–2371.

Messing, R. A. (1970*a*). *Enzymologia*, **38**, 39–42.

Messing, R. A. (1970*b*). *Enzymologia*, **38**, 370–372.

Messing, R. A. (1974). *Biotechnol. Bioeng.*, **16**, 897–901.

Messing, R. A. (1975*a*). In *Immobilized Enzymes for Industrial Reactors*, ed. R. A. Messing. Academic Press, New York, pp. 63–77.

Messing, R. A. (1975*b*). In *Immobilized Enzymes for Industrial Reactors*, ed. R. A. Messing. Academic Press, New York, pp. 79–98.

Messing, R. A. (1978). *Advances in Biochemical Engineering*, Vol. 10: 'Immobilized Enzymes I', eds T. K. Ghose, A. Fiechter & N. Blakebrough. Springer-Verlag, Berlin, pp. 51–73.

Messing, R. A. (1982). *Biotechnol. Bioeng.*, **24**, 1115–1123.

Messing, R. A. & Oppermann, R. A. (1979*a*). *Biotechnol. Bioeng.*, **21**, 49–58.

Messing, R. A., Oppermann, R. A. & Kolot, F. B. (1979*b*). *ACS Symp. Ser.*, **196**, 13–28.

Messing, R. A. & Weetall, H. H. (1970*c*). *Ger. Offen.*, 1,944,418. cf. (1970). *Chem. Abstr.*, **73**, 21 713g.

Miles, M. J., Morris, V. J. & Carroll, V. (1984). *Macromolecules*, **17**, 2443–2445.

Milton, H. J. & Koji, Y. (1989). *J. Bact. Compact. Polym.*, **4**, 281–295.

Miron, T. & Wilchek, M. (1987). In *Methods in Enzymology*, Vol. 135: 'Immobilized Enzymes and Cells, Pt. B', ed. K. Mosbach. Academic Press, Orlando, pp. 84–90.

Mohamed, S. & Salleh, A. B. (1982). *Biotechnol. Lett.*, **4**, 611–614.

Molteni, L. (1985) In *Methods in Enzymology*, Vol. 112: 'Drug and Enzyme Targeting', eds K. J. Widder & R. Green. Academic Press, Orlando, pp. 285–298.

Monsan, P. & Durand, G. (1971). *FEBS Lett.*, **16**, 39–42.

Moore, G. K. & Roberts, G. A. F. (1980). *Int. J. Biol. Macromol.*, **2**, 7377; cf. *ibid.*, **2**, 78–80.

Moo-Young, M., Lamptey, J. & Robinson, C. W. (1980). *Biotechnol. Lett.*, **2**, 541–548.

Morikava, Y., Karube, I. & Suzuki, S. (1980). *Biotechnol. Bioeng.*, **22**, 1015–1023.

Morris, E. R., Powell, D. A., Gidley, M. J. & Rees, D. A. (1982). *J. Mol. Biol.*, **155**, 507–516.

Morris, V. J. & Chilvers, G. R. (1984). *J. Sci. Food. Agric.*, **35**, 1370–1376.

Morris, E. D. (1986). *Br. Polym. J.*, **18**, 14–21 .

Mosbach, K., ed. (1976). *Methods in Enzymology*, Vol. 44: 'Immobilized Enzymes'. Academic Press, New York.

Mosbach, K., ed. (1987). *Methods in Enzymology*, Vol. 135: 'Immobilized Enzymes and Cells, Pt B.' Academic Press, Orlando.

Motozato, Y. & Hirayama, C. (1984). *J. Chromatogr.*, **298**, 499–507.

Munoz, A. M., Pangborn, R.-M. & Noble, A. N. (1986). *J. Texture Stud.*, **17**, 17–36.

Muzzarelli, R. A. A. (1977). *Chitin*. Pergamon Press, Oxford.

Muzzarelli, R. A. A. (1980). *Enzyme Microb. Technol.*, **2**, 177–184.

Muzzarelli, R. A. A. (1984). In *The Polysaccharides*, Vol. 5, ed. G. E. Aspinall. Academic Press, Orlando, pp. 417–450.

Navarro, J. M. & Durand, G. (1977). *Eur. J. Appl. Microbiol.*, **4**, 243–254.

Navarro, J. M., Durand, G., Duteutre, B., Moll, M. & Corrieu, G. (1976). *Ind. Aliment. Agric.*, **93**, 695–703.

Navarro, A. R., Rubio, M. C. & Callieri, D. A. S. (1983). Eur. *J. Microbiol. Biotechnol.*, **17**, 148–151.

Navarro, A. R., Marangoni, H., Plaza, I. M. & Callieri, D. (1984). *Biotechnol. Lett.*, **6**, 465–470.

Neurath, A. R. & Weetal, H. H. (1970). *FEBS Lett.*, **8**, 253–256.

Ngo, T. T. (1986). *Bio/Technology*, **4**, 134–137.

Nguyen, A. L. & Luong, J. H. T. (1986). *Biotechnol. Bioeng.*, **28**, 1261–1267.

Nilsson, K. (1987a). In *Methods in Enzymology*, Vol. 135: 'Immobilized Enzymes and Cells, Pt. B', ed. K. Mosbach. Academic Press, Orlando, pp. 387–393.

Nilsson, K. & Mosbach, K. (1987b). In *Methods in Enzymology*, Vol. 135: 'Immobilized Enzymes and Cells, Pt. B', ed. K. Mosbach. Academic Press, Orlando, pp. 65–78.

Ohmiya, K., Ohashi, H., Kobayashi, T. & Shimizu, S. (1977). *Appl. Envir. Microbiol.*, **33**, 137–146.

Ohtsuka, Y., Kawaguchi, H. & Yamamoto, T. (1984). *J. Appl. Polym. Sci.*, **29**, 3295–3306.

O'Neill, S. P., Wykes, J. R., Dunnill, P. & Lilly, M. D. (1971). *Biotechnol. Bioeng.*, **13**, 319–322.

Oriel, P. (1988). *Biotechnol. Lett.*, **10**, 113–116.

Pedersen, J. K. (1980). In *Handbook of Water-Soluble Gums and Resins*, ed. R. L Davidson. McGraw-Hill Book Co., New York, Chpt 15.

Pelzbauer, Z., Lukáš, J., Švec, F. & Kálal , J. (1979). *J. Chromatogr.*, **171**, 101–107.

Peng, L., Calton, G. J. & Burnett, J. W. (1986). *Enzyme Microb. Technol.*, **8**, 681–685.

Peng, L., Calton, G. J. & Burnett, J. W. (1987). *J. Biotechnol.*, **5**, 255–266.

Peška, J., Štamberg, J., Hradil, J. & Ilavskýr, M. (1976). *J. Chromatogr.*, **125**, 455–469.

Pettit, D. J. (1986). In *Gums and Stabilizers for the Food Industry* **3**, eds G. O. Phillips, D. J. Wedlock & P. A. Williams. Elsevier Applied Science Publishers, London, p. 451.

Pifferi, P. G. & Preziuso, M. (1987a). *Lebensm. Wiss. Technol.*, **20**, 137–142.

Pifferi, P. G., Spagna, G. & Busetto, L. (1987b). *J. Mol. Catal.*, **42**, 137–149.

Pitha, J. (1978). *Eur. J. Biochem.*, **82**, 285–292.

Poehlein, G. W. (1989). In *Encyclopedia of Polymer Science and Engineering*, Vol. 6, ed. H. Mark. J. Wiley, New York, pp. 1–51.

Pommerening, K., Jung, W., Kühn, M. & Mohr, P. (1979). *J. Polym. Sci., Polym. Symp.*, **66**, 185–188.

Pommerening, K., Ristau, D., Rein, H., Dautzenberg, H. & Loth, F. (1983). *Biomed. Biochem. Acta*, **42**, 813–823.

Poncelet de Smet, B., Poncelet, D. & Neufeld, R. J. (1989). *Enzyme Microb. Technol.*, **11**, 29–37.

Popov, G. & Schwachula, G. (1981). *Plaste Kautsch.*, **28**, 312–315.

Porath, J. & Axén, R. (1976). In *Methods in Enzymology*, Vol. 44: '*Immobilized Enzymes*', ed. K. Mosbach. Academic Press, New York, pp. 19–45.

Powell, D. A., Morris, E. R., Gidley, M. J. & Rees, D. A. (1982). *J. Mol. Biol.*, **155**, 517–531.

Pusch, W. (1987). In *Wood Celluloses*, eds. J. F. Kennedy, G. O. Phillips & P. A. Williams. Ellis Horwood, Chichester, pp. 475–482.

Puvanakrishnan, P. & Bose, S. M. (1980). *Biotechnol. Bioeng.*, **22**, 2449–2453.

Rees, D. A. (1969). In *Advances in Carbohydrate Chemistry and Biochemistry*, Vol. 24, eds. M. L. Wolfrom & R. S. Tipson. Academic Press, New York 1969, pp. 267–332.

Rees, D. A. (1972). In *Gellation and Gellating Agents*. British Food Manufacturing Industries Research Association, Symposium Proceedings No. 13. London.

Reese, E. T. (1982). *Process Biochem.*, **17**, 1–10.

Regnier, F. & Noel, R. (1976). *J. Chrom. Sci.*, **14**, 316.

Regnier, F. E. & Maszaroff, I. (1987). *Biotechnol. Progr.*, **3**, 22–26.

Rehg, T., Dorger, C. & Chau, P. C. (1986). *Biotechnol. Lett.*, **8**, 111–114.

Rexová-Benková, L'., Stratilová, E. & Čapka, M. (1986). *Collect. Czech. Chem. Commun.*, **51**, 2924–2932.

Rexová-Benková, L'., Čapka, M. & Stratilova, E. (1990). *Biocatalysis*, **4**, 1–7.

Riley, R. T. (1983). *J. Agric. Food Chem.*, **31**, 202–206.

Robinson, P. J., Dunnill, P. & Lilly, M. D. (1971). *Biochim. Biophys. Acta*, **242**, 659–661.

Rogovin, Z. A., Virnik, A. D., Khomiakov, K. P. & Laletina, O. P. (1972). *J. Macromol. Sci.-Chem.*, **A6**, 569–593.

Rokugawa, K., Fujishima, T., Kuminaka, A. & Yoshino, H. (1980). *J. Ferment. Technol.*, **58**, 423–429.

Romero, C., Sanchez, C., Manjon, S. & Iborra, J. L. (1989). *Enzyme Microb. Technol.*, **11**, 837–843.

Rouxhet, P. G., van Heecht, J. L. Dideler, J., Gerard, P. & Briquet, M. (1981). *Enzyme Microb. Technol.*, **3**, 49–54.

Rovito, B. J. & Kittrell, J. R. (1973). *Biotechnol. Bioeng.*, **15**, 143–161.

Royer, G. P. & Green, G. M. (1971). *Biochem. Biophys. Res. Commun.*, **44**, 426–432.

Royer, G. P. & Andrews, J. P. (1973). *J. Biol. Chem.*, **248**, 1807–1812.

Royer, G. P., Green, G. M. & Sinha, B. K. (1976). *J. Macromol. Sci., Chem.*, **A1C**, 289–307.

Rutenberg, M. W. (1980). In *Handbook of Water-Soluble Gums and Resins*, ed. R. L. Davidson. McGraw-Hill, New York, Chpt. 22.

Rutenberg, M. W. & Solarek, D. (1984). In *Starch: Chemistry and Technology*, eds R. L. Whistler, J. N. BeMiller & E. F. Paschall. Academic Press, New York, pp. 312–388.

Salleh, A. B. & Mohamed, S. (1982). *Biotechnol. Lett.*, **4**, 387–392.

Salter, G. J., Kell, D. B., Ash, A., Adams, J. M., Broun, A. J. & James, R. (1990). *Enzyme Microb. Technol.*, **12**, 419–430.

Scher, B. ed. (1977). In *Controlled Release Pesticides*. American Chemical Society Symposium Series, Vol. 53. Washington.

Schmid, R. (1979). In *Advances In Biochemical Engineering*, Vol. 12: '*Immobilized Enzymes II*', eds T. K. Ghose, A. Fiechter & N. Blakebrough. Springer–Verlag, Berlin, pp. 41–118.

Schnabelrauch, M., Heinze, T. & Klemm, D. (1990). *Acta Polym.*, **41**, 112–120.

Schröder, U. (1985). In *Methods In Enzymology*, Vol. 112: '*Drug and Enzyme Targeting*', eds K. J. Widder & R. Green. Academic Press, Orlando, pp. 116–128.

Schulte, D. M. & Horser, F. A. (1982). *Polym. Bull.*, **7**, 77–81.

Scott, C. D. (1987). *Enzyme Microb. Technol.*, **9**, 66–73.

Scott, C. D., Woodward, C. A. & Thompson, J. E. (1989). *Enzyme Microb. Technol.*, **11**, 258–263.

Scouten, W. H. (1981). In *Affinity Chromatography. Bioselective Adsorption on Inert Matrices*. J. Wiley & Sons, New York, pp. 20–25.

Scouten, W. H., van den Tweel, Delhaes, D., Kranenberg, H. & Dekker, M. (1986). *J. Chromatogr.*, **376**, 289–298.

Scouten, W. H. (1987). In *Methods in Enzymology*, Vol. 135: '*Immobilized Enzymes and Cells, Pt. B*', ed. K. Mosbach. Academic Press, Orlando, pp. 30–65.

Scouten, W. H., van den Tweel, W., Kranenburg, H. & Dekker, M. (1987). In *Methods in Enzymology*, Vol. 135: '*Immobilized Enzymes and Cells, Pt. B*', ed. K. Mosbach. Academic Press, Orlando, pp. 79–84.

Seidl, J., Malinský, J., Dušek, K. & Heitz, W. (1967). *Adv. Polymer Sci.*, **5**, 113–213.

Seitz, U. & Pauly, H. E. (1979). *Angew. Makromol. Chem.*, **76/77**, 319–327.

Shapira, J., Lecocq, J., Furst, A. & Weetall, H. H. (1972). *Experientia*, **28**, 1261–1262.

Shim, J. L. (1985). *US Patent*, 4, 517, 216.

Shimizu, K. & Ishihara, M. (1987). *Biotechnol. Bioeng.*, **29**, 236–241.

Sitton, O. C. & Gaddy, J. L. (1980). *Biotechnol. Bioeng.*, **22**, 1735–1748.

Skjåk-Braek, G., Smidsrød, O. & Larsen, B. (1986). *Int. J. Biol. Macromol.*, **8**, 330–336.

Skjåk–Braek, G., Grasdalen, H. & Smidsrød, O. (1989). *Carbohyd. Polym.*, **10**, 31–54.

Slininger, P. G., Fanta, G. F. & Abbot, T. P. (1988). *Biotechnol. Bioeng.*, **31**, 759–769.

Smidsrød, O. & Haug, A. (1972). *Acta Chem. Scand.*, **26**, 79–88.

Smidsrød, O. (1974). *Farad. Discuss. Chem. Soc.*, **57**, 263–274.

Smith, G. D., Muallen, A. & Hall, D. O. (1982). *Photobiochem. Photobiophysiol.*, **4**, 307–319.

Specht, B.-D. & Brendel, W. (1977). *Biochim. Biophys. Acta*, **484**, 109–114.

Spier, R. E. & Whiteside, J. P. (1976). *Biotechnol. Bioeng.*, **21**, 49–58.

Srivastava, G. P., Agrawal, A. K., Singhal, S. S., Singh, C. & Rao, V. K. M. (1987). *J. Microb. Biotechnol.*, **2**, 58–63.

Stanley, W. L., Watters, G. G., Kelly, S. H. & Olson, A. C. (1978). *Biotechnol. Bioeng.*, **20**, 135–140.

Stevenson, W. T. K. & Sefton, M. V. (1987). *Biomaterials*, **8**, 449–457.

Stjärnkvist, P., Laakso, T. & Sjöholm, I. (1989). *J. Pharm. Sci.*, **78**, 52–56.

Stockwell, A. F., Davis, S. S. & Walker, S. E. (1986). *J. Controlled Release*, **3**, 167–175.

Strandberg, G. W. & Smiley, K. L. (1972). *Biotechnol. Bioeng.*, **14**, 509–511.
Stratilová, E., Čapka, M. & Rexová-Benková, L. (1987). *Biotechnol. Lett.*, **7**, 511–516.
Stratilová, E.,Čapka, M. & Rexová-Benková, L. (1989). *Biocatalysis*, **2**, 317–326.
Sturgeon, C. M. (1982). In *Carbohydrate Chemistry*, Vol. 13, ed. J. F. Kennedy. The Royal Society of Chemistry, London, pp. 633–637; cf. (1983–1986). *ibid.*, Vols. 14, 15, Pt II: *Macromolecules*. The Chemical Society, London.
Sundaram, P. V. & Pye, E. K. (1974). In *Enzyme Engineering*, Vol. 2, ed. E. K. Pye & L. B. Wingard, Jr. Plenum Press, New York, pp. 449–452.
Škachová, H. & Kučera, J. (1983). *Collect. Czech. Chem. Commun.*, **48**, 2874–2878.
Škodová, H., Chaloupka, J. & Škoda, J. (1981). *Biotechnol. Bioeng.*, **23**, 2151–2159.
Štamberg, J. & Peška, J. (1980). *React. Polym. Ion Exch. Sorbents*, **1**, 145–147.
Štamberg, J., Peška, J., Dautzenberg, H. & Phillip, B. (1982). In *Affinity Chromatography and Related Techniques*, eds T. C. Gribnau, J. Visser & R. J. F. Nivard. Elsevier Science Publishing Co., Amsterdam, pp. 131–141.
Štamberg, J. (1988). *Separ. Purif. Method.*, **17**, 155–184.
Štefuca, V., Gemeiner, P. & Báleš, V. (1988). *Enzyme Microb. Technol.*, **10**, 306–310.
Štefuca, V., Gemeiner, P., Kurillová, L'., Danielsson, B. & Báleš, V. (1990). *Enzyme Microb. Technol.*, **12**, 830–835.
Štefuca, V., Gemeiner, P., Kurillová, L'., Dautzenberg, H., Polakovič, M. & Báleš, V. (1991). *Appl. Biochem. Biotechnol.*, **30**, 313–324.
Šťovíčková, J., Turková, J., Urbánek, B., Bažant, V. & Kavalírová, J. (1988). *Biotechnol. Tech.*, **2**, 121–126.
Švec, F., Hradil, J.,Čoupek, J. & Kálal, J. (1975). *Angew. Makromol. Chem.*, **48**, 135–143.
Takata, I., Tosa, T. & Chibata, I. (1977). *J. Solid-Phase Biochem.*, **2**, 225–236.
Tako, M., Sakae, A. & Nakamura, S. (1989). *Agr. Biol. Chem.*, **53**, 771–776.
Tamizawa, K. & Bender, M. L. (1974). *J. Biol. Chem.*, **249**, 2130–2134.
Tampion, J. & Tampion, M. D. (1987). *Immobilized Cells: Principles and Applications*. Cambridge University Press, Cambridge.
Tanaka, H., Kurosawa, H., Kokufuta, E. & Veliky, I. A. (1984*a*). *Biotechnol. Bioeng.*, **26**, 1393–1394.
Tanaka, H., Matsumura, M. & Veliky, I. A. (1984*b*). *Biotechnol. Bioeng.*, **26**, 53–58.
Tarhan, L. & Pekin, B. (1983). *Biotechnol. Bioeng.*, **25**, 2777–2783.
Thibault, J.-F. & Rinaudo, M. (1985) *Br. Polym. J.*, **17**, 181–184.
Toft, K., Grasdalen, H. & Smidsrød, O. (1986). In *ACS Symposium Series* No. 310: '*Chemistry and Function of Pectins*', eds. M. Fishman & J. J. Jen. American Chemical Society, Washington, pp. 117–132.
Torchilin, V. P. (1983). *Targeted Drugs*, ed. E. P. Goldberg. John Wiley & Sons, New York, p. 127.
Tori, T., Sato, T., Tosa, T. & Chibata, I. (1972). *Enzymologia*, **434**, 213–226.
Tosa, T., Sato, T., Mori, T., Yamamoto, K., Takata, I., Nishida, Y. & Chibata, I. (1979). *Biotechnol. Bioeng.*, **21**, 1697–1710.
Tóth, D., Tomašovičová, D., Gemeiner, P. & Kurillová, L. (1989). *Folia Microbiol.*, **34**, 515–524.
Trimnell, D. & Shasha, B. S. (1988). *J. Controlled Release*, **7**, 263–268.

Turková, J., Vajčner, J., Vančurová, D. & Štamberg, J. (1979). *Collect. Czech. Chem. Commun.*, **44**, 3411–3417.

Turková, J., Kozlov, L.V., Bessmertnaya, L. Ya., Kudryatseva, I. V., Krasilnikova, V. M. & Antonov, V. K. (1980). *Bioorg. Khim.*, **6**, 108–115.

Tye, R. J. (1989). *Carbohyd. Polym.*, **10**, 259–280.

Ugelstad, J., Mork, P. C., Nordhuus, I., Mfutakamba, H., Soleimany, E., Berge, A., Ellingsen, T. & Khan, A. A. (1985). *Makromol. Chem. Suppl.*, **10/11**, 215–232.

Unger, K. (1979). *Porous Silica. J. Chromat. Lib.*, Vol. 16, Elsevier, Amsterdam.

Valentová, O., Marek, M., Švec, F., Štamberg, J. & Vodrážka, Z. (1981). *Biotechnol. Bioeng.*, **23**, 2093–2104.

Veber, P., Kuniak, L'., Szanto, J., Hána, L., Antal, M. & Beneš, M. (1984). Czech. CS 234, 367.

Veliky, I. A. & Williams, R. E. (1981). *Biotechnol. Lett.*, **3**, 275–280.

Vijayalakshmi, M. A., Marcipar, A., Segard, E. & Broun, G. B. (1979). *Ann. New York Acad. Sci.*, **326**, 249–254.

Vijayalakshmi, M. A., Marcipar, A. & Cochet, N. (1980). *J. Polym. Sci., Polym. Symp.*, **68**, 57–62.

Voivodov, K. I., Galunsky, B. P. & Dyankov, S. S. (1979). *J. Appl. Biochem.*, **1**, 442–447.

Vojtíšek, V. & Jirků, V. (1983). *Folia Microbiol.*, **28**, 309–340.

Vorlop, K. D. & Klein, J. (1987). In *Methods in Enzymology*, Vol. 135: 'Immobilized Enzymes and Cells, Pt. B', ed. K. Mosbach. Academic Press, Orlando, pp. 259–268.

Walker, J. M. & Gingold, E. B. eds. (1990). *Molecular Biology and Biotechnology*. Royal Society of Chemistry, Cambridge, 2nd edn.

Walters, R. R. (1982). *J. Chromatogr.*, **249**, 19–25.

Weaver, M. O., Bagley, M. O., Fanta, G. F. & Doane, W. M. (1976). *US Pat.* 3 985 616.

Weetall, H. H. (1969). *Science*, **166**, 615–617.

Weetall, H. H. (1970*a*). *Biochim. Biophys. Acta*, **212**, 1–7.

Weetall, H. H. & Baum, G. (1970*b*). *Biotechnol. Bioeng.*, **12**, 399–407.

Weetall, H. H. & Hersh, L. S. (1970*c*). *Biochim. Biophys. Acta*, **206**, 54–60.

Weetall, H. H. (1971). *Nature*, **232**, 473–476.

Weetall, H. H. & Havewala, N. B. (1972*a*). *Biotechnol. Bioeng. Symp.*, **73**, 241–266.

Weetall, H. H. & Messing, R. A. (1972*b*). In *The Chemistry of Biosurfaces*, ed. M. L. Hair. Marcel Dekker, New York.

Weetall, H. H. & Mason, R. D. (1973). *Biotechnol. Bioeng.*, **15**, 455–466.

Weetall, H. H., Havewala, N. B., Pitcher, W. H. Jr., Defar, C. C., Vann, W. P. & Yaverbaum, S. (1974). *Biotechnol. Bioeng.*, **16**, 689–696.

Weetall, H. H. (1976). In *Methods in Enzymology*, Vol. 44: 'Immobilized Enzymes', ed. K. Mosbach. Academic Press, New York, pp. 134–148.

Weibel, M. K. & Bright, H. J. (1971). *Biochem.*, **124**, 801–807.

Weibel, M. K., Dritschilo, W., Bright, H. J. & Humphrey, A. E. (1973). *Anal. Biochem.*, **52**, 402–414.

Weimer, P. J. & Weston, W. M. (1985). *Biotechnol. Bioeng.*, **27**, 1540–1547.

Weliky, N. & Weetall, H. H. (1965). *Immunochemistry*, **2**, 293–322.

White, C. A. (1985). In *Cellulose and its Derivatives. Chemistry, Biochemistry and Applications*, eds. J. F. Kennedy, G. O. Phillips, D. J. Wedlock & P. A. Williams. Ellis Horwood, Chichester, pp. 425–438.

Wikström, P., Mandenius, C. F. & Larsson, P.-O. (1988). *J. Chromatogr.*, **455**, 105.

Widmer, F., Dixon, J. E. & Kaplan, N. O. (1973). *Anal. Biochem.*, **55**, 282–287.

Wierzbicki, L. E., Edwards, V. H. & Kisikovski, F. V. (1974). *J. Food Sci.*, **39**, 338–341.

Wilchek, M., Miron, T. & Kohn, J. (1984). In *Methods in Enzymology*, Vol. 104: '*Enzyme Purification and Related Techniques, Pt. C*', ed. W. B. Jakoby. Academic Press, Orlando, pp. 3–55.

Wiseman, A. (1985). *Handbook of Enzyme Biotechnology*. Ellis Horwood, Chichester.

Wojcik, A., Lobarzewski, J., Blaszczinska, T. & Fiedurek, J. (1987). *Biotechnol. Bioeng.*, **30**, 983–986.

Woodward, J., Krasniak, S. R., Smith, R. D., Spielberg, F. & Zachry, G. S. (1982). *Biotechnol. Bioeng. Symp.*, **12**, 485–489.

Woodward, J. (1988). *J. Microbiol. Methods*, **8**, 91–100.

Yalpani, M. (1986). *CRC Crit. Rev. Biotechnol.*, **3**, 375–421.

Yalpani, M. & Sanford, P. A. (1987). In *Industrial Polysaccharides: Genetic Engineering, Structure/Property Relations and Applications*, ed. M. Yalpani. Elsevier, Amsterdam, pp. 311–336.

Yalpani, M. (1988). *Polysaccharides. Syntheses, Modifications and Structure/ Property Relations*. Elsevier, Amsterdam.

Yamaguchi, R., Arai, Y., Kaneko, T. & Itoh, T. (1982). *Biotechnol. Bioeng.*, **24**, 1081–1091.

Zaborsky, O. R. (1973). *Immobilized Enzymes*. CRC Press, Cleveland.

Zemanová, I., Turková, J., Čapka, M., Nakhapetyan, L. A., Švec, F. & Kálal, J. (1981). *Enzyme Microb. Technol.*, **3**, 229–232.

Zingaro, R. A. & Uziel, M. (1970). *Biochim. Biophys. Acta*, **213**, 371–379.

Zittle, C. A. (1953). *Advan. Enzymol.*, **14**, 319–374.

Zobel, H. F. (1988). *Starch/Stärke*, **40**, 44–50.

Žurková, E., Drobník, J., Kálal, J., Švec, F., Tyráčková, V., Vojtíšek, V. & Zeman, R. (1983). *Biotechnol. Bioeng.*, **25**, 2231–2242.

Chapter 2

IMMOBILIZATION OF PROTEINS FOR BIOSENSOR DEVELOPMENT

ELAINE T. VANDENBERG, R. STEPHEN BROWN AND ULRICH J. KRULL*

Chemical Sensors Group, Department of Chemistry, University of Toronto, Mississauga, Canada

*Corresponding author

CONTENTS

ABBREVIATIONS

ab	antibody (anti-rabbit immunoglobulin G (whole molecule) FITC conjugate, developed from goat)
ag	antigen (rabbit immunoglobulin G)
APTES	3-aminopropyl triethoxy silane
BSA	bovine serum albumin
Con A	concanavalin A
FITC	fluorescein isothiocyanate
glut	glutaraldehyde-derivatized surface
glutaralde	glutaraldehyde-derivatized surface
HABA	2-[4-hydroxyphenylazo]benzoic acid
IgG	immunoglobulin G
ligand	a molecule bound by a receptor
receptor	includes molecular receptors, antibodies, enzymes and lectins
RT-APTES	APTES deposited under room temperature conditions
TEA-APTES	APTES deposited under reflux in the presence of the base catalyst triethylamine
tresyl	2,2,2-trifluoroethane sulfonyl
tresyl chloride	2,2,2-trifluoroethane sulfonyl chloride

2.1   INTRODUCTION

There is considerable interest in the permanent attachment of bio-molecules to solid surfaces because of the many uses of immobilized proteins: for purification of molecules that bind to antibodies, lectins or other binding proteins; in solid-phase analytical assays; for improved biocompatibility of materials; and to provide selectivity to biosensors. In such applications, the amount of protein immobilized and the retention of physiological activity of the immobilized protein (such as binding of a second molecule) are of interest. This work will report the amount and activity of immobilized proteins as a function of the method of immobilization, using instrumental analytical techniques. The results are of interest to all whose work involves immobilized proteins, but the study was inspired by its application to the field of biosensors, as evidenced by the choice of solid surfaces (quartz and silicon), and the types of proteins employed (those which bind a second molecule).

2.2   BIOSENSORS

Biosensors have been defined in many ways, and have been described well by A.P.F. Turner (1987): 'A biosensor is a device incorporating a biological sensing element either intimately connected to or integrated within a transducer. The usual aim is to produce a digital electronic signal which is proportional to the concentration of a specific chemical or set of chemicals. The apparent alien marriage of two contrasting disciplines combines the specificity and sensitivity of biological systems with the computing power of the microprocessor.' Biosensors are classified according to the type of transduction mechanism that is used to quantitate the amount of analyte. There are electrochemical, optical, piezoelectric and thermal devices.

Electrochemical systems represent the oldest and best known family of devices, and operate amperometrically, coulometrically or potentiometrically. A typical electrochemical biosensor would consist of a polymeric sensing membrane containing a selective reagent such as an enzyme, which could be interfaced to a measurement device such as an ion selective electrode or a platinum electrode. Such an enzyme electrode would catalytically convert a target substrate to products, and the development of these molecular or ionic products could then be monitored by subsequent redox or potentiometric methods. A number of useful reviews have been published in this area (Guilbault, 1984; Kobos, 1980, 1987). The enzyme is usually covalently bound or trapped near the electrode surface, and covalent immobilization has often resulted in improvement of enzyme stability. Usually a large excess of enzyme is deposited so that response becomes limited by mass transport and is independent of the concentration of the enzyme. Such an arrangement also serves to extend the life of the electrode, increase the linear range, and reduce susceptibility to low concentrations of interferents such as inhibitors. The signal may be derived at steady-state conditions where the rate of formation of products is equivalent to the rate of loss of the product by diffusion from the zone containing the enzyme. The rate of change of the signal can also be used as a kinetic method of concentration determination.

Coupled reactions of two or more enzymes and/or substrates have been described as both a method for extending the variety of analytes to be measured, or improving the sensitivity of analyses from μM to nM detection levels. In the former case, the product of the first selective reaction is consumed in a subsequent reaction to produce new products amenable to detection by standard ISEs or coulometric electrodes. This makes it possible to convert an electro-inactive species to one that can be detected by the measurement electrode. In the case of amplification of an analytical signal, one enzyme would regenerate the substrate of a second enzyme, leading to an increased analytical signal magnitude. Better reversibility and control can also be achieved by co-immobilization of an enzyme with its co-factor.

Antibody–antigen interactions can be observed if either of the binding species is tagged with an enzyme, in a strategy which involves a sandwich assay or a competitive binding assay (Guilbault, 1984; Kobos, 1980). The localization of the enzyme at the electrode surface by the immuno-chemical complexation provides a source of measurable analyte if sub-strate is also available. Enzyme-linked immunosensors generally are limited by slow response times, and slow reversibility. Immunopotentio-metric systems able to directly observe interactions at an electrode sur-face have been proposed, but are not practically or theoretically functional in complex matrices. Present limitations of electrochemical devices include the necessity for stable reference electrodes, electrical noise susceptibility, non-selective adsorption and surface occlusion lead-ing to drift of surface activity, the evolution of mixed potentials, Debye length restrictions for larger biological 'receptors', and ionic strength, temperature and pH sensitivity.

Semiconductor devices are available for observation of products from catalytic reactions, as well as for observation of temperature alterations and other physical changes caused by biochemical reactions. A representa-tive device is the chemically-selective field-effect transistor (CHEMFET), in which the gate of a field-effect transistor is replaced with a conventional potentiometric membrane supported on an insulator. Modulation of the gate voltage by a biochemical reaction can control an electronic current within the transistor, in a device structure which typically occupies an area of a few square millimetres. The majority of enzyme-based FET devices (ENFET) have used a transduction scheme based on potentio-metric determination of hydronium ion activity. This has set a fun-damental limitation to the type of enzyme that can provide an analytical signal. Efforts to overcome this problem have resulted in the develop-ment of new devices such as the ammonia-sensitive iridium-metal oxide semiconductor system. The use of semiconductor technology has made possible the development of a sensor associated with microelectronic sig-nal processing and control all located in a miniature package on a chip. Semiconductor devices for biosensor applications usually use potentio-metric methods, and are therefore limited by the same problems as previ-ously listed. Further problems originate from device instability in liquid media, and electrical noise sensitivity.

Piezoelectric devices are generally based on specially cut quartz crys-tals which can mechanically oscillate when subjected to an alternating electrical potential. Though other piezoelectric materials are available, the commercial availability of appropriate quartz crystals has led to extensive investigation of two modes of operation. Surface acoustic wave (SAW) devices have been described as gas sensors, but apparently cannot operate simply in liquid media (Ballantine & Wohltjen, 1989). Bulk acoustic wave (BAW) devices have been used extensively as gas sensors, and recently as liquid phase devices with potential for biosensing as determined by studies of antibodies located on one surface of a device (Thompson *et al.*, 1987). The mechanical oscillation of the crystal (usually at MHz frequencies) can be very regular, and can be perturbed by minor mass or microviscosity changes caused by selective binding reactions at the surface of the crystal. Frequency changes smaller than 1 Hz can sometimes be measured reproducibly, providing ng sensitivity with respect to adsorption to the surface of the device. Evolution of products from selective reactions is not necessary for transduction to take place. Remote sensing may be possible with these devices since the frequency response information is not necessarily distorted by transmission through extraneous electrical or magnetic fields. The major limitations of these devices would be governed by non-selective adsorption and surface occlusion. These devices, therefore, offer significant practical advantages in contrast to electrochemical devices.

Optical sensors are perhaps the simplest to understand since they represent an extension of spectrophotometric techniques to miniature systems. Significant innovation has been achieved in this area by the use of fibre-optic technology, where the optical fibres can act as light guides to transmit optical information to/from a remote compartmentalized target reaction. In another strategy, optical fibres can be directly coated along the waveguide surface with selective reagents, thereby permitting optical transduction to ensue by means of evanescent wave stimulation (intrinsic sensor). Sensors which use optical fibres are referred to as optrodes, in analogy to the electrode, and a useful collection of research articles about these systems has appeared recently in *Talanta*, **35**, 1988. The sensitivity of an optical system based on absorption of electromagnetic radiation is greatly limited by short path lengths. However, optical sensors based on fluorescent processes are unique in that they offer the possibility of multi-dimensional analysis by concurrent observation of wavelength, intensity, polarization and event lifetime. In combination, these analytical parameters can be used to define unique solutions to both the qualitative and quantitative aspects of analysis. The multi-dimensional approach allows optical systems to be self-correcting and self-calibrating, and can potentially obviate most of the problems listed for the other device categories. Limitations in such an ideal case will then be driven by the signal-to-noise ratio and reversibility.

Thermistor devices are based on the measurement of minute temperature changes resulting from exothermic analyte–receptor interactions, usually involving enzymes. The transducer may be a thermistor or a pyroelectric device. Although thermistor biosensors are based on well-developed technology, the probe configuration has been developed only in the past five years, and less knowledge about the limitations of these devices is available as compared to the other biosensors.

A major challenge in the creation of a biosensor is to ensure that the selective chemical layer generates a large chemical signal when interacting with the analyte(s) of interest, and that no signal is generated due to other interactions or conditions. Biological molecules have been successfully employed in the selective layer, particularly the following protein types: antibodies, enzymes, lectins and molecular receptors. In living organisms, these proteins bind chemicals in the external and internal environment, resulting in generation of various chemical signals. *In vitro*, these chemical signals may be converted to electronic signals, and are thus highly appropriate for use in biosensors. Molecular receptors, antibodies, enzymes and lectins will be collectively known as 'receptors', and the molecules they bind as 'ligands' for ease of nomenclature in this chapter.

2.2.1 The Use of Proteins as Selective Binding Molecules

2.2.1.1 *Selectivity from Biochemical Systems*

The common feature of the affinity systems is a protein molecule containing a selective binding site. The binding site consists of a combination of functional groups as well as hydrophilic and hydrophobic regions. Selectivity of binding depends on a combination of characteristics of the regions and their orientation in three-dimensional space. Detailed descriptions of the binding mechanisms and energetics will not be given here, but a general description which may be used for all of the affinity systems is:

$$A + BS \underset{k_{-1}}{\overset{k_1}{\rightleftharpoons}} A - BS \tag{1}$$

where A represents the analyte, BS is the binding site and A–BS is the bound complex. k_1 and k_{-1} are the forward and reverse rate constants for the binding process. The equilibrium expression is then:

$$k_{A-BS}-k_{as} = \frac{[A-BS]}{[A][BS]} = \frac{k_1}{k_{-1}} \tag{2}$$

Some characteristics of the binding interaction may be described in terms of the parameters of equation 2. This general treatment may be used to evaluate affinity-based systems for sensor development since the equilibrium constants for many affinity systems are well characterized as commercial assays.

The sensitivity of the binding event is determined by K_{as}, as this is directly related to the amount of complex formed for given analyte and binding site concentrations. The response rate will be determined by k_1, which often includes diffusion and permeation steps in practical systems. Selectivity may be defined as the relative affinity of the binding site for one molecule over any interfering molecules to be considered. This is the relative K_{as} in a reversible system. In an irreversible system, the relative values for k_1 may better describe the selectivity, which would then be dynamically determined at short times and equilibrium determined only after many periods of $1/k_{-1}$.

The reverse rate k_{-1} will determine the reversibility of the complex. Systems for which k_{-1}, is very small ($1/k_{-1}$ is the reversal time, and may be hours, days or longer) are referred to as irreversible, and may provide 'one-shot' or 'probe' devices. In some cases it may be possible to modify the system to increase k_{-1}, though it should be noted that if k_{-1} is increased without a complementary increase in k_1, a loss of sensitivity will result from a decrease in the k_1/k_{-1} term. It is possible to manipulate k_{-1} after a measurement is made, as in the regeneration of an immunochemical binding surface by lowering the pH to dissociate bound complex prior to another analysis cycle.

2.2.1.2 Antibody : Antigen Systems

Antibody : Antigen (Ab : Ag) complexation where one of the species is labelled, is one of the best techniques for determining analyte concentration (generally, the antigen is the analyte) with respect to sensitivity, selectivity and availability for a wide range of compounds. This is mainly because complexation constants are on the order of 10^9, yielding high sensitivity and a high degree of selectivity.

Assays are generally extended to many analytes using competitive binding and sandwich techniques. Competitive binding assay begins with a labelled analogue of the antigen already bound to the antibody. Antigen is then introduced, and displaces the analogue. The change in the amount of free or bound labelled component is measured. In sandwich assay (or immunosorbent assay) antigen is initially allowed to bind to a layer of antibody. A second layer of labelled antibody is added which complexes with the bound antigen forming a 'sandwich' structure. The amount of signal from this second labelled antibody layer is related to the amount of antigen originally present. The sensitivity of sandwich assays has been enhanced using enzyme-linked immunosorbent assay (ELISA), which has extended detection limits to the sub-picomolar regime (Kitamura *et al.*, 1989).

The main deficiencies of immunoassay based on labelled species are heterogenous assays, slow response rate, and formation of irreversible complexes. The heterogenous aspect refers to the need for a 'separation' or 'washing' step to be done during analysis to separate bound analyte from unbound and non-specifically bound components. Incubation of Ab : Ag pairs often requires many minutes or hours to reach a complete response. The irreversibility means that de-complexation, if necessary, would require hours to days to occur. For some assays, it may be possible to induce de-complexation by addition of appropriate reagents, providing an assay which is renewable though not reversible. These deficiencies eliminate the possibility of continuous or on-line analysis, and restrict application to a single analysis, as in a probe device. Further complications arise when it is desirable to do sandwich or competitive binding assays. Sandwich assays require a separate addition of antibody after antigen has reacted, which is difficult to do on-line. In competitive binding assays, it may be necessary to detect the quantity of displaced component in solution. In an on-line or flow-through scheme, this component must be contained in the sensing region, but it is difficult to find a selective containment barrier which is permeable to the analyte but not to an analogue.

Homogenous immunoassay (i.e. the elimination of separation procedures) requires some alternate means of distinguishing the relative amount of bound and unbound components. In some examples of fluoroimmunoassay this is possible because the multiple parameters of the fluorescence measurement may be used to identify separate signals corresponding to each component.

2.2.1.3   *Enzyme : Substrate Systems*

Enzyme : substrate systems are available with a wide range of selectivity, some approaching that of Ab : Ag complexes. There are some important advantages over Ab : Ag systems. The response rates of enzyme reactions are generally much faster, and some measurements may be made in only seconds. The enzyme-derived signal may be described as pseudo-reversible, meaning that although some components are irreversibly consumed in the reaction, the measured parameter varies reversibly with the concentration of the analyte. Since many detected species react with each enzyme in a period of seconds (depending on the 'turnover' number of the reaction) there is an inherent amplification of signal over other affinity complexes.

Michaelis–Menton kinetics indicate a dependence of reaction rate on substrate and enzyme concentration where the enzyme reaction may be written as:

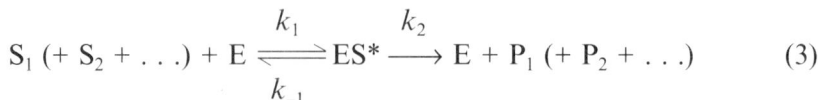

$$S_1 (+ S_2 + \ldots) + E \underset{k_{-1}}{\overset{k_1}{\rightleftharpoons}} ES^* \overset{k_2}{\longrightarrow} E + P_1 (+ P_2 + \ldots) \tag{3}$$

Where S_x represents the possible substrates and E the enzyme, ES^* is the complex, P_x represents possible products, and k_x represents the rate constant for the various steps in the enzymatic process. For the case where there is one substrate the rate of enzyme reaction with substrate may be written as:

$$R = k_2 \times \frac{[E]\,[S]}{([S] + k_m)} \tag{3}$$

where k_m is the Michaelis constant and $k_m = (k_1 + k_2)/k_1$.

Selectivity of the enzyme for one substrate over another (generally the analyte is a substrate) may be approximated by the relative values for k_1/k_{-1}, assuming all complexed species have the same probability of reacting to give products (k_2 is identical for different substrates). This selectivity differs greatly from one enzyme to another, and in part depends on the classification of analytes. In some cases, an enzyme may react with many different compounds which possess a common functional group, as in protease reactions which cleave amide bonds of proteins. In others, the enzyme selectivity is similar to that of antibodies.

Sensitivity of a reaction may be defined as the slope of a plot of R vs [S], which is $k_2 \times [E]/k_m$ in the case where $[S] \ll k_m$. Sensitivity of a device further depends on the relationship between R and the measured signal.

Sensing schemes based on enzymes commonly operate in three modes. (1) The rate of reaction may be measured by determining substrate or product concentration as a function of time. This is usually extrapolated to an initial rate at time zero. This takes advantage of the rapid response observed with some enzyme reactions, and avoids problems encountered with the build-up of products which may have deleterious effects on the enzyme resulting in dynamic alteration of reaction rates. (2) The final concentration of product (or secondary substrate where applicable) may be measured and related to the total amount of substrate initially present. This allows for signal enhancement by chemical integration, though at the expense of a long analysis time and possible product interferences. (3) The concentration of substrate or product may be measured after reaching a steady-state level dependent on substrate concentration. The steady-state condition is usually achieved when the enzyme reaction is much faster than diffusion of the measured species, resulting in a diffusion-limited concentration in the measurement region.

2.2.1.4   Enzyme Availability for Analytes

The most significant limitation of enzyme-based sensors is that enzymes are known for only a limited number of potential analytes, and not all of these are easily isolated or immobilized. One method in which this is overcome is the ELISA technique, in which an antibody is isolated or designed for a target analyte, and then the enzyme is coupled to that antibody. ELISA is usually done as a sandwich assay, which introduces all of the limitations of the immunoassay techniques. It allows for determination of any antigen using an enzyme-based measurement, and can enhance the sensitivity of the immunoassay by chemical integration of the signal as the enzyme reaction proceeds over an extended period of time.

Synthetic catalysts which function like enzymes have been designed, and are referred to as catalytic antibodies or synzymes (Schulz, 1988). This is based on the principle that some enzymes catalyse reactions by stabilizing one or more target molecules in a conformation which favours the reaction. An example is ester hydrolysis, which may be catalysed by fixing the ester molecule so that the ester linkage is exposed to solution and steric hindrance from substituents is minimized. The preparation of a synzyme begins with the selection of an appropriate molecule (not a target) to which an antibody can be raised. This molecule is chosen to have a structure that closely resembles the electrostatic and steric configuration that is required in a transition state complex to catalyse reactions of the target molecule. The antibody is then prepared and tested for complexation and catalysis of the original target molecule. Using this method, molecules with selectivity and binding constants similar to enzymes have been synthesized. Hydrolysis of substrates by synzymes has been increased by a factor of 10^3 or 10^4 over hydroxyl-induced reaction in bulk solution. Catalytic antibodies usually do not approach the activity of true enzymes, and therefore detection in an assay more closely resembles immunoassay, but the catalytic behaviour does serve to provide reversibility of the signal by eliminating the complex. Systems have been reported (Tramontano *et al.*, 1988) which do approach the activity of enzymes, and therefore represent true synthetic enzymes designed to detect a substrate of choice.

2.2.1.5   Lectin : Saccharide Systems

The interactions of lectin and saccharides are attractive for sensor applications as they are reversible, even though sensitivity and selectivity are generally less than for enzymes and antibodies. The lectin concanavalin A (Con A) will form reversible complexes with different saccharide residues. Selectivity is limited as many different molecules have saccharide functionalities and will therefore complex with Con A. Sensitivity is characterized by a range of association constants, such as $2\cdot06 \times 10^4$ M^{-1} for the polysaccharide methyl α-D-mannopyranoside and $5\cdot88 \times 10^2$ M^{-1} for the monosaccharide D-glucose.

2.2.1.6   Molecular Receptor : Agonist Systems

The molecular receptor protein : agonist complexes are the affinity system which is most directly designed for biosensing in natural systems. The complexes are formed with reasonably high sensitivity ($K_{as} \approx 10^6$) and selectivity, fast formation rates, and are reversible. The proteins themselves, however, are the most difficult to manipulate. Molecular receptors are membrane-bound, and the presence and composition of the surrounding lipid bimolecular layer has a dramatic effect on the overall membrane/protein response. This prevents normal characterization using methods such as recrystallization, and only one such protein, the acetylcholine receptor (AChR), has been isolated and characterized to any great extent. It has been reported that the selectivity of AChR may be retained in a biosensor when not using a lipid membrane support (Eldefrawi, 1989).

2.2.2   Desirable Characteristics of a Biosensor

Table 2.1 summarizes desirable characteristics of a biosensor (Krull & Thompson, 1985). The device should be selective for analytes closely related in structure or function and give no response to other compounds in a complex mixture. Accuracy to ±1% sensitivity of millivolts to volts in the concentration range of interest, and a linear range of many decades of concentration are challenging to obtain in a biosensor, as clinically useful concentrations are often in the nanomolar range. The response should be stable to better than ±1% throughout minor variations in conditions, e.g. temperature changes of a few degrees, agitation of the device, and the device should be sufficiently rugged that it will not be damaged by normal use. In keeping with its use as a probe device, a biosensor should be as easy to use as a pH electrode, small enough to be inserted intravenously (i.e. maximum dimension 1 to 2 mm) and able to

Table 2.1
Desirable Characteristics for Biosensors

---

selectivity
sensitivity
fast response
reversibility with respect to determinant-device interaction
small size
rugged structural characteristics
inexpensive (throw-away construction for medical purposes)
facile use by non-expert personnel
large linear range
facile interfacing with computers and other configurations for telemetric purposes
biocompatibility

---

respond within seconds. It would be best to have available low-cost, throw-away devices for diagnostic purposes, as well as reversible devices for continuous monitoring of analyte. When used for *in vivo* monitoring, the biosensor must be biocompatible. The device should have a lifetime of weeks or more. Biological molecules are ideally suited to attain some of these characteristics, while other characteristics are more difficult to attain, as discussed below.

2.2.2.1 *Selectivity*

Selectivity is the key feature of biological receptors that has brought about their popularity for use in biosensors. Receptors range in degree of selectivity from recognizing a whole class of compounds (e.g. subtilisin binds and cleaves any protein or peptide) to recognizing only a few closely related compounds (e.g. antibodies). Selectivity is determined by the number and strength of interactions in three dimensions between receptor and ligand. If a suitably selective receptor for the analyte does not exist, in many cases an antibody can be made for it.

2.2.2.2 *Sensitivity and Response Time*

The reaction time of receptor and ligand is governed by the time required for the diffusion of the two together. For minimum response time, the receptor should be coupled directly to the device with no barriers to the analyte solution. In practice, this may not be possible, since the analyte solution may contain species that interfere with the device or degrade the receptor, requiring a selective membrane barrier to be employed. Since each receptor can generate a signal from one molecule of analyte, sensitivity is dependent on the efficiency of signal transfer from receptor to device, and response characteristics of the device. Again, the receptor should be coupled to the device for maximum sensitivity. Receptors that are not permanently attached to the device may diffuse away from the device, causing a decrease in sensitivity with time, drift in response and a need for frequent calibration.

2.2.2.3 *Reversibility*
Receptors bind ligands with a broad range of affinities, and binding may be reversible or irreversible. If a biosensor uses a high affinity receptor–ligand combination it must be considered a throw-away sensor, unless the ligand can be dissociated and active receptor regenerated, for example, by increasing ionic strength, changing pH or heating. Reusable devices generally employ receptor–ligand combinations of lower affinity which exhibit reversible binding. One danger of using biological molecules is that there usually exist irreversible binding receptor inhibitors, synthetic or natural, which may be present in the analyte solution and subsequently render the device ineffective. At present some receptors are very expensive making them unsuitable for throw-away devices. With cloning technology and the economics of supply and demand, this consideration should be minimized in the next ten years.

2.2.2.4 *Lifetime and Ruggedness*
The lifetime of biosensors containing proteins is limited because of the susceptibility of biological molecules to degradation, as previously discussed. In addition, the receptors will be exposed to proteases, which exist in all bodily fluid samples. Invasion of the biosensor by microorganisms will also eventually occur since manufacturing the device under sterile conditions and sterilizing all samples is impossible or impractical at this time. The transducers described above are sufficiently rugged within themselves to withstand normal handling and a wide range of chemical environments, but when biological molecules are added to the device, faults often develop at the receptor–transducer interface, or within the biological layer. For example, chemical field effect transistors are plagued by leakage of water and charged species into the gate/chemically sensitive layer which causes a change in potential, and most proteins cannot be exposed to organic solvent or extremes of pH. Improved device design and protein immobilization methods are required to minimize these problems.

2.2.2.5 *Biocompatibility*
Biosensors may experience impaired function when exposed to biological fluids. Proteins and lipids adsorb to most surfaces, and blood compo-

nents will form a clot over most surfaces with time. The problem is exacerbated *in vivo* because clotting will be more extensive, the immuno logical system will deliberately attack the device, primarily with polymor-phonuclear leukocytes (Eldefrawi, 1989), and the biosensor will be surrounded by a fibrous tissue sheath if implanted for more than a few days. These reactions may be prevented or lessened in severity if the biosensor surface is coated with an anticoagulant such as heparin, or with a common protein such as albumin that is not recognized by the body as being foreign (Thompson & Vandenberg, 1986). In addition, the device may be corroded by salt and water of the body.

2.2.2.6 Biosensor Design

Selective interactions of ligand with receptor must be differentiated from non-selective adsorption of other compounds. The design of many biosensors assumes that all interactions will be selective. Where this assumption proves false, signal generation from non-selective adsorption will result. Partial compensation for non-selective adsorption can be made by subtraction of the signal from a reference device in which the selective chemical layer is either absent, or is altered, for example, by irreversibly denaturing the receptor. Other variables which affect biosensor function, such as temperature and pH, can be co-analysed, and the biosensor calibration curve can be corrected for changing conditions (Caras & Janata, 1980).

Amplification of the signal generated by the analyte, either in the device or the selective chemical layer, can increase precision, sensitivity and signal-to-noise. Chemical field effect transistors have an internal gain mechanism in the device due to the high input impedance. It is also poss-ible to amplify the signal in the selective layer by coupled enzyme reac-tions, catalytic antibodies or enzyme amplification. An example of the latter is a biosensor for biotin, wherein lipoate (covalently attached to the sensor surface) and biotin (in the analyte solution) compete for the binding of enzyme-linked avidin. The avidin bound to the lipoate on the sensor is determined by adding the substrate to the enzyme (Ikariyama *et al.*, 1985). This creates many molecules of chemical signal (enzyme prod-uct) per molecule of analyte. Device design also determines biosensor size, facility of interfacing, and ease of use by non-expert personnel.

Despite the plethora of research papers and the money invested into biosensor research, there are few viable commercial devices on the mar-ket. Table 2.2 lists some commercially available biosensors. Enzyme elec-trodes are the most successful devices and are available for a variety of analytes.

Table 2.2
Examples of Commercially Available Biosensors

Manufacturer	Biosensor type	Analytes detected
Ajinomoto	microbial electrode	biol. oxygen demand
Analytical Instr. (Toyo Jozo)	pot. electrode	glucose, alcohol glycerol, lactate
Daiichi Kagaku Genetics Int. Mitsubishi Toa Electric	amp. electrode	glucose
Gambro	pot. electrode	glucose
Lithuanian Acad. Sci. Radelkis	electrode	glucose
Fuji Electric	enz. electrode	alpha-amylase, acetic acid ethanol, glucose, uric acid
Leeds & Northrup	electrode	sugars
Midwest Res. Inst	electrode	pesticides
Omron Toyobo Co.	amp. electrode	glucose, lactate
Oriental Electric	pot. electrode	fish, freshness
Owens-Illinois (Kimble)	amp. electrode pot. electrode	glucose urea
Seres	pot. electrode	glucose, lysine, alcohol
Universal Sensors	amp. electrode	alcohol, L-amino acids creatinine, glucose, sucrose lactate, lactose, uric acid oxalate, salicylate
	pot. electrode	aspartate, tryptophan, urea aspartame, leucine, lysine
Univ. of Lund	enz. thermistor	sugars, alcohol, lactate penicillin, urea, oxalate triglycerides
Yellow Springs Instrument Co.	electrode	fructose, lipase, amylase glucose, triglycerides lactate, alcohol, sucrose

If a protein is to be used in the chemically selective layer, it should ideally cover the surface fully, be coupled directly and permanently to the device, and retain its properties of selective binding. Much effort has been directed towards attainment of these goals, and the following sections will summarize some general information and indicate the current state of research regarding biological molecules immobilized onto surfaces. The adsorption of proteins to surfaces will be discussed because adsorption of protein is fast and easy, and has been used successfully in certain instances. It is necessary to understand adsorption stabilization to compare and contrast covalent stabilization. Also, covalent binding of receptor to biosensor begins with the adsorption of receptor to the surface.

2.2.3 Protein Adsorption

2.2.3.1 *Energetics*

The three-dimensional conformation of a protein is not very stable – the native state of a globular protein is only 21–63 kJ/mol more stable than the unfolded state under physiological conditions (Pace, 1975). This marginal stability is due to opposing forces in optimizing coulombic, hydrogen bond, hydrophobic and van der Waals interactions as well as rotational mobility, and bond lengths and angles for the protein (Norde, 1986). The three-dimensional structure of a protein in a particular environment is the result of intramolecular interactions and interactions between the protein and the environment (Norde, 1986).

Proteins always adsorb to some extent to a solid surface and adsorb to different surfaces by different mechanisms (Andrade *et al.*, 1984). This adsorption is energetically favoured because of the decrease in interfacial tension and resultant decrease in Gibb's free energy of the system. Proteins are ideal for decreasing the free energy of a system due to their ability to change conformation and provide low energy surfaces to each phase in the system. For example, there are two types of non-interacting adsorption sites for bovine serum albumin, gamma-globulins and fibrinogen on most surfaces: (a) hydrophilic, exothermic, reversible sites with enthalpy of adsorption –42 kJ/mol, and (b) hydrophobic, endothermic, tight binding sites with enthalpy of adsorption 21–84 kJ/mol (Dillman & Miller, 1973). Proteins often change conformation on adsorption to a hydrophobic surface in order to expose hydrophobic residues to the surface, possibly leading to surface denaturation of the protein (Saderquist & Walton, 1980). For most applications this is undesirable since orientation and tertiary structure are critical for receptor–ligand interactions (Andrade *et al.*, 1984). Hydrophilic surfaces are somewhat less likely to denature proteins since the water-soluble protein has many polar and charged residues facing the environment which can interact with the surface.

Adsorption is heavily entropic (Saderquist & Walton, 1980). The difference in the electrokinetic charge density before and after protein adsorption shows that low molecular weight ions participate in the adsorption process (Norde, 1986). For each adsorbed protein segment, several solvent molecules (and possibly other components) at the interface are released. The resultant gain in translational entropy of the solvent is large, thus the loss of translational entropy of the protein is of minor importance (Norde, 1986). The enthalpy of adsorption for albumin on silica was found to be +136 kJ/mol·K as measured by calorimetry, demonstrating the reaction to be entropy driven; the entropy gain from a small change in secondary structure being large enough to compensate for the unfavourable enthalpy change (Norde *et al.*, 1986).

Thermal motion is a major factor in the total free energy of adsorption of a protein. Although the free energy of each segment of the protein is only 1 kT, since many segments of the protein are adsorbed, the sum is great (Norde, 1986). For adsorption at the air–water interface of bovine serum albumin, it was found that 3–5 amino acid residues were needed to anchor albumin at the air–water interface assuming the average surface area per amino acid is about 15 Å^2 (Damodaran & Song, 1988). This is a surprisingly small contact area for a protein of molecular weight 67 000 and dimensions 140×40 Å (Hendrickson *et al.*, 1975).

2.2.3.2 *Electrostatics*

The electrostatic considerations of protein adsorption to surfaces has been thoroughly discussed in a review by Norde (1986). The important points are summarized below. Electrostatic conditions often favour adsorption of protein at a solid surface. The interface can be viewed as an electrical double layer; the electrokinetic charge is neutralized by counterions diffusely distributed in the buffer solution. When protein is added to the buffer solution, some will randomly diffuse near the surface, whereupon the electrical double layers of the protein and interface will overlap and a redistribution of charge will result. If the protein and the surface have like charges, an energy barrier exists which retards adsorption, and the protein must rearrange its structure to a more favourable conformation for adsorption. Upon adsorption, titratable groups are likely to undergo environmental alterations, which affect titration behaviour, and the electrokinetic charge density can be monitored to follow adsorption. In addition, interactions between charges can be screened by electrolyte, but adsorption is sensitive to the type of ion (McLaren, 1954; Dulm *et al.*, 1981), perhaps due to differing sizes and charge densities. The role of ions is more intricate than simple screening.

The electrical charge of the protein affects its structural stability, which may affect its adsorption properties. The electrostatic Gibb's free energy of a protein is greater for a compact protein than a loose, solvent-permeated one, and is at a minimum at the isoelectric point of the protein since all charges are paired and there are no repulsions. For example, the surface concentration of albumin on negatively charged polystyrene decreases as the pH strays from the isoelectric point, due not to charge repulsions but to structural changes in the molecule (Norde & Lyklema, 1978*a*,*b*,*c*). However, albumin is adsorbed onto hydrophobic and hydrophilic surfaces even if the protein and the surface have the same charge, demonstrating that charge–charge interactions are not always the most important force.

2.2.3.3 Hydrophobic Interactions

Hydrophobic forces are usually more important than polar or ionic attractions between a protein and a surface. This is due to the large entropy gain during dehydration of hydrophobic parts of the protein and sorbent. Hydrophobic hydration water is highly ordered, thus has low entropy, but has a labile structure in contrast to bulk water (shown by anomolous heat capacity) (Norde, 1986). This drives adsorption to occur spontaneously (Norde, 1986).

Furthermore, there may be a strong configurational entropy driving force for an adsorbed globular protein to assume an extended configuration, especially if the extended state can be accommodated by maintaining a high degree of hydrophobic interaction comparable to that provided by the globular state (Andrade & Hlady, 1987; Andrade *et al.*, 1987). Dispersion (van der Waals) forces favour a compact structure but the magnitude of this effect is small (Norde, 1986). The new extended configuration can be stabilized by hydrogen bonding between peptide units, which is of minor importance in a solubilized protein, but becomes an important stabilizing force in a protein shielded from water (Norde, 1986). The intrinsic solution stability of a protein is very important in understanding adsorption behaviour (Andrade *et al.*, 1987). For example, a distorted or extended protein is less soluble than a globular protein, thus it has an increased tendency to bind to, and denature at, a surface (Andrade & Hlady, 1987).

In general, the strength of protein adsorption increases with the hydrophobicity of the surface (Dezelic *et al.*, 1971; Jonsson *et al.*, 1982; MacRitchie, 1972). The minimum adhesion is suggested to occur on surfaces having a critical surface tension of 20–30 mN/m (Baier, 1975; Baier & Glantz, 1978; Dexter, 1979). In aqueous solution, non-polar interactions are more important for adsorption than polar interactions, but bond strength is at a maximum if both are involved (Norde, 1986).

2.2.3.4 Kinetics

Protein adsorption is dynamic, and the proteins continually reorient themselves on the surface until they finally reach an equilibrium (Andrade &
Hlady, 1986). If the surface is deformable, as with a polymer surface, it
may also be conformationally changed on interaction with the protein
(Andrade *et al.*, 1987). Due to the plastic characteristics of protein and
surface, the kinetics of adsorption of proteins at interfaces is highly complex and poorly understood (Damodaran & Song, 1988). Adsorption is
governed by diffusion, the energy barrier at the interface, and the conformation of the protein (Damodaran & Song, 1988), and depends on the
nature of the protein and surface, the solvent, and their interdependencies
(Andrade *et al.*, 1984). General conclusions about the interfacial behaviour
of proteins are difficult to make because the protein itself may be difficult
to describe chemically. Not all impurities (e.g. fatty acids, ions) are eliminated in the purification process, and the structural stability of the protein can be varied by preparation and storage conditions (Norde, 1986).

Different researchers have analysed the events in protein adsorption to
surfaces, and have categorized them into several stages. One study
describes three stages: (1) a brief period of reversible adsorption, (2) a
period during which the protein is essentially irreversibly adsorbed and
undergoes slow conformational changes, and (3) a stage where denatured
protein is desorbed slowly and adsorption is unlikely (Soderquist &
Walton, 1980). Another interpretation assumes five stages: (1) transport
towards the interface, (2) attachment of the protein to the interface, (3)
structural rearrangement during adsorption, (4) detachment, and (5)
transportation away from the interface (Norde, 1986). Several mathematical models of the kinetics of protein adsorption are compared in Norde
(1986). Each model is lacking in fully describing protein behaviour due to
the heterogeneity of the adsorbed protein layer with respect to desorption,
and a poor understanding of the mechanism underlying reduced desorption of a fraction of the protein layer. It should be stressed that none of
the Langmuir premises is satisfied for protein adsorption (i.e. adsorption
is not reversible, adsorption is not at fixed sites, molecules may change
form on adsorption and lateral interaction is possible), therefore protein
adsorption should not be modelled to fit the classical Langmuir isotherm.

At low concentrations of protein, the rate-limiting step in protein
adsorption is transport towards the interface, as:

$$\Gamma(t) = 2c \sqrt{(D/3 \cdot 14)} \sqrt{t} \qquad (5)$$

where Γ = surface concentration, c = concentration, D = diffusion constant
and t = incubation time (Norde, 1986). At high concentration, the rate-

limiting step is attachment at the interface because less surface area is available for adsorption. However, simple diffusion theory often cannot adequately explain protein adsorption at fluid interfaces. For example, the measurement of apparent diffusion coefficients (calculated from surface concentration versus time curves) of a series of proteins were found to decrease as the protein structure became more compact, in conflict with well-established diffusion theory (Damodaran & Song, 1988; Andrade & Hlady, 1987).

Partially denatured bovine serum albumin takes two hours to reach a steady state at the air–water interface, whereas native bovine serum albumin takes 15 hours (Damodaran & Song, 1988). Two hours are required for saturation of antibody onto dimethyldichloro silane-treated surfaces, whereas seven to eight hours are needed for saturation of glutaraldehyde-activated surfaces (Lin *et al.*, 1988). Immunoglobulin G adsorbs more quickly onto methylated (hydrophobic) surfaces than onto silicon oxide (hydrophilic) surfaces in low ionic strength buffer. In high ionic strength buffer, the amount of immunoglobulin G adsorbed is less for hydrophilic surfaces than hydrophobic surfaces, but immunoglobulin G adsorbs to both surface types with the same kinetics (Jonsson *et al.*, 1985*b*).

2.2.3.5 *Conformation at the Interface*

A good example of how conformational differences and amino acid composition and distribution affect the rate of adsorption of protein is given in Damodaran & Song (1988). The effect of conformation, not chemistry, on adsorption was studied at the air–water interface using bovine serum albumin that was denatured to different extents. The initial adsorption was independent of conformation, but the rearrangement of the protein, detected by the surface area/molecule, was conformation-specific. Native protein, containing more α-helix, occupied more surface area than denatured protein, and the rate of adsorption was slower. In another study, it was demonstrated that native and denatured proteins have different affinities to polyvinyl chloride (Golander & Kiss, 1988). Inspection of adsorption isotherms for compact molecules reveals well-defined plateaus at low concentrations, indicating monolayer formation (Norde, 1986). Flexible, highly-solvated polymers do not exhibit such plateaus, indicating that adsorbed molecules detach segments to create surface area to which new molecules may attach (Norde, 1986).

The extent of protein structural rearrangement at an interface depends on the relative rates of arrival and structural changes. More structural changes will occur if there is more room at the surface (Andrade *et al.*, 1987). Rearrangement of the protein on the surface may sometimes be detected by adsorption isotherms. Low plateaus in adsorption isotherms signify incomplete monolayer coverage or spreading of the protein on the surface. Globular proteins adsorb in a compact fashion and do not expand, but they may flatten under some conditions (Norde, 1986). Albumin on platinum decreases in thickness when the platinum is positively charged, indicating that the protein collapses to form a dense monolayer (Morrisey *et al.*, 1976). Steps in adsorption isotherms signify two different modes of binding; multilayer adsorption, or structural transitions (Norde, 1986). Proteins have been observed to shrink and swell on substrates with time (Cuypers *et al.*, 1977, 1978).

If the dimensions of the protein are known and the surface density is measured, one can calculate whether the protein has adsorbed with sub-monolayer, monolayer or multilayer coverage, and if the smallest or largest dimension is in contact with the surface. For example, observed thicknesses indicated that bovine serum albumin, immunoglobulin G and fibrinogen adsorb to polyvinyl chloride side-on at low concentration and end-on at higher concentration (Golander & Kiss, 1988; Morrisey & Ham, 1978; Vroman, 1967). Albumin is adsorbed side-on to copolypeptide and silicone surfaces (Soderquist & Walton, 1980). Adsorption of albumin usually occurs in a close-packed monolayer side-on or end-on (Norde, 1986). Bovine serum albumin onto hydrophilic surfaces has a coverage of $2 \cdot 1 \ \mu g/mm^2$ (Hlady *et al.*, 1987), demonstrating side-on adsorption (maximum coverage $2 \cdot 5 \ \mu g/mm^2$ side-on, $6 \cdot 6 \ \mu g/mm^2$ end-on) (Golander & Kiss, 1988). Albumin on platinum forms a $2 \ \mu g/mm^2$ monolayer, 20 nm thick (Morrisey *et al.*, 1976). A brief literature review in Jonsson *et al.* (1985) concludes that immunoglobulin G, adsorbing to a variety of surfaces at 2–$7 \ ng/mm^2$, contacts the surface mostly side-on (maximum coverage $2 \ ng/mm^2$) and partly end-on (maximum coverage 14–$19 \ ng/mm^2$) (Golander & Kiss, 1988; Morrisey *et al.*, 1976). The shortest dimension of concanavalin A contacts mica surfaces giving a coverage of $2 \cdot 9 \ \mu g/ mm^2$ (Afshar *et al.*, 1987).

Coverage of protein may not be uniform – islands may form. Studies on ferritin have shown that protein adsorption is initiated when a single molecule adsorbs, followed by protein–protein adsorption and cluster growth (Nygren, 1988). This uneven sub-monolayer coverage indicates that cohesive forces between molecules are stronger than protein-surface interactions (Afshar *et al.* 1987).

2.2.3.6   *Examples of Protein-Sorbent Interactions*
Literature pertaining to interaction of proteins with solid surfaces is extensive but somewhat confusing (Jonsson *et al.*, 1985*b*). A few examples

of systematic studies are given below. Adsorption isotherms (concentration of protein versus amount protein adsorbed to various surfaces) are often in conflict since minor changes in pH or ionic strength create major changes in the isotherm (Jonsson *et al.*, 1985*b*). Silicon wafers have been derivatized by dichlorodimethylsilane to give a continuous gradient of hydroxyl and methyl groups which is excellent for the study of protein adsorption to substrates of different polarities (Hlady *et al.*, 1987; Elwing *et al.*, 1987). Bovine serum albumin adsorbs most to hydrophilic substrates (Hlady *et al.*, 1987). Polyethylene glycol was found to decrease adsorption of various proteins to silicon wafers by 25–33%, thus nonspecific adsorption is attributed to hydroxyl groups (Fusek *et al.*, 1987). Adsorption of bovine serum albumin, immunoglobulin G and fibrinogen are reduced on hydrophilic polymethacrylate/methacrylic acid and polyethylene oxide as compared to polyvinyl chloride (Golander & Kiss, 1988). However, there is a very low amount of adsorption when the polyethylene oxide chains are covalently bound to the surface (Golander & Kiss, 1988).

Denaturation of protein upon interaction with surfaces is well documented. Short literature reviews of protein conformational changes on surface binding are given in Andrade *et al.* (1984) and Andrade *et al.* (1987). Surface-bound protein is expected to be in a different conformation than protein in solution (Andrade *et al.*, 1984). Human fibrinogen eluted after adsorption to Pyrex glass shows a 50% loss of α-helicity due to interactions with the glass (Chan & Brash, 1981). Albumin experiences a 10% decrease in α-helix (Norde *et al.*, 1986) and fibrinogen loses most of its β-sheet after desorption from copolypeptide and silicone surfaces (Soderquist & Walton, 1980). For albumin this corresponds to 60 amino acids being liberated from the α-helix, an entropy gain of 691 J/mol.K, and a Gibb's free energy of adsorption of –205 J/mol (Norde *et al.*, 1986). Surface binding can change conformation on albumin and fibronectin (Andrade *et al.*, 1984). Fibrinogen aggregates on silica, but no structural change is seen for albumin or prothrombin (Morrisey & Stromberg, 1984).

Antibodies have been known to recognize antigenic determinants on proteins only when adsorbed (not in solution), thereby demonstrating a change in conformation on adsorption (Mierendorf & Dimond, 1983). This has caused erroneous results in solid phase immunoassays (Miller *et al.*, 1983). Antibodies adsorbed to aminopropyl triethoxy silane (APTES) are suspected to change conformation since antigen binding to immobilized antibody is inhibited by antibodies in solution (Lin *et al.*, 1988). Adsorption of antigen onto polydimethyl siloxane (relatively hydrophobic) results in the loss of some antigenic determinants (Darst *et al.*, 1988).

A contrasting report states that antibody–antigen complexes formed at the solid phase are more stable than if formed in solution (Nygren, 1988).

2.2.3.7   Prevention of Adsorption

Nonspecific adsorption on the solid phase may in some applications be minimized by incorporation of a detergent into buffer solutions (Jonsson *et al.*, 1985*a*). Detergent greatly decreases immunoglobulin G adsorption onto methylated surfaces (Jonsson *et al.*, 1985*a*). Concentrations of the non-ionic detergent Tween 20 ranging from 0·05 to 0·5% give the identical degree of adsorption inhibition for various proteins on hydrophobic/hydrophilic gradients, and the degree of protein coverage depends on the type of protein (Elwing *et al.*, 1987). γ-globulin adsorption to slightly hydrophobic surfaces decreases when Tween 20 is used (Elwing *et al.*, 1987). Ionic strength can also prevent non-specific adsorption and its influence should be the greatest on surfaces with a net charge at the pH used (Jonsson *et al.*, 1985*b*). High ionic strength buffer decreases the amount of immunoglobulin G bound to polar surfaces (Jonsson *et al.*, 1985). In contrast, no effect of ionic strength is seen with immunoglobulin G adsorption onto methylated surfaces, indicating that electrostatic forces are not important in this interaction (Jonsson *et al.*, 1985*b*).

2.2.3.8   Desorption

Some salient features of protein desorption may be summarized as follows: (1) none or a small fraction of protein desorbs on dilution with solvent, (2) additional adsorption may occur by changing the external conditions (pH, ionic strength, temperature) but desorption is at most only partially realized by reversing these conditions, (3) exchange of protein between the surface and solution occurs, (4) adsorbed protein may be displaced by other protein, and (5) protein that changed conformation during adsorption may not revert back to the native structure (Norde, 1986).

The molar free energy of adsorption is positive and large, and increases further on structural rearrangement (Norde, 1986). Thus, desorption is not the reverse of adsorption – the sum of the energies of adsorption and rearrangement must be overcome (Norde, 1986). The activation free energy of adsorption is smaller than for desorption so the latter process is much slower, and may take hours, days, or even longer (Norde, 1986). Distorted or extended protein is less soluble in solution, and thus has a decreased tendency to desorb (Andrade & Hlady, 1987). As protein–sorbent contact time increases, the probability of desorption decreases because more segments of the protein attach, each with a free energy of adsorption of 1 kT (where k = Boltzmann constant and T = temperature) (Andrade *et al.*, 1987; Andrade & Hlady, 1987; Norde, 1986). The adsorption affinity is greater on hydrophobic surfaces due to a greater number of contacts, thus desorption is less than from hydrophilic surfaces (Dezelic *et al.*, 1971; Jonsson *et al.*, 1982; MacRitchie, 1972; Norde, 1986).

After 16 hours only a small portion of adsorbed protein on various surfaces desorbs (Jonsson *et al.*, 1982). Clusters of ferritin adsorbed to quartz over 24–55 hours do not desorb after two hours of rinsing (Nygren, 1988). Adsorbed antibody can be desorbed by buffer or bovine serum albumin solutions if adsorption times are short, but for longer adsorption times, desorption is less (Darst *et al.*, 1988). Methylated surfaces lose 18% of adsorbed antibody in 10 hours, but only 2% more in the next 120 hours in phosphate-buffered saline (Lin *et al.*, 1988). Some adsorbed albumin is removed from silicon wafers by changing pH and ionic strength, but more is removed by low molecular weight displacers (Norde *et al.*, 1986). A medium which reduces surface tension and increases ionic strength elutes most immunoglobulin G from alkylamino surfaces (Jonsson *et al.*, 1985*a*).

2.2.3.9 *Exchange of Protein*
Where two or more proteins are present, one is generally more labile or conformationally adaptable to the interface, and therefore it spreads more effectively and displaces the other (Andrade *et al.*, 1987). Exchange of protein involves the sequential detachment of a segment of the adsorbed protein, and attachment of the new molecule (Norde, 1986). The free energy required for protein exchange is that of a single segment, rather than that of the whole molecule. Displacement occurs much faster than desorption, the latter requiring all attached segments to detach simultaneously (Norde, 1986). The shear rate of flow at a hydrophobic polyethylene surface plays an important role in the exchange of albumin due to increased collisions between adsorbed and dissolved protein, but shear rate has no effect on fibrinogen exchange (Brash *et al.*, 1974; Brash & Samak, 1978; Chan & Brash, 1974). One study on albumin showed that even after 40 hours, only part of the monolayer was exchangeable (Brash *et al.*, 1974; Brash & Samak, 1978; Chan & Brash, 1974).

2.2.4 Non-covalent Immobilization

Literature surveys of non-covalent methods of immobilization are found in Scouten (1987) and Messing (1987). These methods include physical adsorption, ionic binding and metal chelation, and are the simplest, fastest and least expensive methods of immobilizing proteins (Messing, 1987). A few examples follow.

The classic example of non-covalent immobilization is that of invertase immobilized on activated carbon. The discovery was made in 1916 by Nelson and Griffin, but was patented in 1955 by Stone using the enzyme α-amylase. The enzyme and activated carbon were stirred at 10°C for one hour, then packed into a column. Passage of starch through the column resulted in glucose formation.

Glucose oxidase was incubated with an organic, ionic crystal for several hours to effect the adsorption of the enzyme to the crystal. The crystal was then packed in a cavity of a silver wire coated with Teflon. A reference electrode made with glucose oxidase that had been boiled for three minutes corrected for interferences. The response decreased 5% per day over four days, and after 28 days response was 31% of the original (Boutelle *et al.*, 1986). A similar system was created when ubiquinone, graphite powder and liquid paraffin were packed in the end of a glass tube, and dipped into a solution of D-gluconate dehydrogenase (Ikeda *et al.*, 1987). Approximately 1 μg/cm^2 enzyme was immobilized in this fashion (Ikeda *et al.*, 1987).

2.2.5 Covalent Stabilization

Covalent stabilization of proteins onto solid supports is described primarily in biosensor and chromatography literature. Almost all the papers describe a particular receptor for a particular application, stressing optimization of the specific system output, rather than details of protein binding and activity. An exception is the 'Methods in Enzymology' series in which four volumes (vols 44, 135, 136, 137) have been devoted to the immobilization of enzymes and cells, with primary emphasis on the chemistry of binding, along with characterization procedures and applications of immobilized enzymes. Volume 137 features analytical applications with emphasis on biosensors.

The biosensor literature dealing with covalent stabilization of receptors for the purpose of selective reaction with ligand frequently lacks documentation as to the conditions under which the reaction was done, and often reports only cursory characterization of the bound receptor. Indeed, many of the described biosensors are 'quick and dirty' devices which display proof of principle, rather than any detailed science. However, there are some rigorous studies which discuss frankly the problems associated with deposition of active protein, interferences, and the limits on the lifetime of the device. These papers will be discussed below, with emphasis placed on the reactions that were used in this work.

There are numerous reviews dealing with enzyme electrodes (Czaban, 1984; Mullen & Vadgama, 1986; Pinkerton & Lawson, 1982), field effect transistor devices (Caras *et al.*, 1985), fibre optic sensors (*Talanta*, **35**, 1988) and immunosensors (Pace *et al.*, 1985). As stated in a review on chemical sensors, there is one review paper for every four research papers in the field (Thompson *et al.*, 1987).

Differentiation must be made between spherical supports, used in most chromatography and flow-injection systems, and planar supports. The former can be expected to immobilize more receptor and react with more ligand, because the curvature of the support ensures good accessibility of reagents to the surface. Planar and large diameter spherical supports force more lateral interactions between immobilized receptors, increasing the probability of denaturation. Hindered access of the ligand to the active site of the receptor is also more likely due to the proximity of the surface or adjacent protein, especially when the ligand is large.

For inorganic supports to be coupled with organics, the most common coupling agents are the silanes, which simply require free hydroxyl groups on the support. The most commonly used silanes are 3-aminopropyl triethoxy silane (APTES) and glycidoxypropyl trimethoxy silane (GOPS), but others such as 3-mercaptopropyl triethoxy silane (Mannens *et al.*, 1988) and 3-(2-aminoethylamino)propyl trimethoxy silane (Genshaw, 1988) have been used. Deposition of the silanes is from either aqueous solution (Garcia, 1989; Masoom & Townshend, 1984, 1986; Sutherland, 1984; Watanabe *et al.*, 1988) or anhydrous solvent (Garcia, 1989; Williamson *et al.*, 1989; Moody *et al.*, 1986*a*). Reaction in aqueous solution gives very thick multilayers which may be subject to hydrolysis, while reaction in anhydrous solvent gives stable sub-monolayer to multilayer coverage, depending on the reaction time (Ishida & Koenig, 1980). The silane and support are heated or refluxed for 45 minutes (Ishida & Koenig, 1980) to six hours (Moody *et al.*, 1986*a*), or may be reacted at low pH in aqueous solution (Masoom & Townshend, 1986; Sutherland *et al.*, 1984).

In order to covalently bind protein, the silane may be treated with one of a number of reagents, of which glutaraldehyde, carbonyldiimidazole and cyanogen bromide are the most common. The choice of activating reagent usually seems to be arbitrary since rationale for use of the particular activating agent is rarely given. It has been found that different receptors retain varying amounts of activity depending on the activating reagent used. Taking the example of glutaraldehyde, it is usually reacted at a concentration of 2·5% (v/v) from aqueous solution (Kiba *et al.*, 1986; Masoom & Townshend, 1984, 1986; Moody *et al.*, 1986*a*; Watanabe *et al.*, 1988), although concentrations from 0·1% (Williamson *et al.*, 1989) to 12·5% (Arnold, 1985; Moody *et al.*, 1986*b*) are documented. The pH varies from 6·8 (Sutherland *et al.*, 1984) to 8·5 (Arnold, 1985; Girotti *et al.*, 1986; Moody *et al.*, 1986*b*), and the time of reaction from 15 minutes to two hours. In one case glutaraldehyde was vapour deposited from a 50% (w/v) solution at 0·5 torr, 80°C for 30 minutes (Karube *et al.*, 1986). During or after protein incubation, the resultant Schiff bases may be reduced with sodium borohydride, or the gentler reagent sodium cyanoborohydride (Matsumoto *et al.*, 1986). In one case, sodium cyanoborohydride was added simultaneously with the glutaraldehyde (Williamson *et al.*, 1989). In all cases, the activation was apparently successful, since the protein was subsequently bound to the activating agent.

The specifics of the conditions of protein incubation are often poorly described, which is surprising considering the importance of protein immobilization. The concentration of protein ranges from 5 to 25 mg/ml (Karube *et al.*, 1986; Sutherland *et al.*, 1984) where given, but is usually not stated. Incubation is generally at 4°C overnight, but the incubation time may vary from 30 minutes (Matsumoto *et al.*, 1986) to 48 hours (Sutherland *et al.*, 1984; Williamson *et al.*, 1989), and this may be preceded by a two (Moody *et al.*, 1986*b*; Williamson *et al.*, 1989) to five hour (Garcia *et al.*, 1989) room temperature incubation. In one case (Watanabe *et al.*, 1988), the only incubation was at room temperature for one hour. The pH of this reaction varies in literature from 5 to 9·3 (Comfort *et al.*, 1988) and is a compromise between the required pH for the coupling chemistry and the pH of greatest stability of the enzyme. Buffer type and concentration likewise vary greatly in response to pH requirements. Addition of detergent or salt to reduce non-specific adsorption is common.

Receptor has been bonded to an inorganic sensor surface without a spacer group in a few cases. Cyanuric acid was used to couple glucose oxidase and L-amino acid oxidase to a graphite electrode (Ianello & Tacynych, 1981), with efficient coupling of the chemical signal to the device surface. Cyanuric chloride coupling of glucose oxidase to a tin oxide electrode (Watanabe *et al.*, 1988) retained more enzyme activity than coupling to platinum or gold electrodes. Glucose oxidase was linked to an oxidized glassy carbon electrode (Narasimban & Wingard, 1986) with carbodiimide and aminophenyl boronic acid allowing good electron transfer to the electrode. The effect of spacer molecules was investigated by activating quartz by p-toluenesulfonyl chloride or p-nitrophenyl chlorotormate, and then binding protein to these reagents either directly, or via the spacer molecule hexanediamine. This resulted in varying degrees of protein density and activity (Williamson *et al.*, 1989).

Protein binding to organic or biomolecule surfaces such as agarose and dextran is routine in chromatography. Sepharose (agarose) resins are commercially available in unactivated form or activated by cyanogen bromide or tresyl chloride. Unactivated resins may be activated in the laboratory by other agents, for example carbodiimidazole (Comfort *et al.*, 1988). Nylon has been used after activation by diaminohexane or adipic acid dihydrazide, followed by glutaraldehyde or dimethylpimelimidate (Girotti *et al.*, 1986). Nylon has also been derivatized with dimethylsulfate, lysine and glutaraldehyde, then reacted with the receptor of choice (Arnold, 1985; Moody *et al.*, 1986*b*). Chloromethylated polystyrene has been activated with 1,8-diamino-4-aminooctane then glutaraldehyde (Kiba *et al.*, 1986). Other materials, such as aminated controlled pore glass (Hayashi *et al.*, 1986; Narasimban & Wingard, 1986) and aminocellulofine (Girotti *et al.*, 1986) have been glutaraldehyde activated.

Between each reaction step, appropriate rinsing steps are required to prevent carryover of reagent, but again, details of these procedures are usually lacking. This is an especially serious omission following the protein incubation step, because the inevitable non-specific adsorption of protein may be reduced by thorough rinsing with detergent, high salt, or protein-containing buffers. Also, unreacted sites may still be exposed after receptor incubation, possibly allowing ligand to bind covalently. Blocking of reactive sites is usually accomplished by incubation with a small ligand such as lysine (Matsumoto *et al.*, 1986), or with Tris buffer in the case of tresyl chloride activation ('Tresyl-Activated-Sepharose', Pharmacia Fine Chemicals, Uppsala, Sweden).

Studies of protein binding to solid supports often do not report the density of protein coverage. This may be due to the fact that the surface area is undefined, as with agarose and dextran resins and controlled pore glass, or because the researcher has no reliable way of measuring the amount of protein deposited. In many cases, the amount deposited is taken as the difference between the original amount and the amount measured after binding. This method may be acceptable when large amounts of protein are bound, relative to the concentration of protein and sensitivity of the protein assay, but may be prone to error, especially when dilute amounts of protein are measured in large volumes of wash solution. Succinylated calmodulin on aminopropyl triethoxy silane-derivatized silicon oxide was measured at $89 \cdot 6 \, \text{ng/cm}^2$. Anti-T antibody was estimated to be $181–231 \, \text{ng/cm}^2$ on nitrophenyl chloroformate-activated quartz, 212–260 ng/cm^2 on toluenesulfonyl chloride-activated quartz, and 233–297 ng/cm^2 on glutaraldehyde-APTES quartz (Williamson *et al.*, 1989). Anti-immunoglobulin G antibody was estimated at 0·2–0·5 μg/cm^2 on glutaraldehyde-APTES quartz (Sutherland *et al.*, 1984).

A high surface coverage of receptor is of no value if the receptor has lost its activity in the immobilization process. Therefore, it is important to quantitate activity after immobilization, and also as a function of time and device usage. The initial density of coverage inevitably includes adsorbed as well as covalently bound protein. Loss of activity of the receptor-bound surface can be measured by exposure to ligand which is labelled in some fashion. However, it may not be clear whether activity decreases are due to protein desorption or to inactivation of bound receptor. Receptors usually have several potential points of attachment, and little or no control can be exercised over which point attaches to the surface. Therefore, a multiple population of immobilized receptor exists and different types of vulnerability to storage are expected (Williamson *et al.*, 1989). It has been noted that the apparent increase in lifetime and stability reported for some immobilized enzymes may actually be due to overloading of the support with receptor, so that the prevailing denaturation is not evident (Mosbach, 1976). A more recent report asserted that the immobilization of enzymes may confer greatly increased resistance to various denaturing agents: extreme pH and temperature, high ionic strength, denaturing reagents, and proteases (Monsan & Combes, 1988).

Glucose oxidase linked by cyanuric chloride to tin oxide showed a 50% decrease in activity after 24 hours, but the response level was then stable for three months (Watanabe *et al.*, 1988). Cellulase bound to iron oxide via polyethyleneglycol-glutaraldehyde-APTES showed two stages of in-activation at 40°C, a fast component for the first 30 hours, and a slow component for the remaining 70 hours (Garcia *et al.*, 1989). The rapid activity loss was attributed to either desorption or inactivation of one component, and the slow inactivation to a slower denaturation of another enzyme component (Garcia *et al.*, 1989). In one case where many different supports and chemistries were compared, the leaching profiles for all binding situations were similar, but the absolute amount of receptor lost was a function of the binding chemistry. The loss of receptor was greatest from cyanogen bromide-activated supports, less from tresyl chloride-activated supports, and least from epoxy-activated supports (Comfort *et al.*, 1988).

In many cases, researchers state the lifetime of the device or column, but do not state whether the end of the lifetime represents 0% activity, the last day of 100% activity, or an intermediate level of activity. Glucose oxidase and L-amino acid oxidase are stated to be active on an enzyme electrode for 20–30 days (Ianello & Tacynych, 1981). On a platinum electrode, glucose oxidase was sufficiently active for 10 days or nine hours of continuous use (Moody *et al.*, 1986), and on a glassy carbon electrode, the enzyme was stable for several days at room temperature (Narasimban & Wingard, 1986). When bound to nylon through lysine and glutaraldehyde, glucose oxidase could be stored for four months, or used continuously for 24 hours (Moody *et al.*, 1986). In contrast, a controlled pore glass-aminopropyl triethoxy silane-glutaraldehyde-glucose oxidase column had a lifetime of over 10 months at room temperature (Masoom & Townshend, 1984).

Urease on a glutaraldehyde-aminopropyl triethoxy silane-ion selective field effect transistor demonstrated only a 10% loss in sensitivity over three weeks (Karube *et al.*, 1986). A luciferase column showed no loss of activity in two months at 4°C, but at room temperature lost 50% of all activity in four days (Worsfold & Nabi, 1986). A column of glycerol dehydrogenase was stable for three months in storage, but lost 80% of its activity when used for two months with 50 analyses per day (Kiba *et al.*, 1986). A column of controlled pore glass–glutaraldehyde–yeast alcohol dehydrogenase did not decrease in activity over two months (Masoom & Townshend, 1986). Fructose dehydrogenase attached to aminocellulofine by glutaraldehyde lost 60% activity over 30 days (Matsumoto *et al.*, 1986). A lactate dehydrogenase column with an aminated controlled pore glass–glutaraldehyde support retained 95% of the original activity after six months (Zaitsu *et al.*, 1987). One study found that although 75–100% of available enzyme of various types was bound by various methods to organic and inorganic supports, only 0·3–23·2% was active. Of the active amount, after 15 days at 4°C the activity decreased to 40–75% of the original amount. At a higher temperature (37–40°C), activity decreased by the same amount in five hours (Mannens *et al.*, 1988). Elevated temperature was responsible for an 80% loss of activity of antibody within 80 hours when bound to quartz by glutaraldehyde–aminopropyl triethoxy silane, whereas at 4°C, 15–45% was lost in 55 days, depending on the method of immobilization (Mannens *et al.*, 1988).

Spacer molecules have been used to determine whether insulation of the receptor from the solid support had a beneficial effect on receptor activity. A 76% increase in activity was observed when polyethylene glycol or polyvinylalcohol was used as a spacer between iron oxide-silane and cellulase (Garcia *et al.*, 1989). However, when a hexanediamine spacer was used between quartz and an antibody, a decrease in activity was observed (Williamson *et al.*, 1989). A spacer is expected to be beneficial when the substrate is large by providing better availability of the enzyme to the substrate (Garcia et al., 1989), but obviously this is not always the case. Another method of increasing receptor activity is to crosslink the immobilized receptor with a bifunctional reagent in order to minimize losses due to desorption. This was found to increase activity for most but not all systems (Comfort et al., 1988).

It is common to see changes in the intrinsic properties of receptors immobilized on surfaces. Shifts in the pH optimum and Michaelis constant of an enzyme reaction are routinely observed, and antibody-antigen binding constants may be altered. The shift in pH on immobilization can be acidic (Karube et al., 1986; Worsfold & Nabi, 1986) or basic (Garcia et al., 1989; Ianello & Tacynych, 1981). This is due to the creation of a microenvironment around the immobilized enzyme that differs from the conditions in the bulk solution, usually due to proton consumption or generation by the enzymic reaction (Mattiasson & Mosbach, 1976), or to the interaction of the protein with the surface, resulting in salt bridge or hydrogen bond formation. In addition, the pK_A values of acidic or basic residues may be altered by interaction with the surface, causing more complex interactions (Garcia et al., 1989). The Michaelis constant for enzymes is equal to the concentration of enzyme at which the reaction rate is half its maximal value (Stryer, 1981). This constant may be unaffected by immobilization as in the case of alcohol dehydrogenase and a nicotinamide oxidoreductase on nylon-diaminohexane-glutaraldehyde (Girotti et al., 1986), or altered as in the case of glucose oxidase on graphite-cyanuric acid where the Michaelis constant is an order of magnitude greater than in the free enzyme (Ianello & Tacynych, 1981).

An additional consideration when using biological molecules in biosensors, is that non-specific interactions between biological molecules can occur. Unfortunately, very few tests of the devices for non-specific interaction have been reported. In the case of immobilized 2-[(4-hydroxyphenyl)azo]benzoic acid (HABA) which selectively interacts with avidin, the HABA–avidin complex could not be dissociated on addition of the competitive receptor biotin (Ikariyama et al., 1985). Thus, non-specific interaction of avidin with HABA was estimated to constitute 50% of the total interaction even in the presence of bovine serum albumin or ovalbumin which should compete for non-specific binding sites (Ikariyama et al., 1985). Antibody fragments (F_{ab}') were found to bind more non-specific antigen than whole antibody (Alarie et al., 1990). In another case, radiolabelled T2 demonstrated only minute amounts of non-specific binding with bare surfaces or surfaces with non-specific antibody (Valdes et al., 1988).

Table 2.3
Important Factors in the Optimization of Protein Function in Biosensors

pH	interfacial tension
temperature	ultraviolet radiation
ionic strength	viscosity
metal ions	detergents
shear forces	pressure
organic solvents	chemical agents (e.g. urea, guanidine salts)

It is the consensus of most workers in the field that optimization of reaction conditions for a particular method are needed with each new receptor (Alarie *et al.*, 1990; Wienhausen & DeLuca, 1982). Different proteins have different environmental requirements, and may be affected by the factors listed in Table 2.3.

2.2.6   Other Immobilization Methods

After adsorption and covalent coupling, the next most popular type of immobilization is entrapment/encapsulation of receptor inside a polymer, liposome, fibrous matrix or semi-permeable membrane. These methods are popular since the receptor is protected from many interferents, and its freedom of movement is retained. The major drawbacks of this technique are leakage of the receptor from the matrix, rupture of the encapsulating sphere of material, and the increased diffusion time necessary for the analyte to reach the receptor. Copolymerization of chemically-modified receptors with polymers is another popular immobilization technique as the protein/polymer layer can be cast directly onto the device. A variation of this method is to polymerize receptor with a bifunctional reagent (e.g. glutaraldehyde) that can attach to functional groups already present. An inexpensive inert protein such as bovine serum albumin may be used to effectively dilute the polymer film. Affinity immobilization is useful for those receptors which have a highly selective absorbance to another biomolecule, for example avidin and biotin, protein A and immunoglobulin G, and lectin and carbohydrate.

2.2.7   Comparison of Common Methods of Covalent Immobilization

Literature reveals that immobilization of proteins is a very individual process, with no one best method of protein immobilization for all proteins. Therefore, there is a need to evaluate different immobilization processes with regard to binding density, conformation and function. This work deals with the use of instrumental methods of analysis to evaluate these processes, and illustrates the strengths and weaknesses of each instrumental method by case studies of the immobilization of certain proteins by several widely used and successful methods. These case studies also provide specific information on the systems studied, and are valuable since characterization of each step of the immobilization by multiple techniques is rarely done. In addition, the work described herein assumes no specific morphology or transduction mechanism, as is usually the case in biosensor publications. It requires only that surface hydroxyl groups be available. This is a very common surface chemistry for both inorganic and organic substrates, as almost all metals experience surface oxidation, and hydroxyls are found in abundance on sugars and many hydrogels and polymers.

This work will compare and contrast immobilization of various proteins onto silicon and quartz substrates by adsorption onto hydroxylated silicon oxide and 3-aminopropyl triethoxy silane, and by covalent binding via glutaraldehyde, tresyl chloride, glutaric anhydride/thionyl chloride and chloroacetic acid. The total binding densities of bovine serum albumin, acetylcholinesterase, beta-glucosidase, and immunoglobulin G-anti-immunoglobulin G (antigen-antibody) were investigated, along with the biological activities of the latter three systems. The activated surfaces and protein binding were characterized by the following methods: contact angle measurements, ellipsometry, fluorescence microscopy, radio-isotope labelling, X-ray photoelectron microscopy, scanning tunnelling microscopy and scanning electron microscopy.

2.3   EXPERIMENTAL PROCEDURES

2.3.1   Materials and Chemicals

Fluorescein-di-β-D-galactopyranoside (FDG) (F1179, lot #9E) was obtained from Molecular Probes (Eugene, OR). Tween 20, packed in 10 ml ampoules under nitrogen, was purchased from Pierce (Rockford, IL). Aldrich (Milwaukee, WI) was the supplier for 3-aminopropyl triethoxy silane (APTES) (11.339-5), and 2,2,2-trifluoroethane sulfonyl chloride (tresyl chloride) (32,478-7). Scintillation fluid was obtained from Beckman (Mississauga, ON) under the names of Ready-Solv Hp/B or Ready-Protein (158727). All other solvents were obtained from BDH (Toronto, ON) and were of analytical grade or were purified as described below. Most biological reagents and buffers were obtained from Sigma

(St Louis, MO). All other reagents were of analytical grade and were purchased from BDH.

Quartz was purchased in the form of ground and polished square plates of commercial grade (TO8) quartz, 5·08 × 5·08 × 0·42 cm or 6·20 × 6·20 × 0·42 cm, from Heraeus Amersil (Sayerville, NJ), and plates were cut to 0·7 cm width and various lengths. A hole of 0·5–1 mm diameter was made near one end by ultrasound. Silicon wafers were type N(100)C , with low resistance (2 or 4 ohm) and thickness 0·017 inch, and were obtained from Aurel (St Charles, MO).

2.3.1.1 *Purification of Solvents*

Methanol – 50 ml methanol, 5 g magnesium turnings and 0·5 g iodine were refluxed until the iodine colour disappeared and hydrogen gas evolved. One litre of methanol was added and the solution was refluxed 30 minutes and distilled.

Acetone – Acetone was dried over potassium carbonate overnight, and distilled from a small amount of potassium permanganate.

Chloroform – Chloroform was dried over calcium chloride overnight and distilled from phosphorus pentoxide. The distilled solvent was stored in the dark.

Toluene – Sodium metal (about 30 g) was freshly cut into small pieces and added to 2 l toluene and 2 g benzophenone. The mixture was refluxed until it turned a deep purple colour, then was distilled.

Water – Water was purified by reverse osmosis, then was passed through a Milli-Q filtration system (Millipore, Mississauga, ON) and had a specific resistivity of 18 Mohm/cm.

Pyridine – Pyridine was dried over KOH for several days, then distilled from BaO. This procedure was repeated when the solvent showed traces of colour.

2.3.2 Procedures

2.3.2.1 *Cleaning of Substrates*
Silicon wafers were cleaned either by refluxing for two hours in dried and distilled chloroform, washing with dried and distilled acetone and methanol, and vacuum-drying for 15 minutes, or by the following procedure (Mandenius *et al.*, 1984): wafers were added to a 1 : 1 : 5 (volume) mixture of 25% ammonia : 30% hydrogen peroxide : water heated to 80°C (for five minutes, then to a 1 : 1 : 5 (volume) mixture of concentrated hydrochloric acid : 30% hydrogen peroxide : water at 80°C for 5 minutes.

Quartz was cleaned by refluxing two hours in dried and distilled chloroform, washing with dried and distilled acetone and methanol, and vacuum-drying for 15 minutes or by the following procedure (adapted from Tscharner & McConnell, 1981): quartz slides were washed in warm Alconox detergent and sonicated in a bath sonicator for 30 minutes, rinsed in water, soaked 15 minutes in chromic acid, rinsed thoroughly in water, sonicated in 3 mM sodium hydroxide in an 80 watt bath-type sonicator (Bransonic 12, Branson Cleaning Equipment Company, Shelton, CN) for 15 minutes, rinsed for 20 minutes with water, then dried at least 30 minutes in a 110–150°C oven.

2.3.2.2 *Derivatization of Silicon or Quartz with Silanes*

For the reflux reaction of the silane 3-aminopropyl triethoxy silane (APTES) with quartz, about 10–12 slides of 0.7×1.2 cm with a small hole drilled at one end were strung on polyethylene thread through a three-necked flask in order to keep the slides separated.

For the reflux reaction of the silane with silicon, a 7·5 cm diameter wafer was hand cut into strips of about 0·5 cm wide and varying lengths, and no more than half the wafer reacted at a time. Separation of the wafers was accomplished by slow stirring with a magnetic stir bar. About 200 ml of dried toluene and 1 ml of APTES was added with stirring. If base catalysis was desired, 160 μl to 1 ml of triethylamine was added. Refluxed reactions proceeded for 1·5 hours, and room temperature re-actions were incubated for five hours with occasional swirling in an octadecyl trimethoxy silane-coated petri dish (Tscharner & McConnell, 1981). If anhydrous conditions were desired, the reflux took place under a nitrogen atmosphere. APTES-coated wafers or slides were washed with warm chloroform, acetone and methanol, and vacuum-dried for 30 minutes. The substrates were stored in a desiccator under vacuum until use. The various methods of APTES deposition are referred to as follows: RT-APTES if the reaction was performed at room temperature, TEA-APTES if base catalysis was used, and APTES if reacted under reflux conditions without base.

2.3.2.3 *Protein Coupling*

The tresyl chloride coupling method was performed as follows: about four APTES-coated surfaces were added to a mixture of 0·1 ml 2,2,2-trifluoroethane sulfonyl chloride (tresyl chloride), 0·2 ml pyridine and 20 ml acetone at 4°C for 20 minutes with occasional gentle swirling. The surfaces were rinsed with acetone, 1 : 1 (volume) acetone : 1 mM hydrochloric acid, 1 mM hydrochloric acid, and water (Mandenius *et al.*, 1984). These will be referred to as tresyl surfaces.

The glutaraldehyde coupling method was performed as follows: APTES slides or wafers were incubated with 6% glutaraldehyde in 10 mM sodium phosphate buffer pH 9·0 for three hours at 4°C, and then rinsed thoroughly with water. These will be referred to as glutaraldehyde surfaces.

2.3.2.4 Protein Incubation

Activated, APTES and plain surfaces were incubated overnight at 4°C in a solution containing: protein at a concentration of 0·01–1 mg/ml, 10 mM sodium phosphate, pH 7·4–7·6, and in some cases 0·01% Tween 20. Thin test tubes were used for incubation so that diffusion alone would suffice for protein mixing. The substrates were usually subsequently incubated in buffered solutions of 1 mg/ml bovine serum albumin, then 0·1% (v/v) Triton-X 100 for 24 hours each.

2.3.2.5 Chloramine T Iodination of Protein

Protein was radiolabelled by a scaled-up version of the chloramine T technique ('Radioiodination Techniques', Review 18, Amersham, Oakville, ON, Canada). In a polystyrene tube 10 µl Na^{125}I (100 mCi/ml) and 40 µl of 0·25 M sodium phosphate, pH 7·5 were mixed with a glass-coated magnetic stirrer. The following were added in rapid succession: 20 µl of 100 mg/ml protein in 50 mM sodium phosphate, pH 7·5, and 4 µl of two mg/ml chloramine T in 50 mM sodium phosphate, pH 7·5. The reaction proceeded with stirring for two minutes, then was quenched by the addition of 2 ml of 0·2 mg/ml tyrosine or 0·17 mg/ml cysteine in 50 mM sodium phosphate, pH 7·5 and 10 ml of 2 mg/ml sodium iodide in the same buffer. After mixing for two minutes, the entire tube was counted for radioactivity. Meanwhile, a 0·9 × 12 cm Sephadex G-75 column was conditioned by passing through 1 ml of 100 mg/ml bovine serum albumin. Aliquots of reaction mixture were applied to the column and eluted with 50 mM sodium phosphate, pH 7·5. The eluate was monitored for absorbance at 280 nM by a flow-through detector (Gilson HM Holochrome Variable Wavelength Detector with a 10 mm, 40 µl cell, Mandel Scientific, Rockwood, ON), and two peaks were observed. The peaks were collected, counted and assayed for protein. The first peak, consisting of radiolabelled protein, was saved, and the second peak, consisting of free label, salt and amino acids, was discarded.

2.3.2.6 Flourescein Labelling of Protein

Fluorescein labelling of proteins was a modified version of Rinderknecht (1962). Protein was dissolved in sodium carbonate, pH 9·0 and fluorescein isothiocyanate on Celite (10%) was added to a concentration of 1–5 mg/ml. The mixture was stirred at 4°C for 15 minutes, then passed through a 0·45 μm filter (product E04-WP047-00, Fisher, Don Mills, ON), and applied to a Sephadex G-75 column equilibrated in 10 mM sodium phosphate, pH 7·5. The protein was eluted with the same buffer, and the void volume was collected.

2.3.2.7   Protein Assay
Protein was determined by using the Bio-Rad Microassay, as per the manufacturer's instructions ('Bio-Rad Protein Assay', Lit-33, Bio-Rad Laboratories, Bio-Rad Press, San Francisco, CA), using bovine serum albumin as a standard.

2.3.2.8   Beta-Galactosidase Activity Assay
Substrates with immobilized enzyme or various amounts of enzyme in solution were added to a solution of 7 μl of 2 mM fluorescein-di-β-galacto-pyranoside and 10 mM sodium phosphate, pH 7·5 to a volume of 1·5 ml, and incubated five or eight minutes. Fluorescence at 520 nm was measured (Pavari *et al.*, 1983*b*).

2.3.3   Equipment

2.3.3.1   Fluorescence Spectrophotometer
Fluorescence spectra of solid substrates were obtained using a fluorescence spectrometer consisting of: an atmospheric pressure nitrogen laser (Model LN 103, PRA) for excitation at 337·1 nm; a Bentham M300 monochromator with SMD 3B stepper motor controller (Optikon, Waterloo, ON, Canada); a Hamamatsu R928 photomultiplier tube (PMT) (Hamamatsu, Bridgewater, NJ) in a linearly-optimized housing (PRA) operated at –1250 V; and an SR 250 Gated integrator/Boxcar averager (SRS, Stanford, CA) operated by an IBM personal computer through an SR 245 interface.

2.3.3.2   Radiation Counters
Radioactive samples were counted in a gamma counter (Gamma Trac 1191, Tracor Analytical, Elk Grove Village, IL) equipped with a programmed cartridge for ^{125}I, or in a β-counter (Beckman LS7000 Liquid Scintillation System, Beckman Instruments, Fullerton, CA) using library program 9 for ^{125}I. Counts were taken for one minute with the appropriate background subtracted.

2.3.3.3   X ray Photoelectron Spectroscopy

X-ray photoelectron spectroscopy was carried out at the Surface Science Laboratory, University of Western Ontario, London, ON, Canada, on a Surface Science Laboratory Model SSX-100 spectrometer attached to an analytical chamber (vacuum 5×10^{-9} torr). Survey spectra were obtained with a monochromatized Al K(alpha) X-ray source with a spot size of 600 μm. Two scans of a 0–1000 eV region were taken and added. High resolution spectra were obtained with either a 150 or a 300 μm spot size and a pass energy of 50 eV, adding 5–20 scans. The XPS was calibrated to give 83·9 eV for the Au(4f7/2) peak. The sample X-ray source geometry was such that the take-off angle was 55°. High resolution peaks were deconvoluted on a Hewlett-Packard computer 9836 and plotted on an HP 7470A plotter.

2.3.3.4   Contact Angle Measurements

Contact angle determinations were made with the kind permission of Prof. Graham Nairn (Department of Pharmacy, University of Toronto) using his NRL contact angle goniometer model 100–00. Static contact angles were taken within 15 seconds of application of a large drop of doubly distilled water.

2.3.3.5   Ellipsometry

Ellipsometry was performed with an Auto EL II Ellipsometer (Rudolph Research, Flanders, NJ) with a 1 mW continuous wave helium-neon laser source (632·8 nm), an angle of incidence of 70°, and a phototransistor detector. The data were analysed on a Hewlett-Packard 85 computer (Fisher Scientific, Toronto, ON) using the 'Film 85' software package, version 3·0, program 10 (Rudolph Research). Three silicon oxide standards (455 Å, 1133 Å and 2000 Å) were used for calibration purposes.

2.3.3.6   Scanning Tunnelling Microscopy

The scanning tunnelling microscopy images were collected by Dr Wolfgang Heckl (IBM Research, Physics Group Munich, West Germany) using an in-house assembled scanning tunnelling microscope (Losche & Mohwald, 1984).

2.3.3.7   Infra-red Spectroscopy

The reflection infra-red spectra were recorded in a Bruker IFS 113v Fourier transform spectrometer operated at *ca.* 10 Torr. The spectrometer was equipped with a reflection accessory at an angle of incidence of 85° and a single reflection. Each spectrum was recorded by averaging 1500 interferograms at 4 cm^{-1} resolution.

2.4   RESULTS AND DISCUSSION

2.4.1   Substrate Derivatization Reactions – Theory

2.4.1.1   Cleaning Procedures
The ammonia/peroxide and acid/peroxide cleaning procedure for silicon wafers is a standard practice in the semiconductor industry (Kern & Puotinen, 1970). Acid treatment of the silicon wafers increases the density of surface hydroxyl groups, thereby increasing the number of possible points of attachment for the silanes. However, when the APTES reaction was performed under strict anhydrous conditions this cleaning procedure was not used, in order to prevent exposure of the substrates to water. Instead, the substrates were refluxed in chloroform to remove organic matter. Similarly, quartz slides were cleaned with chloroform when the APTES reaction was performed under anhydrous conditions, but were otherwise cleaned with chromic acid to increase the surface hydroxyl density.

Ammonia/hydrogen peroxide solutions remove organic contaminants by the solvating action of ammonia, the powerful oxidizing action of peroxide, and the complexation of some group I and II metals by ammonium hydroxide (Kern & Puotinen, 1970). At low pH, peroxide solutions desorb metals primarily by complexation (Kern & Puotinen, 1970). The hydrogen peroxide prevents etching of the wafers, presumably by the formation of a thin (less than 50 Å) continuous layer of hydrous oxide which is very resistant to chemical attack by ammonium hydroxide under these conditions (Kern & Puotinen, 1970). The two solutions are also attractive because they are completely volatile (Kern & Puotinen, 1970).

Once cleaned, the quartz pieces were suspended separately to ensure full access of APTES to the surface during the reflux. Since it was not feasible to make holes in the silicon due to its brittleness, pieces of different sizes were cut, and were gently stirred during the reflux. This did cause some scratching of the surfaces, but was unavoidable if the wafers were to be fully exposed to the reagents and reacted in batches.

2.4.1.2   3-Aminopropyl Triethoxy Silane Reactions
3-aminopropyl triethoxy silane (APTES) reacts with hydroxyl groups on the silicon surface resulting in the loss of ethanol (Fig. 2.1a). This reaction can be catalyzed by triethylamine, or self-catalysed by the APTES amine. In either case, the base reacts with the silane to form a pentacoordinate structure, subsequently undergoing hydrolysis by nucleophilic replacement of ethoxy by surface water. The resulting silanol then reacts with the surface hydroxyl groups to form a bond (Kinkel & Unger,

A

B

C

$$R = -(CH_2)_3-NH_2 \quad \text{in APTES}$$
$$= -(CH_2)_3-CH-CH_2 \quad \text{in GOPS}$$
$$\qquad\qquad \backslash_O{}^{\diagup}$$

$$R' = -OC_2H_5 \quad \text{in APTES}$$
$$= -OCH_3 \quad \text{in GOPS}$$
$$= -OH \quad \text{after } H_2O \text{ treatment}$$

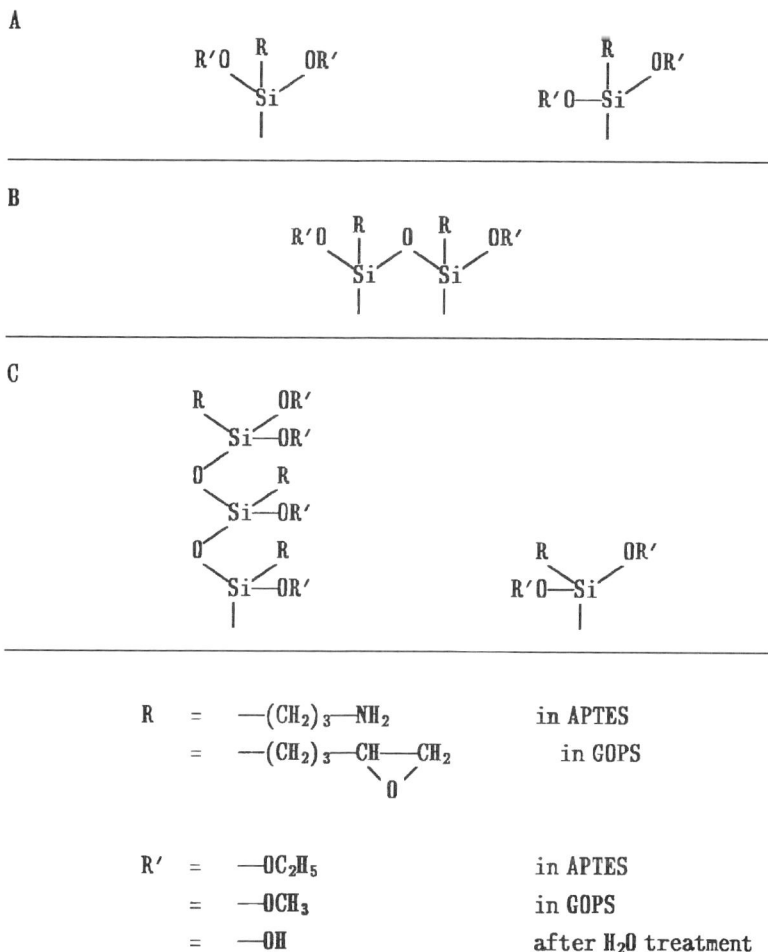

Fig. 2.1 Schematic diagram of silane-treated surfaces (*a*) deposition under anhydrous, non-catalysed conditions, (*b*) polymerization of immobilized silanes, (*c*) multilayer silane formation. APTES = 3-aminopropyl triethoxy silane; GOPS = glycidoxypropyl trimethoxy silane.

1984). Uncatalysed conditions were routinely used in this work since catalysed reactions are less certain to have extensive covalent attachment to the surface.

Conditions for the reaction of APTES with surfaces vary in the literature from an anhydrous environment to aqueous solutions. Water catalyses hydrolysis of APTES ethoxy groups, leading to dehydration and the formation of a siloxane bond between (a) APTES and a surface hydroxyl group, constituting multilayer formation, or (b) between two APTES molecules, which constitutes polymer formation (Fig. 2.1b). Much of the polymerized APTES is adsorbed to the surface rather than covalently bonded, so may slough off, taking with it any subsequently bound or adsorbed reagents (Ishida & Koenig, 1980). In order to obtain a covalently attached APTES layer, APTES was deposited under anhydrous conditions in this experiment. Note that the glassware used for APTES reactions should be alkylated (for example, with octadecyl trichloro silane) to prevent covalent reaction of APTES with the glassware.

Differences in reaction conditions of APTES deposition may result in varying surface chemistries (i.e. proportion of protonated, uncharged and hydrogen bonded amines) and extent of siloxane formation between adjacent APTES residues. Nevertheless, as discussed below, there was no significant difference in the APTES surfaces with respect to contact angles, protein adsorption to APTES, or covalent protein binding to activated APTES, even though differences in thickness and surface coverage were observed. Taken altogether, the results point to very similar surface chemistry for APTES deposition, although the number of layers and extent of surface coverage may vary. Therefore, this line of research was not pursued further.

2.4.1.3 Activation of Silane with Tresyl Chloride

2,2,2-trifluoroethane sulfonyl chloride (tresyl chloride) reacts with APTES amine groups forming a sulfonamide, with loss of hydrochloric acid (Fig. 2.2). The reaction is catalysed by pyridine which removes the hydrochloric acid from solution and helps drive the reaction to completion (Nilsson & Mosbach, 1981). The sulfamide is stable to hydrolysis, but is readily substituted by nucleophiles and is an excellent leaving group due to the high electron withdrawing properties of the fluorines (Scouten *et al.*, 1986). Several residues on the protein can displace the sulfonate, such as amines (found in lysine), thiols (cysteine), imidazole (histidine), and the hydroxyl groups of tyrosine. A stable nitrogen–carbon or sulfur–carbon bond is formed (Scouten *et al.*, 1986) between the protein and APTES with concurrent displacement of 2,2,2-trifluoroethanesulfonamide.

The literature states that coupling of protein to surfaces that are activated by tresyl chloride treatment is normally complete after 16 hours at 40°C ('Tresyl-Activated Sepharose', Pharmacia Fine Chemicals, Uppsala, Sweden). The optimum pH for coupling depends on the amino acid composition of the protein, as proteins with free thiols show better protein loading at a higher pH ('Tresyl-Activated Sepharose', Pharmacia Fine Chemicals, Uppsala, Sweden). However, immobilization at pH 7·8 is recommended so that fluorine displacement is avoided, and maximum protein activity is retained (Nilsson & Mosbach, 1981). This pH range was

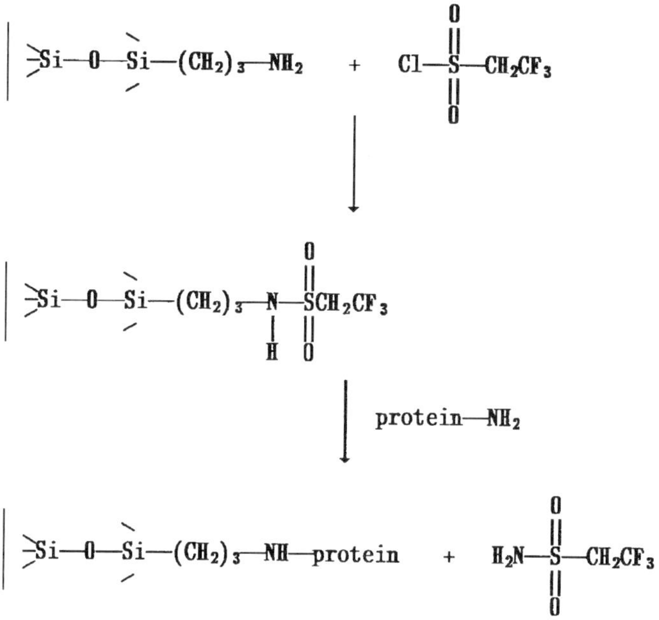

Fig. 2.2 Schematic diagram of tresyl chloride activation of 3-aminopropyl triethoxy silane and immobilization of protein.

maintained for all experiments. Residual active groups were deactivated by incubation with Tris buffer, as the nucleophilic nitrogen on the Tris can displace the sulfonamide ('Tresyl-Activated Sepharose', Pharmacia Fine Chemicals, Uppsala, Sweden). The only drawbacks of the tresyl chloride reaction are the cost and volatility of this reagent (Nilsson & Mosbach, 1981).

2.4.1.4 *Activation of Silane with Glutaraldehyde*

Glutaraldehyde reacts directly with APTES by formation of a Schiff base. This reaction requires base catalysis, as the APTES amine must be unprotonated to react. Incubation of the substrate with a protein solution at basic pH conditions results in the formation of a second Schiff base (Fig. 2.3). Commercial solutions of glutaraldehyde contain proportions of polymer ranging from a few per cent to nearly 100% (Gillett & Gull, 1972; Monsan 1977/78; Monsan *et al.*, 1975), which creates the possibility of side-reactions. With time, the unstable Schiff bases undergo complex reactions (Miller & Robyt, 1983). For example, the formation of aldol condensation products has been observed, leading to the creation

$$\left| \begin{array}{l} \\ \end{array} \right. \!\!\! \ge\! Si\!-\!O\!-\!Si\!-\!(CH_2)_3\!-\!NH_2 \quad + \quad \overset{\overset{\textstyle O}{\|}}{HC}\!-\!(CH_2)_3\!-\!\overset{\overset{\textstyle O}{\|}}{CH}$$

$$\downarrow$$

$$\left| \begin{array}{l} \\ \end{array} \right. \!\!\! \ge\! Si\!-\!O\!-\!Si\!-\!(CH_2)_3\!-\!N\!=\!CH\!-\!(CH_2)_3\!-\!\overset{\overset{\textstyle O}{\|}}{CH}$$

$$\downarrow \text{protein}\!-\!NH_2$$

$$\left| \begin{array}{l} \\ \end{array} \right. \!\!\! \ge\! Si\!-\!O\!-\!Si\!-\!(CH_2)_3\!-\!N\!=\!CH\!-\!(CH_2)_3\!-\!CH\!=\!N\!-\!protein$$

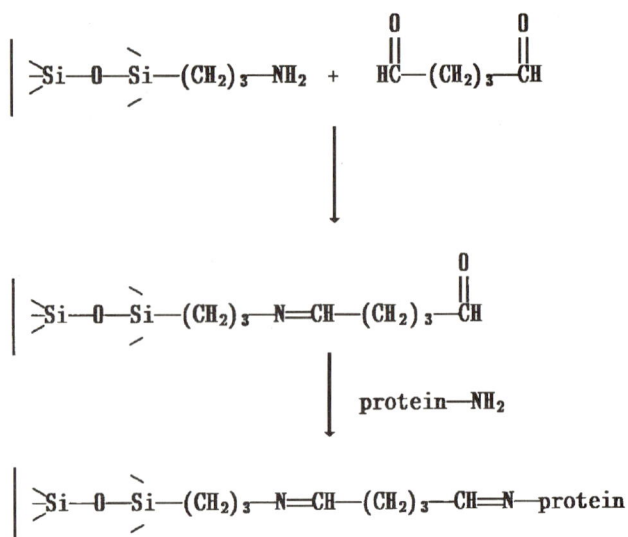

Fig. 2.3 Schematic diagram of glutaraldehyde activation of 3-aminopropyl triethoxy silane and immobilization of protein.

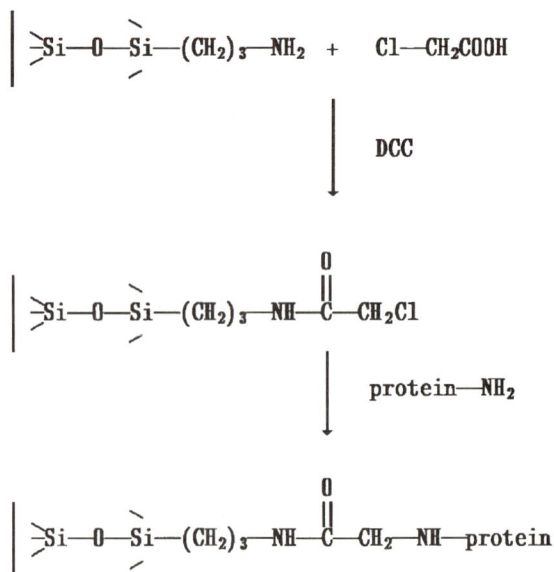

$$\left| \begin{array}{l} \\ \end{array} \right. \!\!\! \ge\! Si\!-\!O\!-\!Si\!-\!(CH_2)_3\!-\!NH_2 \quad + \quad Cl\!-\!CH_2COOH$$

$$\downarrow \text{DCC}$$

$$\left| \begin{array}{l} \\ \end{array} \right. \!\!\! \ge\! Si\!-\!O\!-\!Si\!-\!(CH_2)_3\!-\!NH\!-\!\overset{\overset{\textstyle O}{\|}}{C}\!-\!CH_2Cl$$

$$\downarrow \text{protein}\!-\!NH_2$$

$$\left| \begin{array}{l} \\ \end{array} \right. \!\!\! \ge\! Si\!-\!O\!-\!Si\!-\!(CH_2)_3\!-\!NH\!-\!\overset{\overset{\textstyle O}{\|}}{C}\!-\!CH_2\!-\!NH\!-\!protein$$

Fig. 2.4 Schematic diagram of chloroacetic acid activation of 3-aminopropyl triethoxy silane, and immobilization of protein.

of straight chain or cyclic α, β-unsaturated aldehydes with loss of water (Gillett & Gull, 1972). Further reaction with free amines gives Michael-type adducts, which are stable even to acid hydrolysis (Richards & Knowles, 1968). The α, β-unsaturated aldehyde may also conjugate with ethylenic double bonds (Monsan, 1977/78), and amines may react with aldehydes to give imines, stabilized by resonance with adjacent ethylenic double bonds (Monsan, 1977/78). Therefore, reduction with reagents such as sodium borohydride or sodium cyanoborohydride to increase stability is unnecessary when long incubation times are used (Miller & Robyt, 1983). Figures 2.4 and 2.5 show some further commonly used reactions.

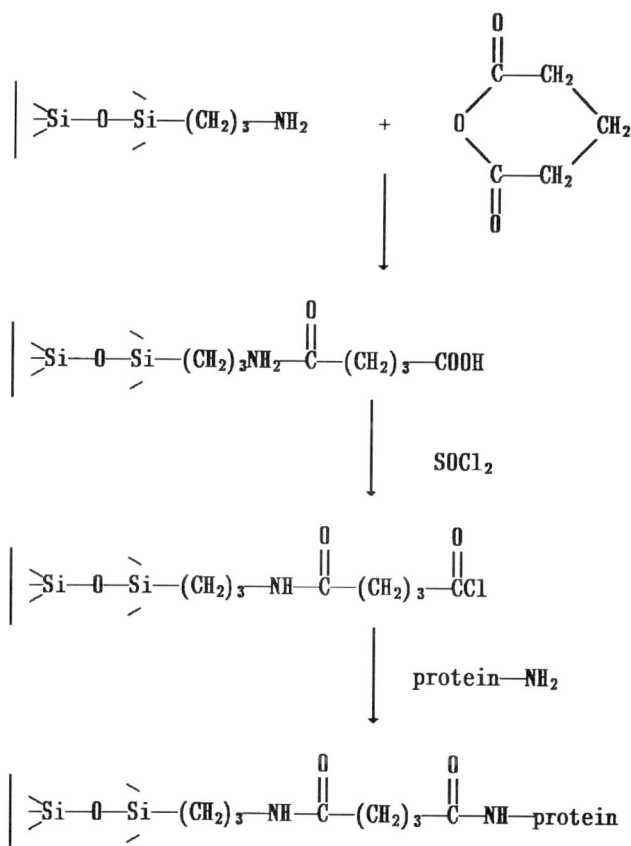

Fig. 2.5 Schematic diagram of acid chloride activation of APTES. Glutaric anhydride reacts with 3-aminopropyl triethoxy silane, then is activated by thionyl chloride. Protein is then immobilized.

2.4.2   Substrate Derivatization Reactions – Instrumental Analysis

2.4.2.1   Contact Angle Measurements

Contact angle measurements indicate the surface free energy of the substrate. The relationship between contact angle and free energy is stated in Young's equation:

$$\gamma_{SV} - \gamma_{SL} = \gamma_{LV} \cos \theta \qquad (6)$$

where γ is interfacial free energy, S refers to solid, L to liquid and V to vapour, and θ is the experimentally measured contact angle. Interfacial free energy can be defined as the reversible work required to create a unit area of surface at constant temperature, volume and chemical potential. Contact angles are most simply measured with a goniometer, which consists of a magnifying lens equipped with a protractor. The point of intersection of the three phases is viewed, and the tangent of the liquid–vapour interface with the liquid–solid interface is measured (Fig. 2.6). When the contact angle is near 90°, the surface is hydrophobic (low wettability) and when the contact angle is near 0°, the surface is hydrophilic (high wettability).

In order to interpret contact angle measurements quantitatively the surfaces should be smooth and non-deformable (Andrade *et al.*, 1985), however, the surfaces studied here are not ideal. 3-aminopropyl triethoxy silane coverage of the surface may not be homogeneous, and APTES may swell and reorient on exposure to water. Nevertheless, contact angles of each reaction step were measured to see if qualitative differences could be distinguished between the derivatized surfaces.

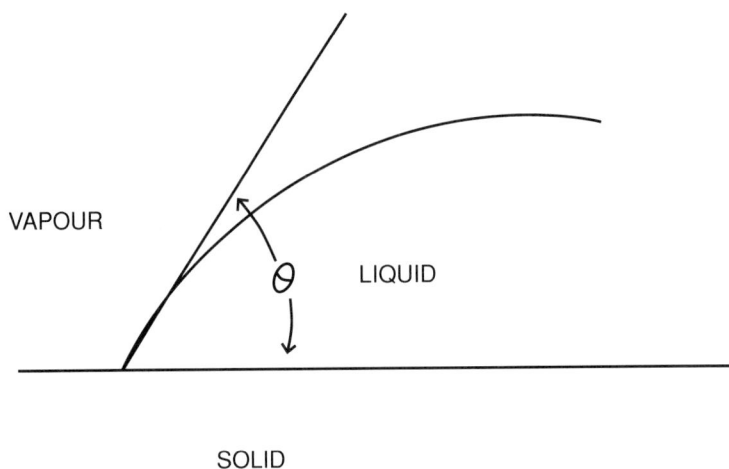

Fig. 2.6 Schematic diagram of the contact angle at a solid/liquid/vapour interface.

The effect of cleaning silicon substrates is distinctly seen by contact angle measurements (Table 2.4). There is a significant decrease in contact angle between uncleaned silicon (49 ± 3°), and silicon treated with ammonia/acid/peroxide (27 ± 2°). Similarly, chromic acid cleaning of quartz decreased the contact angle from 47 ± 7° to 20 ± 3°. These decreases indicate removal of surface organic contaminants which are hydrophobic, and an increase in the number of surface hydroxyls due to the reduction of surface oxides by acid. Note, however, that all samples suffered contamination, as the contact angles of hydrophilic silicon and quartz should be close to 0°. Surface contamination is a function of time between preparation of the sample and measurement, as hydrophobic carbon compounds are present in air. Samples in the same set were prepared and read at the same time to minimize differences in contamination level.

APTES derivatization of substrates increased the measured contact angle by 31–46° (Table 2.4). APTES is moderately hydrophobic due to the methylene groups, which are exposed to the surface when APTES amines are hydrogen bonded to surface hydroxyls (discussed further below). There is greater variation in the contact angle on the silicon supports than on the quartz supports. The results may reflect differences in surface conformations of the APTES, in densities of surface hydroxyls between batches of silicon wafers, in extents of polymerization of APTES, in degrees of attachment of APTES polymers to the surface, or simply amounts of surface contamination.

Table 2.4
Static Contact Angles of Water on Derivatized Silicon and Quartz Substrates

Surface chemistry	Silicon substrate contact angle (degrees) ± std dev	n	Quartz substrate contact angle (degrees) ± std dev	n
uncleaned	49 ± 3	12	47 ± 7	10
hydrophilic	27 ± 2	6	20 ± 3	10
APTES	58 ± 12	85	66 ± 6	49
tresyl	64 ± 4	5		
glutaraldehyde	71 ± 3	4		
APTES/BSA	50 ± 8	6	61 ± 4	16
tresyl-BSA	57 ± 8	20	67 ± 5	30
glutaraldehyde-BSA	66 ± 4	17	60 ± 4	16
FITC-concanavalin A	71 ± 7	24		

APTES = 3-aminopropyl triethoxy silane; BSA = bovine serum albumin; FITC = fluorescein isothiocyanate; *std dev* = standard deviation; *n* = number of measurements.

Table 2.5
Static Contact Angles of Water on APTES-Derivatized Quartz Substrates

APTES reaction conditions		*Average contact angle (degrees) ± std dev*	*n*
toluene	*base catalysis*		
distilled	no	70 ± 10	14
distilled	yes	71 ± 6	6
undistilled	no	65 ± 3	12
undistilled	yes	66 ± 4	13

APTES = 3-aminopropyl triethoxy silane
std dev = standard deviation
n = number of measurements

The contact angles of quartz cleaned in the same batch and reacted with APTES under various conditions (base catalysed versus non-catalysed, and anhydrous versus non-anhydrous) were not significantly different (Table 2.5). Contact angles measure the macroscopic properties of the surface, but cannot detect the presence of microdomains of different chemistry. Because of this fact, and the possibility of contamination, no conclusion about the similarity of these APTES surfaces can be made.

Upon activation of APTES-silicon by tresyl chloride or glutaraldehyde, changes in the contact angle were not significant considering the error in the APTES contact angle (Table 2.4). Therefore, the activated surfaces are either of similar surface free energy to APTES surfaces (the non-ideality of the APTES surface prevents accurate measurement of the change in surface free energy), or the surfaces are similarly contaminated.

In summary, this technique gives a good indication of the chemistry of the cleaning steps and APTES coating of the quartz and silicon surfaces, but reveals no detailed information about the activation reactions. Other methods must be used to determine the degree of success of these reaction steps.

2.4.2.2 *Ellipsometry*

Ellipsometry is a technique for measuring the thickness and refractive index of thin films coating a reflective surface by using elliptically polarized light. Basic principles of ellipsometry are discussed comprehensively in Azzam & Basharah (1977). Under the best conditions, coatings of several angstroms to over 1000 Å thick may be analysed for thickness to less than one angstrom resolution, and refractive index to 0·002 resolution. As shown in Fig. 2.7, laser light elliptically polarized by a polarizer and compensator reflects off the sample where it is further polarized, then

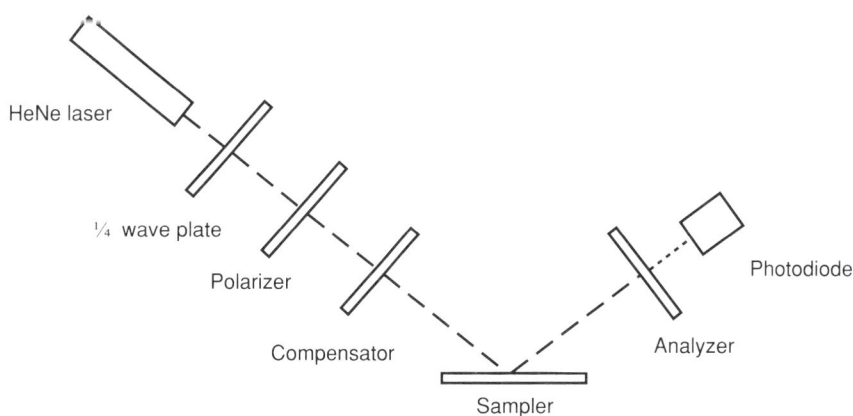

Fig. 2.7 Schematic diagram of the optics of a null reflection ellipsometer.

passes through an analyser and detector. The polarizer and analyser are rotated to extinguish the light emerging from the system, and their positions are used to calculate the ellipsometric parameters Δ and ψ. Δ represents the change in amplitude of light upon sample reflection, and ψ is the phase change of light upon sample reflection. For films under 100 Å, Δ and ψ values are calculated using the Drude linear approximation, and are converted to thickness and refractive index by the McCrackin (1969) programs. These have a number of solutions, and an estimate of the approximate thickness of the coating is required.

Samples should exhibit homogeneous thickness and refractive index, planar parallel interfaces between layers and non-interaction between layers. Calculated thickness and refractive index may not be identical to true thickness and refractive index in cases of submonolayer coverage, mixed multilayers with differences in refractive index between layers, and where interaction between layers occurs. Although the absolute thickness and refractive index calculations may include a variety of errors, relative changes in these parameters can be viewed with more confidence.

The near plane-parallel interface boundary between bulk crystalline silicon (refractive index 3·8) and its oxide layer (refractive index 1·5), and the homogeneous refractive index and thickness of these layers make this substrate ideal for ellipsometric analysis. Light reflects at the silicon/ silicon dioxide interface, and the thickness of the oxide layer can be measured. After cleaning, the measured thickness decreased by up to 21 Å and in no case increased significantly (maximum increase 4 Å). This demonstrates that the substrates were contaminated by a layer of dust, organic deposits or metal deposits that were removed by cleaning. Quartz

Table 2.6
Ellipsometrically Measured Apparent Thicknesses of APTES on Silicon

Substrate cleaning method	*APTES deposition conditions*	*APTES thickness* (Å)	*n*
A	refluxed	35	12
	refluxed	23	12
A	refluxed	107	4
A	catalysed	213	10
B	refluxed	70	3
B	refluxed	153	3
B	refluxed	150	18
A	refluxed	68	3
	catalysed	73	4

A = ammonia/HCl cleaning, B = chloroform cleaning
APTES = 3-aminopropyl triethoxy silane. APTES was deposited on silicon using either reflux conditions, or by refluxing with the catalyst triethyl-amine.
n = number of measurements

was unsuitable as a substrate since its refractive index (1·5) is close to that of organic and biological matter.

The refractive index of organic and biological material is close to that of silicon dioxide. Little reflection of light occurs at the organic/silicon dioxide interface, and addition of such material to silicon wafers appears to increase the thickness of the oxide layer. The thickness of the oxide/organic/biological material will be referred to as the 'coating' to indicate that it is difficult to assign proportions of the measured thickness to individual layers.

The coating thickness increased greatly after 3-aminopropyl triethoxy silane (APTES) deposition (Table 2.6). Upon APTES deposition, an increase in thickness was always observed, amounting to 41–153 Å for non-catalysed reactions, and 77–213 Å for base-catalysed APTES layers. The APTES layer may actually be much thicker and an alternative solution to the Drude equations indicating a thicker coating may be the correct one, especially when large decreases in ψ (by 3–5 angular degrees) are observed. For example, this is expected in the case of base catalysed APTES deposition, which is known to deposit in a very thick layer.

Addition of activating agents to APTES unexpectedly decreased the calculated thickness of most coatings (Table 2.7). All activating reagents should increase the coating thickness, and glutaraldehyde-activated surfaces in particular should exhibit significant increases since glutaraldehyde exists largely in polymer form. This result could be due to an alteration of the refractive index of all or part of the film, or a change in microstructure of the APTES. All activated surfaces were rinsed with water which can cause polymerization of APTES, thereby changing the refractive index and producing inconsistent thickness values. Table 2.7 shows that when APTES is exposed to water, changes in Δ are observed. This could correspond to formation of silane bonds in the APTES layer or swelling of the substrate with hydration. Since the activated supports were rinsed and dried similarly, it can be concluded that ellipsometry cannot detect a significant difference in thickness between any of the activating agent layers.

In summary, ellipsometry was a useful method of monitoring the cleaning procedure quantitatively, and the APTES deposition and

Table 2.7
Ellipsometric Analysis of Activated Silicon Substrates

Surface chemistry	Ellipsometric parameters			d (Å) + std dev	Change in d
	delta	psi	n		
APTES	272 ± 5	26 ± 4	10	2005 ± 130	0
tresyl	278·4 ± 0·3	46·3 ± 0·6	5	1774 ± 7	−231
glutaraldehyde	278·6 ± 0·1	44·7 ± 0·3	5	1802 ± 6	−203
RT-APTES	272·3 ± 0·4	62 ± 1	4	1611 ± 9	0
glutaraldehyde	274 ± 2	58 ± 4	4	1641 ± 35	30
tresyl	272·8 ± 0·1	61·9 ± 0·6	3	1603 ± 7	−8
TEA-APTES	277·2 ± 0·4	47 ± 1	4	1775 ± 16	0
glutaraldehyde	277·2 ± 0·2	47·4 ± 0·2	3	1773 ± 6	−2
APTES	277·5 ± 0·1	50·0 ± 0·4	3	1719 ± 9	0
glutaraldehyde	278·2 ± 0·2	48·7 ± 0·6	4	1729 ± 8	10
APTES	277·1 ± 0·2	48 ± 1	3	1770 ± 20	−5
tresyl	277·97 ± 0·08	47·2 ± 0·2	3	1763 ± 2	−12
APTES	167 ± 2	10·59 ± 0·03	6		
APTES/rinsed	171·5 ± 0·5	10·54 ± 0·02	6		
APTES/water	170 ± 3	10·59 ± 0·05	6		

APTES = 3-aminopropyl triethoxy silane; RT-APTES = room temperature-deposited APTES; TEA-APTES = base catalysed APTES; APTES/rinsed = APTES surface rinsed with water; APTES/water = APTES surfaces incubated in water for 20 minutes
value ± standard deviation
n = number of measurements
d = ellipsometrically measured coating thickness
std dev = standard deviation
The APTES surface is set to 0 thickness in order to emphasize the apparent thickness of the activating reagent layer.

activation semi-quantitatively, but after exposure of APTES to water, the samples exhibit non-ideal characteristics and are not suitable for ellipsometric analysis.

2.4.2.3 *X-Ray Photoelectron Spectroscopy*

X-Ray photoelectron spectroscopy (XPS), also known as electron spectroscopy for chemical analysis (ESCA), is a surface-sensitive method useful for determining the elemental composition of a surface. The surface is bombarded with X-rays, which causes individual atoms to eject inner core electrons. The kinetic energy of an ejected electron is indicative of the atom from which it came, which can be identified from published data, and also of its chemical environment. Analysis of electron energies gives both elemental and chemical speciation information about the surface layer, but the information may be ambiguous and requires careful interpretation, especially for complex surfaces.

X-ray photoelectron spectroscopy can be described by the following equation:

$$KE = h\nu - BE\ \Phi \tag{7}$$

where KE is the kinetic energy, BE is the binding energy of the ejected electron, h is Planck's constant, and Φ is the work function of the spectrometer. As all samples were taken on the same spectrometer, the work function and source energy are the same for all spectra. The kinetic energy of the electrons (measured by the instrument) therefore varies with the binding energy of the electrons, which can be identified from published data. Principles of X-ray photoelectron spectroscopy are discussed comprehensively in Baker & Brundle (1977, 1978).

The samples analysed by X-ray photoelectron spectra all had a certain degree of contamination. All were contaminated by airborne organics, which contain mostly carbon. Thus, the carbon signal is a mixed signal and must be interpreted with caution. Some samples contain unexpected elements. Chlorine signals may indicate the presence of adsorbed chloroform used in the rinsing of the substrates, whereas chlorine and sodium together in the protein spectra indicate that salt from the post-incubation wash has adsorbed to the sample. The origin of the phosphorus, sulfur and calcium signals may be detergent contamination and water residue in the glassware used to rinse the sample. Aluminum may come from the spatulas used to weigh (and occasionally to stir) protein, and fluorine is from the Teflon-coated stirring bars used in the APTES reaction.

During X-ray photoelectron spectroscopy samples lose electrons, and therefore gain positive charge as the analysis progresses. Conducting *n*-doped silicon samples were used rather than insulating quartz, so that the sample could be neutralized by grounding. Fewer artifacts are present in spectra of grounded samples, as opposed to samples neutralized by flooding with electrons. Nevertheless, peaks of identical atoms in different samples may not appear at the same energy because of charging in samples that have a thick insulating organic or biological coating. Therefore, as is common in the field, the carbon 1s peak was used as a calibration standard, and was set to a 285·00 eV (Ratner, 1983).

The curve fitting routines for deconvoluting high resolution elemental signals in the X-ray photoelectron spectra are computerized. The deconvoluted spectra are optimized for best fit, but there still remains a great deal of subjectivity regarding the choice of peak width, peak separation and number of peaks. Each elemental signal was deconvoluted to the minimum number of peaks which gave a good fit ($\chi^2 < 2·5$) while providing reasonable peak widths and separations as determined by past data sets on the instrument and literature values. It is difficult to determine exactly the uncertainty in absolute peak placement and area, so conservative estimates of $\pm10\%$ in area and $\pm0·2$ eV in absolute peak placement should be used, respectively. For comparisons of peak separation, a 0·1 eV difference should be significant.

The silicon substrate had large proportions of silicon (32·11 atom %) and oxygen (53·98 atom %), indicative of the *ca.* 1000 Å silicon oxide layer (Table 2.8 and Fig. 2.8A). Note that the atom % composition has been compensated for differing capture cross-sections of different atoms. One source of silicon, deconvoluted to 2s and 2p peaks, was observed (Fig. 2.9A). Considerable carbon (13·90 atom %) contamination from the atmosphere exists. The pattern of carbon peaks shows a 53% contribution from hydrocarbon (285·00 eV), 27% from hydroxyl or amide carbon (285·84 eV), 13% from carbonyl (287·20 eV) and 6% from carboxyl carbons (288·96 eV) (Table 2.9 and Fig. 10A). For the derivatized samples, only significant deviations from this pattern of carbon contamination will be discussed.

The proportion of silicon signal from 3-aminopropyl triethoxy silane (APTES) coated silicon (20–24 atom %) is considerably less than the 32 atom % observed on plain silicon wafers, and can be deconvoluted into two sources of silicon in a 70 : 30 ratio (Table 2.9 and Fig. 2.9B). These two signals arise from the two sources of silicon – 2s and 2p from silicon oxide appearing at the same energy as in plain silicon, and 2s and 2p from APTES seen as a shoulder at lower energy. The shifts agree with results in the literature (Kallury *et al.*, 1988). Room temperature and base-catalysed APTES deposit with more multilayers or better surface

E. T. Vandenberg, R. S. Brown & U. J. Krull

Table 2.8

Atomic Composition of X-Ray Photoelectron Spectra of Derivatized Silicon

Surface chemistry	Atomic composition (atom %)					
	oxygen	carbon	silicon	nitrogen	chlorine	other
silicon	53·98	13·90	32·11			
APTES #1	34·25	40·82	20·21	3·24	1·03	F-0·45
APTES #2	39·85	32·03	24·04	4·07		F-0·25
TEA-APTES	19·67	55·28	13·24	9·02	2·80	
RT-APTES	20·11	56·87	13·02	8·72	1·28	
glutaralde #1	31·31	47·55	18·66	2·48		
glutaralde #2	23·25	61·62	10·44	4·69		
tresyl #1	38·42	34·99	21·81	2·74	0·53 P-0·70,	F-0·57 Na-0·24
tresyl #2	35·98	38·58	21·63	3·80		
silicon/FITC-BSA	24·29	59·41	10·70	4·83	0·76	
APTES/BSA Tween	30·78	44·11	18·41	3·68	1·54	Na-1·13 P-0·35
APTES/BSA	24·88	52·89	12·20	4·99	2·53	Na-2·52
tresyl-BSA Tween	32·35	44·10	17·42	2·99	1·33	Na-1·25 S-0·57
tresyl-BSA	28·76	45·36	15·24	7·34	0·93 Al-l.19,	Na-0·90 P-0·29
glutaralde-ab-ag	10·99	80·38	2·99	5·37	0·27	
glutaralde-con A	18·84	64·52	6·02	8·46		Na-1·47
FITC-con A #1	14·56	74·37	1·88	9·19		
FITC-con A #2	19·82	64·70	8·32	7·16		

APTES = 3-aminopropyl triethoxy silane; RT-APTES = room temperature-deposited APTES; TEA-APTES = base catalysed APTES; glutaralde = glutaraldehyde; BSA = bovine serum albumin; FITC = fluorescein isothiocyanate; ab = anti-immunoglobulin G; antigen = immunoglobulin G; con A =concanavalin A; Tween = protein incubation with Tween 20

coverage than refluxed APTES, as seen by the greater nitrogen content (Table 2.9). Nitrogen constitutes 3–4 atom % of refluxed APTES-silicon (Table 2.8 and Fig. 2.8B), 8·72 atom % of room temperature deposited APTES-silicon and 9·02 atom % of base catalysed APTES-silicon. The nitrogen signal can be deconvoluted into two peaks (Fig. 2.11A and B), corresponding to protonated or hydrogen bonded (higher binding energy) and neutral (lower binding energy) forms of the amine function (Kallury et al., 1988). Under all reaction conditions, the proportion of higher energy nitrogen is much less in this work than in a previous study

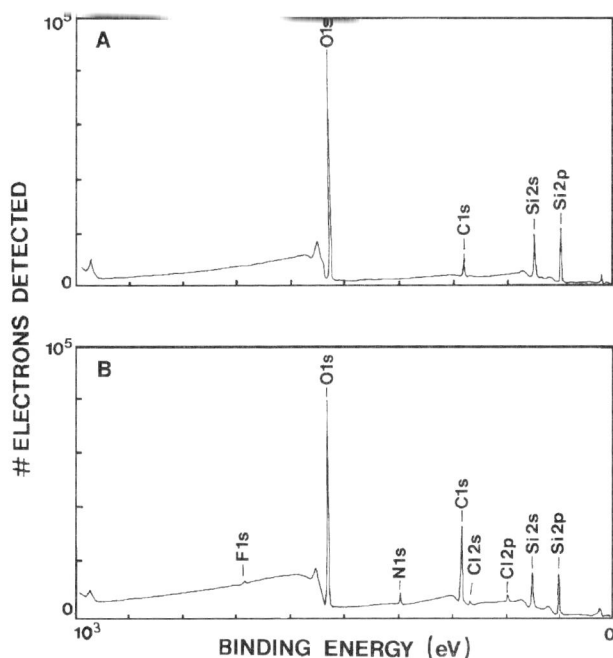

Fig. 2.8 Broad scan X-ray photoelectron spectra from (A) plain silicon, (B) silicon-3-aminopropyl triethoxy silane.

(Kallury *et al.*, 1988). This indicates conformational differences or less protonation of the amines.

Since APTES has three carbons to one nitrogen in a fully hydrolysed and silanized molecule, the nitrogen : carbon ratio is expected to be 0·3 for a complete monolayer, but will be decreased if unhydrolysed ethoxy groups or carbon contamination are present. This ratio may also be affected by APTES conformation at the surface, as atoms closer to the surface will produce a stronger signal than those beneath the surface. Free amines, constituting 76–82% of the nitrogen signal, are at the surface, but amines participating in hydrogen bonding and electrostatic interactions are several angstroms below the surface. Carbons attached to free amines will be buried, but those attached to bonding amines are exposed to the surface. The effect of varying atom placement on the nitrogen : carbon ratio is difficult to quantitate, therefore the results will be qualitative. In no sample did the nitrogen : carbon ratio approach 0·3; the highest observed was 0·163 for base catalysed APTES-silicon (Table 2.10). Making the crude assumption that the amount of contaminating carbon was 13·90 atom % for all samples (as observed for plain

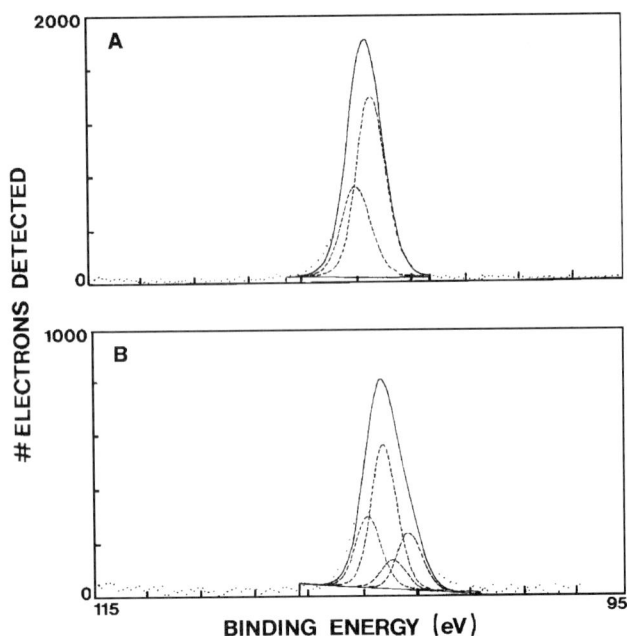

Fig. 2.9 High resolution X-ray photoelectron spectroscopy silicon signals from (A) plain silicon, (B) silicon-3-aminopropyl triethoxy silane.

silicon), nitrogen : carbon ratios for base catalysed and room temperature APTES would increase to only 0·22. Since it is known that base cata-lysed APTES deposition is many multilayers thick, unreacted ethoxy groups or amine bonding interactions are present, and full monolayers are presumed to have formed for room temperature and base catalysed APTES-silicon. Therefore, refluxed APTES clearly demonstrates sub-monolayer coverage, or formation of large islands.

Similarly, an oxygen : nitrogen ratio of 2 (complete siloxane bond formation) to 3 (no siloxane bond formation), and a silicon : nitrogen ratio of 1 are expected for a complete monolayer. However, some sub-strate silicon and oxide will contribute to the signal since electron escape depths are usually 20–50 Å and a monolayer of APTES is about 7·5 Å. Table 2.10 shows that for refluxed APTES oxygen : nitrogen is about 10, silicon : nitrogen is about 6, and nitrogen : carbon is about 0·1, suggesting that submonolayer coverage has occurred. For base catalysed and room temperature APTES–silicon, silicon : nitrogen is about 1·5 and oxygen : nitrogen is 2·2–2·3, indicating good coverage.

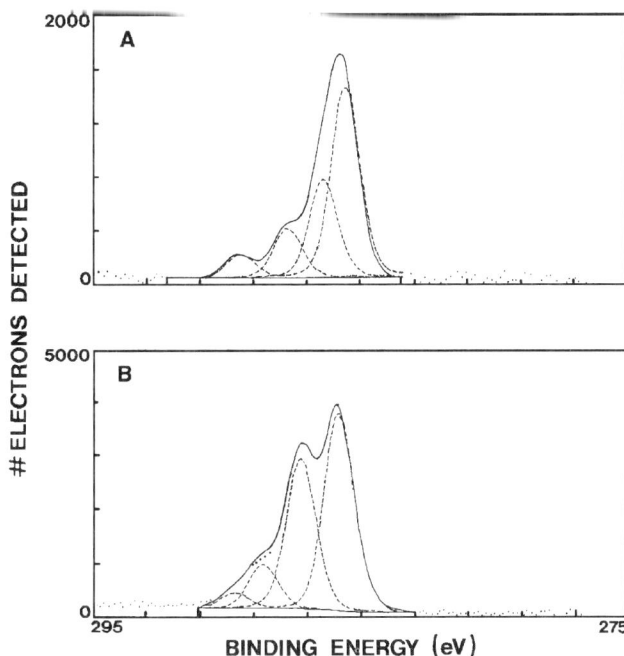

Fig. 2.10 High resolution X-ray photoelectron spectroscopy carbon signals from (A) plain silicon, (B) silicon-3-aminopropyl triethoxy silane.

Therefore, atomic ratios and atomic composition of the X-ray photoelectron spectra reveal that refluxed APTES-silicon has less than a monolayer of APTES, while base catalysed and room temperature deposition of APTES onto silicon gives mono- or multi-layer coverage.

Protonated amine nitrogens have a formal positive charge, therefore less electron density than hydrogen bonded amine nitrogens, which are only partially electron deficient. As the electron density around the nitrogen decreases, the binding energy per electron increases, and the ejected electron will have less kinetic energy and appear at higher binding energy. In systems where protonated and hydrogen bonded nitrogens exist, the nitrogen peak near 401 eV will be shifted to higher energy with respect to the free amine nitrogen peak near 400 eV as the proportion of protonated amines increases. The separation of the two nitrogen peaks is 1·34 eV for refluxed APTES, 1·43 eV for room temperature APTES, and 1·56 eV for base catalysed APTES (Table 2.9). The increasing peak separation indicates a shift to higher energy of the higher binding energy peak due to a greater contribution from protonated amines than from hydrogen bonded amines.

Table 2.9

Detailed Analysis of Individual High Resolution Signals from X-Ray Photoelectron Spectra of Derivatized Silicon

Surface chemistry (eV shift)	Deconvoluted peak energies (% area)		
	silicon	carbon	nitrogen
silicon (−0·66)	103·79 (66·7) 104·39 (33·3)	285·00 (53·1) 285·84 (27·2) 287·20 (13·4) 288·96 (6·3)	
APTES #1 (−0·66)	103·62 (47·5) 104·22 (23·8) 102·71 (19·1) 103·31 (9·6)	285·00 (68·3) 286·43 (21·6) 288·31 (10·0)	400·03 (81·5) 401·31 (18·5)
room temperature APTES (−1·1)			399.71 (75.8) 401·14 (24·2)
base catalysed APTES (−1·1)			399·72 (77·6) 401·28 (22·4)
glutaraldehyde #1 (−0·65)	103·37 (56·3) 102·10 (10·3) 103·97 (28·2) 102·70 (5·2)	285·00 (68·5) 286·28 (19·3) 287·72 (9·2) 289·00 (2·9)	399·52 (64·6) 400·99 (18·9) 402·08 (16·5)
glutaraldehyde #2 (−1·1)			399·60 (72·3) 400·79 (16·0) 402·17 (11·7)
tresyl #1 (−0·71)	103·49 (51·0) 102·45 (15·6) 104·09 (25·5) 103·05 (7·8)	285·00 (65·5) 286·20 (22·8) 287·41 (8·0) 289·00 (3·7)	399·34 (24·3) 400·54 (35·2) 401·94 (40·5)
tresyl #2 (−1·1)			398·90 (69·3) 401·59 (30·7)
APTES/bovine serum albumin (−0·79)	103·44 (54·4) 102·34 (12·3) 104·04 (27·2) 102·94 (6·1)	285·00 (48·6) 286·43 (37·1) 287·85 (10·6) 288·94 (3·6)	400·10 (79·7) 401·68 (20·3)
APTES/bovine serum albumin Tween 20 (−0·88)	103·41 (49·6) 102·48 (17·1) 104·01 (24·8)	285·00 (52·5) 286·34 (35·5) 287·83 (8·7)	399·61 (40·5) 400·52 (47·1) 402·04 (12·3)
tresyl-bovine serum albumin (−0·82)	103·24 (53·3) 101·92 (13·4) 103·84 (26·6) 102·52 (6·7)	285·00 (52·7) 286·33 (31·6) 287·99 (12·2) 288·96 (3·4)	400·16 (76·7) 401·80 (23·3)

continued

Table 2.9 – *contd.*

Surface chemistry	Deconvoluted peak energies (% area)		
(eV shift)	silicon	carbon	nitrogen
tresyl-bovine serum albumin Tween 20 (−0·98)	103·30 (54·6) 102·18 (12·1) 103·90 (27·3) 102·78 (6·0)	285·00 (51·9) 286·34 (32·8) 287·60 (9·5) 289·05 (5·9)	399·23 (16·3) 400·16 (58·5) 401·60 (20·7) 402·94 (4·6)
glutaraldehyde-antibody-antigen (±1·50)		285·00 (61·5) 286·40 (11·4) 288·32 (5·8) 283·49 (21·3)	398·16 (69·7) 398·80 (22·1) 401·22 (8·3)
concanavalin A (−0·33)		285·00 (56·0) 286·29 (26·7) 287·87 (12·0) 288·85 (5·3)	400·15 (75·7) 401·12 (24·3)
fluorescein iso-thiocyanate concanavalin A #1 (±4·68)	111·23 (66·7) 111·83 (33·3)	285·00 (40·7) 286·08 (31·1) 287·27 (15·6) 288·60 (12·6) 289·16 (2·1)	400·14 (55·6) 401·07 (35·8) 402·31 (8·6)

APTES = 3-aminopropyl triethoxy silane
#1 samples use reflux deposited APTES-silicon as a substrate
#2 samples use room temperature deposited APTES-silicon as a substrate

Refluxed APTES has the lowest protonated amine : hydrogen bonded amine ratio. APTES is known to participate in hydrogen bonding and electrostatic interactions with the surface when at submonolayer coverage (Caravajal *et al.*, 1988; Chiang *et al.*, 1982; Sudholter *et al.*, 1985). Furthermore, the proportion of protonated form is known to increase with the amount of water at the silica surface (Caravajal *et al.*, 1988). A high percentage of protonated amines is not expected for the refluxed APTES, since the solvent was dried, distilled toluene. Thus, refluxed APTES amines interact with the surface mainly through hydrogen bonds. The larger proportion of the nitrogen peak near 401·2 eV reported in the literature (Kallury *et al.*, 1988) could indicate the presence of water in that system. Note also that post reaction exposure to water from the atmosphere may also protonate the amines.

The room temperature reaction of APTES was not performed in a closed system, therefore water vapour may have entered the system. APTES deposited under room temperature conditions is more densely packed than if deposited under reflux conditions, so the former surface

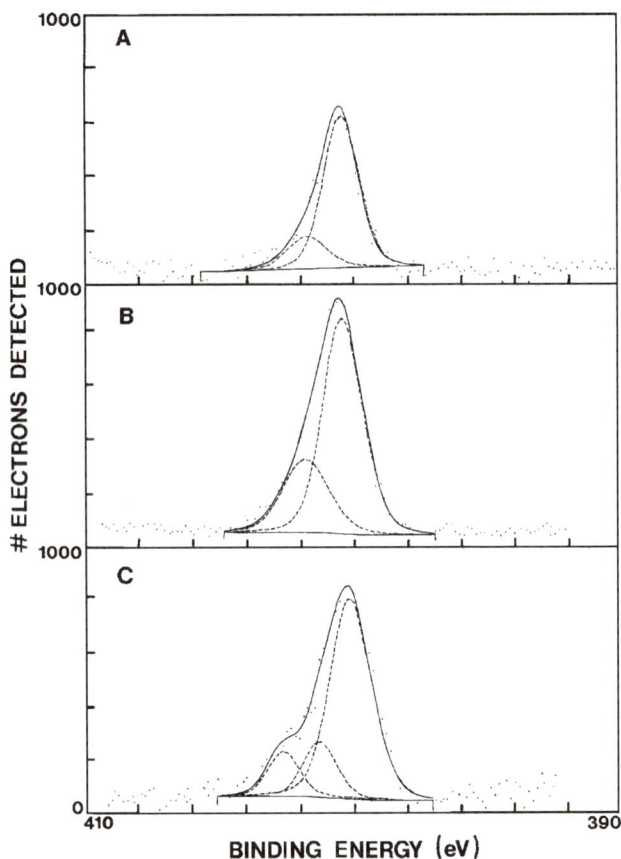

Fig. 2.11 High resolution X-ray photoelectron spectroscopy nitrogen signals from (A) silicon-3-aminopropyl triethoxy silane, (B) as (A) with room temperature silane deposition, (C) silicon-3-aminopropyl triethoxy silane-glutaraldehyde.

has fewer amine–silanol interactions than the latter. Therefore, more free amines exist which are protonated by exposure to water, increasing the proportion of 401 eV to 400 eV nitrogen peak, and increasing the peak separation. Protonated amines may be stabilized by entrapped chloroform. Chlorine is not observed in the refluxed APTES sample, but room temperature-deposited APTES has 1·28 atom % and APTES deposited with base catalysis has 2·08 atom % chlorine, which supports this hypothesis. The electronegative chloride atoms in chloroform will have a partial negative charge, and may stabilize the positively charged nitrogen.

APTES deposited with base catalysis has the largest energy separation between nitrogen peaks. Therefore, a higher proportion of protonated

Table 2.10
Relative Atomic Composition Ratios of X-Ray Photoelectron Spectra

Surface chemistry	Atom % ratios		
	silicon: nitrogen	oxygen: nitrogen	nitrogen: carbon
APTES #1	6·24	10·57	0·079
APTES #2	5·90	9·79	0·127
base-catalysed APTES	1·47	2·18	0·163
room temperature APTES	1·49	2·30	0·153
glutaraldehyde #1	7·52	12·63	0·052
glutaraldehyde #2	2·23	4·96	0·076
tresyl #1	7·96	14·02	0·078
tresyl #2	5·69	9·47	0·098
APTES/BSA	2·44	4·99	0·094
APTES/BSA Tween 20	5·00	8·36	0·083
tresyl/BSA	2·07	3·92	0·162
tresyl/BSA Tween 20	5·81	10·82	0·068
silicon/FITC-BSA	2·22	5·03	0·081
concanavalin A	0·71	2·23	0·13
FITC-con A #1	0·20	1·58	0·124
FITC-con A #2	1·16	2·77	0·11
glutaraldehyde-ab-ag	0·56	2·05	0·067

See notes in Table 2.11; also BSA = bovine serum albumin; FITC = fluorescein isothiocyanate; con A = concanavalin A; ab = anti-immunoglobulin G (antibody); ag = immunoglobulin G (antigen)

amines stabilized by chloroform is proposed to explain the energy separation between the nitrogen peaks. It is surprising that a reaction taking place in the presence of base leaves any amines protonated. It is likely that exposure of the sample to atmosphere after the reaction resulted in protonation of amines, and the high amount of protonated amines is due to the high amount of free amines in the original sample surface.

The X-ray photoelectron spectrum of tresyl-activated samples shows only a trace of fluorine (0·57 atom %) and no sulfur (Table 2.8). It is therefore likely that tresyl group sublimed when placed in the vacuum. This would leave the APTES electron deficient, resulting in the observed greater proportion of the high energy nitrogen peak (30·7%), and the large separation between nitrogen peaks (2·69 eV). The bond between the tresyl group and APTES is therefore not of covalent strength. There is good evidence that the tresyl chloride reagent actually is present in non-vacuum conditions because of the significantly greater amount of bovine serum albumin immobilized on tresyl-activated substrates as compared to APTES substrates, as discussed in section 2.4.4.5.

Glutaraldehyde activation of APTES results in an additional nitrogen peak at $402 \cdot 1$ eV constituting 12–17% of the nitrogen signal (Fig. 2.11C). The source of the peak at $402 \cdot 1$ eV is the imine linkage that forms between the APTES and the glutaraldehyde. The imine nitrogen signal appears at higher energy. The two lower energy peaks are at the same energies as observed in the spectrum of refluxed APTES (within experimental error).

The activated samples displayed increased silicon : nitrogen and oxygen : nitrogen ratios, and a decreased nitrogen : carbon ratio (Table 2.10). This indicates that there is either an alteration of the APTES layer on exposure to the solvents or solutions used in the activation procedures, or a desorption of adsorbed APTES. Since great changes in ellipsometric parameters were observed after exposure of APTES to water, indicating some type of refractive index change, the former explanation is most probable. APTES amines can hydrogen bond with water, and the addition of water to the APTES layer could be largely responsible for the refractive index change and rearrangement of APTES conformation.

X-ray photoelectron spectroscopy differs from the other methods of analysis in that microenvironments, such as sections of uncovered substrate, can be partially resolved by deconvolution of the high resolution elemental signals. This technique, unlike ellipsometry and contact angle measurements, demonstrated that surface coverage of APTES varied as reaction conditions changed. It also provided a means by which the reaction of APTES with the activating agents could be verified. The results indicate that glutaraldehyde reacted with APTES. The APTES layer was demonstrated to be altered upon exposure to solvents or water.

2.4.2.4   *Reflection Infra-red Spectroscopy*

The infra-red spectrum of APTES on aluminum oxide is shown in Fig. 2.12A, and peak assignments were made according to Naviroj *et al.* (1984) and Boerio *et al.* (1980). The broad peak centred at 3400 cm^{-1} corresponds to the oxygen–hydrogen stretch of silanols, formed when the surface was exposed to the water vapour of the atmosphere. Nitrogen–hydrogen stretching also contributes to this peak. The carbon–hydrogen stretching peak of the APTES propylene groups can be seen at 2940 cm^{-1}. The broad region at 1600 cm^{-1} indicates a mixture of free amines, hydrogen-bonded amines, and protonated amines. The peak at 1100 cm^{-1} indicates siloxane bonds, and the peak at 960 cm^{-1} is due to silanols and silicon–oxygen–carbon groups.

The spectrum of APTES that was heated to 120°C for two hours, and not exposed to the atmosphere for prolonged periods, is shown in Fig. 2.12C. Mixed sources of nitrogen are indicated by the broad doublet at 1600 cm^{-1}. The hydroxyl peak at 3400 cm^{-1} is weaker than in the atmosphere-exposed sample, since no water vapour was present to hydrolyse APTES ethoxy groups. The peak at 1100 cm^{-1} is less intense than the atmosphere-exposed APTES sample, indicating that fewer siloxanes had formed, again, due to the lack of water to catalyse this reaction.

The spectrum of an oven-treated APTES sample reacted with tresyl chloride is shown in Fig. 2.12B. The peak at 1600 cm^{-1}, corresponding to nitrogen–hydrogen stretching, is absent. Furthermore, no evidence of the carbon–fluorine bond of tresyl chloride is seen at 1200 cm^{-1}, even though this group strongly absorbs. This result correlates well with the X-ray photoelectron spectrum of the tresyl-activated surface, where no atoms typical of the tresyl group were observed, but the APTES nitrogen signal had an upwardly shifted high energy peak (Table 2.9). Thus, the tresyl group sublimed, even in the low vacuum of the infra-red spectrometer, and left an altered APTES nitrogen.

There are no positive clues in the tresyl-activated sample spectrum as to the structure of the altered APTES surface, other than the absence of the nitrogen–hydrogen bond. No nitrogen–carbon double bond was formed since there is no imine peak at 1670 cm^{-1}.

When finely divided silica was derivatized with tresyl chloride and allowed to dry, the container became cold and the odour of tresyl was detected, giving evidence that an endothermic reaction occurred with the

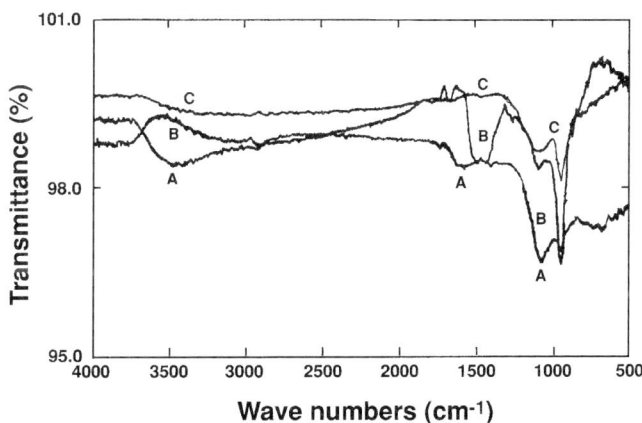

Fig. 2.12 Reflection infrared spectra for (A) 3-aminopropyl triethoxy silane on aluminum, exposed to atmosphere for two hours, (B) as (A) except heated in an oven for two hours, (C) oven-treated 3-aminopropyl triethoxy silane reacted with tresyl cholide.

release of the tresyl group. Characterization of the dried tresyl-activated surface warrants investigation, but since it is a complex problem and peripheral to the covalent stabilization of protein on surfaces, it was not pursued further. Care was taken to maintain hydration of the tresyl-activated surfaces used for protein binding at all times.

2.4.2.5 Summary

Cleaning of the substrates was observed by contact angle measurements and ellipsometry. APTES was observed to react with the surfaces under reflux, base-catalysed and room temperature deposition conditions as seen by changing contact angles and ellipsometric parameters. X-ray photoelectron spectroscopy showed that refluxed deposition of APTES gave sub-monolayer coverage, whereas base-catalysed and room temperature depositions covered the surface. Base-catalysed APTES had a different surface morphology than room temperature deposited APTES, as seen by scanning electron microscopy. APTES chemically reacted with water and changed conformation, as seen by ellipsometry and X-ray photoelectron spectroscopy.

Glutaraldehyde reacted successfully with silicon, whereas tresyl chloride addition was not seen directly with X-ray photoelectron spectroscopy or infra-red spectroscopy, but showed changes in the APTES amines, indicating that the tresyl group may have sublimed on exposure to the atmosphere.

Addition of protein to activated, APTES and plain substrates is studied in the following sections.

2.4.3 Protein Immobilization – Surface Analysis

2.4.3.1 Contact Angle Measurements

Proteins may denature on drying, therefore an accurate measurement of the wettability of protein in the native conformation would best be made on either wet or freshly dried surfaces. However, since there are technical problems associated with reproducing hydration conditions, measurements were taken on dried surfaces. No significant change in contact angle was seen between activated or APTES surfaces before and after incubation with bovine serum albumin and concanavalin A (Table 2.4). No dramatic differences in surface free energy are expected, since protein has both hydrophobic and hydrophilic residues, and should produce an intermediate wettability similar to that of activated surfaces. The contact angle of $60 \pm 6°$ for bovine serum albumin reported here agrees with the literature value for the contact angle of dried human serum albumin surfaces (63·5°), which is expected to be similar to bovine serum albumin surfaces due to the sequence homology of the two proteins (Behrens *et al.*, 1975; Brown, 1975).

Contact angle measurements could not confirm the presence of protein on an activated surface, due either to similarity in wettability between activated substrates, or because the non-ideality of the surface does not allow accurate measurements to be taken.

2.4.3.2 Ellipsometry

When protein was added to APTES or activated surfaces, thickness increases and decreases were observed in comparison to the activated surface thickness (Table 2.11). The irreproducibility indicates a complex system where proteins may adsorb or bind in sub-monolayer, monolayer or multilayer coverage, may adsorb with different orientations on substrates having different surface chemistry, or may interact with the underlying layer, thereby changing the conformation of both layers. Before ellipsometric measurements were taken, the samples were dried

Table 2.11
Ellipsometric Parameters of Protein Deposition onto Derivatized Silicon

Surface chemistry	Measured ellipsometric parameters				Change in d
	Δ	ψ	d (Å)	n	
protein adsorption to silicon					
clean silicon	$275·7 \pm 0·2$	$56·2 \pm 0·5$	1647 ± 6	3	0
silicon-con A	$273·2 \pm 0·3$	$63·0 \pm 0·5$	1588 ± 4	4	-59
protein adsorption to APTES					
APTES	$277·2 \pm 0·5$	$48·5 \pm 0·4$	1753 ± 17	3	153
APTES/IgG	$282·1 \pm 0·3$	38 ± 2	1844 ± 13	2	244
APTES	$276·1 \pm 0·5$	47 ± 2	1797 ± 42	18	0
APTES/con A	$277·1 \pm 0·3$	$48·2 \pm 0·9$	1760 ± 13	4	11
APTES/con A	$276·2 \pm 0·4$	$53·0 \pm 0·7$	1689 ± 7	4	-60
protein covalent binding					
glut	$275·1 \pm 0·2$	$57·6 \pm 0·5$	1635 ± 4	4	0
glut-con A	276 ± 1	54 ± 2	1666 ± 22	2	31
glut-ab-ag	$279·2 \pm 0·3$	38 ± 1	1908 ± 38	3	308

APTES = 3-aminopropyl triethoxy silane; con A = concanavalin A; IgG = immunoglobulin G; glut = glutaraldehyde; ab-ag = antibody-antigen
value ± standard deviation
n = number of measurements
d = ellipsometrically measured coating thickness.
The reference surface (cleaned silicon, APTES, or activated surface) is set to 0 to emphasize the protein thickness.

fully to avoid differences between samples arising from varying amounts of entrapped water. However, it is not possible to regulate strictly the amount of water hydrogen bonded to the protein, which affects protein conformation and refractive index. When protein is dried, the conformation will change to varying degrees depending on the orientation on the substrate, the density of neighbouring molecules, and the conditions of drying (e.g. temperature and humidity). Variations in these conditions are difficult to control, resulting in inconsistent changes in the ellipsometric parameters. Ellipsometry can be used quantitatively to study protein deposition and conformation at interfaces under well-controlled conditions (Azzam *et al.*, 1977; Mandenius *et al.*, 1984; Mathot & Rothen, 1969; Poste & Moss, 1972; Vroman, 1967). However, under the conditions of this experiment, ellipsometry could not confirm the presence of protein on activated substrates.

2.4.3.3   X-Ray Photoelectron Spectroscopy

Surfaces composed of bovine serum albumin bound to 3-aminopropyl triethoxy silane and tresyl-activated silicon were examined by X-ray photoelectron spectroscopy. The amounts of carbon, oxygen, nitrogen and silicon observed were similar to the amounts on APTES and tresyl-activated surfaces (Table 2.8). The silicon signals of the albumin samples should be reduced in intensity, since a monolayer of bovine serum albumin should be thicker than the escape depth of silicon (*ca.* 50 to 150 Å (Hendrickson *et al.*, 1975) versus 40 Å (Powell, 1974)). Quantitative calculations are difficult for this heterogeneous surface, but qualitatively, the addition of 50 Å of material should decrease the silicon signal by about half. Therefore, it is unlikely that monolayer coverage of bovine serum albumin was achieved.

The deconvoluted carbon spectrum for bovine serum albumin is distinct from APTES and from general carbon contamination. There is a significant increase in the peak at about 286·4 eV (Fig. 2.10B and Table 2.9) due to amide carbons, which are the second most numerous type of carbon in protein. The nitrogen peaks of bovine serum albumin samples deposited in the absence of detergent can be deconvoluted to two peaks at 400·1 and 401·7 eV of ratio 78 : 22 (Fig 2.13A and Table 2.9). The higher energy peak is shifted by 0·4–0·5 eV from its position in the APTES spectrum, reflecting the contribution of protonated and amide carbons in the protein. The spectrum is the same for bovine serum albumin covalently bound to tresyl-activated surfaces or adsorbed to APTES surfaces, since covalent binding does not involve sufficient numbers of albumin nitrogens to alter the observed pattern. The silicon signal deconvolutes into two sources of silicon, from the APTES and from the surface, similar to the pattern observed with APTES samples.

Bovine serum albumin samples deposited in the presence of the non-ionic detergent Tween 20 have irregularly shaped nitrogen signals, making deconvolution of the signals difficult (Figs. 2.13B and C). The low energy end of the peak was broad, but this was not due to sample charging as high resolution signals of silicon (Fig. 2.14) and carbon were sharp. The nitrogen peak was deconvoluted into four peaks for bovine serum albumin adsorbed to APTES (Fig. 2.13B) and three peaks for covalently bound albumin (Fig. 2.13C). Bovine serum albumin is a transport protein and has binding sites for a variety of amphiphilic anions and for aliphatic alcohols (Makino *et al.*, 1973). Tween 20 has hydroxyl groups on the end of long polymer chains, and thus is able to bind to

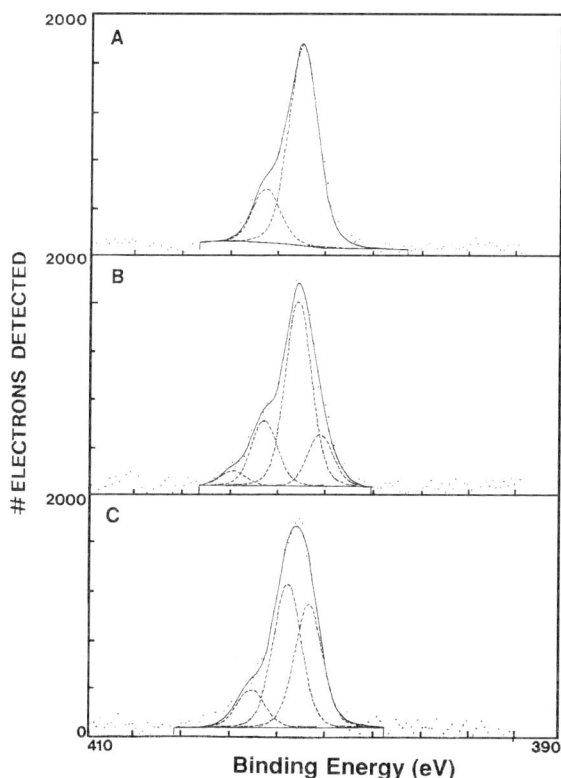

Fig. 2.13 High resolution X-ray photoelectron spectroscopy nitrogen signals from (A) silicon-3-aminopropyl triethoxy silane/bovine serum albumin, (B) as (A) with Tween 20 present during protein incubation, (C) silicon-3-aminopropyl triethoxy silane-tresyl-bovine serum albumin with protein incubation in the presence of Tween 20.

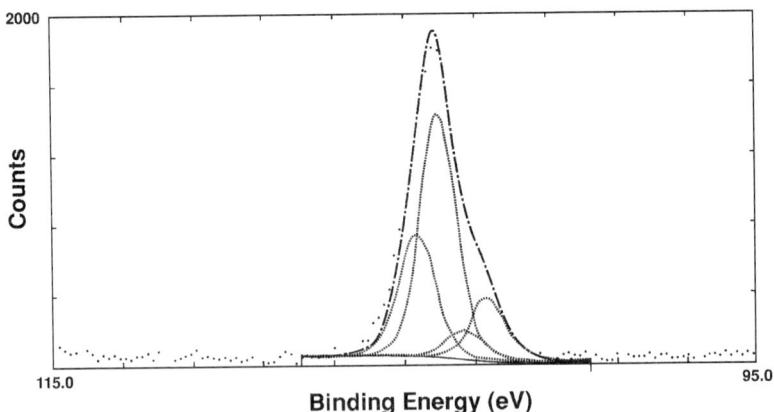

Fig. 2.14 High resolution X-ray photoelectron spectroscopy silicon signal from silicon-3-aminopropyl triethoxy silane-tresyl-bovine serum albumin with protein incubation in the presence of Tween 20.

bovine serum albumin. Judging from the magnitude of change of the signal, the aliphatic alcohol binding site may consist of part of the protein backbone or be rich in nitrogen-containing amino acid side chains (lysine, arginine, histidine, tryptophan, asparagine and glutamine).

Concanavalin A and fluorescein-labelled concanavalin A were incubated with glutaraldehyde-activated surfaces. The samples showed low oxygen (15–20 atom %) and silicon (2–8 atom %), and high carbon (65–74 atom %) and nitrogen (7–9 atom %) when compared to the atom % composition of APTES or activated substrates (Table 2.8). The small silicon signal demonstrates that monolayer or greater coverage was achieved, since the protein thickness (40 Å) is the same as the escape depth of silicon (Powell, 1974). It was necessary to flood the surface with electrons to produce spectra free of charging peaks, again indicating a thick protein coating. Concanavalin A is well known to adsorb non-specifically to most surfaces (Krull *et al.*, 1989), and given its molecular weight of 100,000 under these pH conditions, a 40 Å thick coating of protein (Reekes *et al.*, 1975) is expected. This correlates well with the fact that fluorescein-labelled concanavalin A samples were intensely fluorescent.

As with the albumin samples, an increased contribution of the carbon peak at 286·1 eV was seen for concanavalin A samples, due to amide carbons (Fig. 2.15A). It was not possible to deconvolute the weak silicon signal. The nitrogen signal of the concanavalin A sample deconvoluted to three peaks at 400·1, 401·0 and 402·3 eV of ratio 56 : 36 : 9 (Fig. 2.16A). The high energy peak can be attributed to the nitrogen in the isothiocyanate of the fluorescent label.

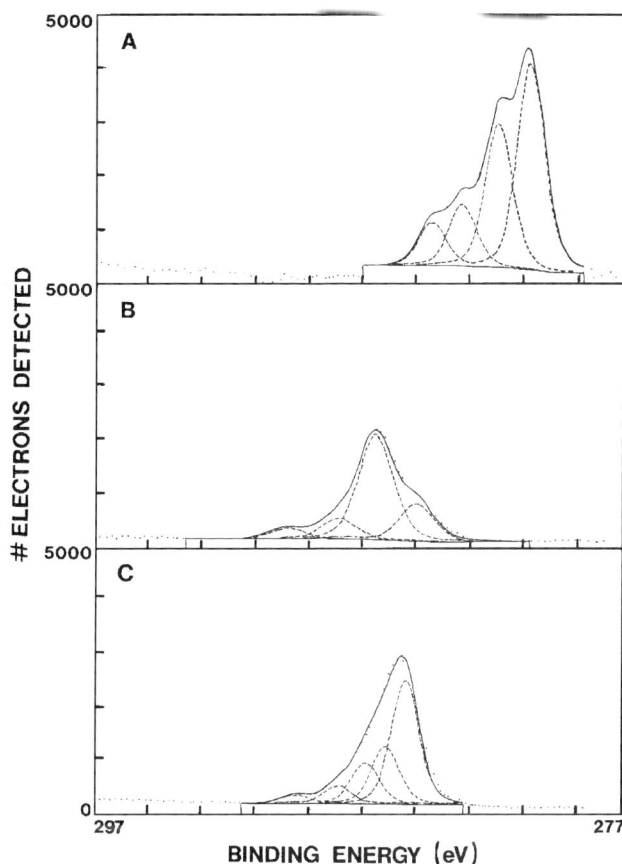

Fig. 2.15 High resolution X-ray photoelectron spectroscopy carbon signals from (A) immobilized concanavalin, (B) immobilized antibody-antigen, (C) immobilized acetylcholine receptor.

Glutaraldehyde-activated surfaces were incubated with fluorescein-labelled anti-rabbit immunoglobulin G (antibody), then with rabbit immunoglobulin G (antigen), to create a covalently attached antibody–antigen complex. There is ample evidence of a thick protein coating; sample charging (seen as broad, low energy peaks) comprised 21–22 % of the carbon (Fig. 2.15B) and nitrogen (Fig. 2.16B) signals despite flooding of the sample with electrons, the oxygen and silicon contributions were very low (10·99 and 2·99 atom % respectively, see Table 2.8), the carbon signal was very strong (80·38 atom %), and the silicon signal was very weak. A thick coating is expected for a double layer of protein.

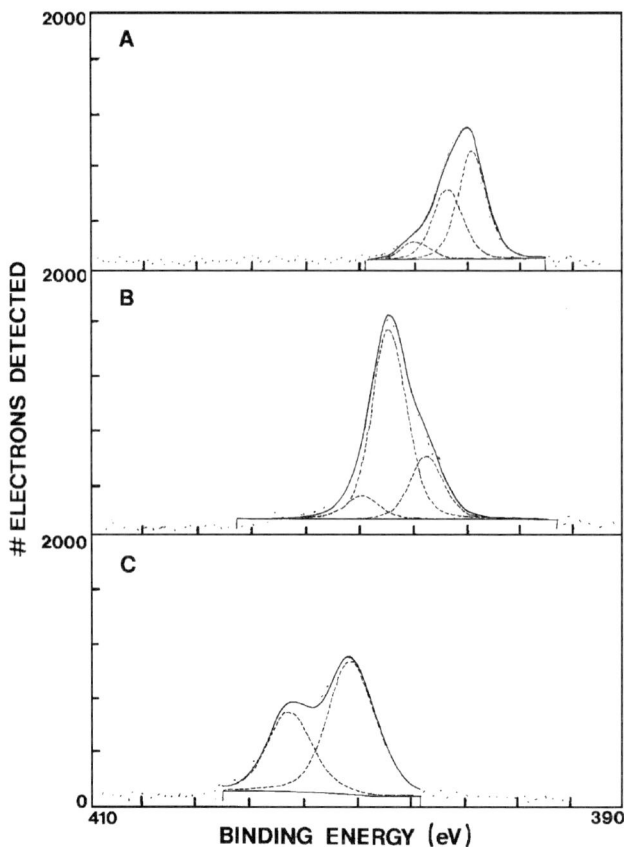

Fig. 2.16 High resolution X-ray photoelectron spectroscopy nitrogen signals from (A) immobilized concanavalin, (B) immobilized antibody-antigen, (C) immobilized acetylcholine receptor.

X-ray photoelectron spectroscopy showed definite differences in the low and high resolution signals of bovine serum albumin, concanavalin A, and antibody–antigen supports. Therefore X-ray photoelectron spectroscopy can be used to detect the presence of protein qualitatively. The information from high resolution atomic signals decreases as the protein coverage and thickness increases, due to sample charging. Nevertheless, charging peaks are informative to the extent that they indicate good protein coverage. A broad scan of the sample gives the most useful information for estimation of protein coverage, since signals from the substrate (silicon) and the protein (nitrogen) can be compared. High resolution spectra may give useful information when the protein is rich in a different chemistry compared to the substrate (e.g. the isothiocyanate of fluorescein-labelled concanavalin A), or where the protein has bound an identifiable ligand or is chemically altered by it (e.g. Tween 20 binding to bovine serum albumin).

2.4.3.4 Scanning Tunnelling Microscopy

Scanning tunnelling microscopy (STM) works on the following principles: an electric field is applied between two electrodes, one being flat and positioned under the sample, and the other being fashioned into a point, and placed above the sample. The point of the latter electrode is constructed so that only one atom is present at the tip of the point. An overlap in the wavefunctions between the atom on the tip of the electrode and the sample atom nearest to the tip occurs, allowing tunnelling of an electron from the sample to the electrode tip. Scanning tunnelling microscopes are operated either under constant voltage or constant current, and the tip is moved to compensate for deviations from the set parameter. The displacement of the tip indicates the surface position of the atoms, with a vertical resolution of $0.1\,\text{Å}$, and horizontal resolution of $6\,\text{Å}$ under the best circumstances. Principles of scanning tunnelling microscopy are discussed in Hansma & Tersoff (1987).

Fluorescein-labelled bovine serum albumin was attached via glutaraldehyde and 3-aminopropyl triethoxy silane to mica and graphite substrates. These substrates were chosen because they are conductive and molecularly smooth, which is necessary to obtain clear images. Total surface coverage of graphite by APTES was not achieved since areas of bare graphite, distinguishable by the regular crystalline structure (Fig. 2.17a), were observed. Since APTES requires surface hydroxyls in order to bind, and since graphite does not readily oxidize, monolayer coverage was not expected. The mica samples, which are rich in surface hydroxyls, showed qualitative differences between uncoated and APTES coated mica, indicating APTES coverage. However, the mica was not sufficiently conductive to obtain clear images of immobilized bovine serum albumin.

A high resolution image (Fig. 2.17b) of the covalently bound bovine serum albumin to graphite shows albumin molecules aligned end to end and several rows wide. This peculiar alignment may be indicative of oxidation of the graphite surface at ridges or faults in the graphite, and subsequent deposition of APTES, glutaraldehyde and protein occurring only at these ridges or faults. High resolution images (Fig. 2.17c) clearly show the globular structure of bovine serum albumin bound side-on, and the molecular dimensions are estimated as $100 \times 40\,\text{Å}$ (error $20\,\text{Å}$), in good agreement with the literature value of $140 \times 40\,\text{Å}$ (Hendrickson *et al.*,

⊢——⊣   a   ⊢——⊣   b
1 nm   10 nm

⊢——⊣   c
2·5 nm

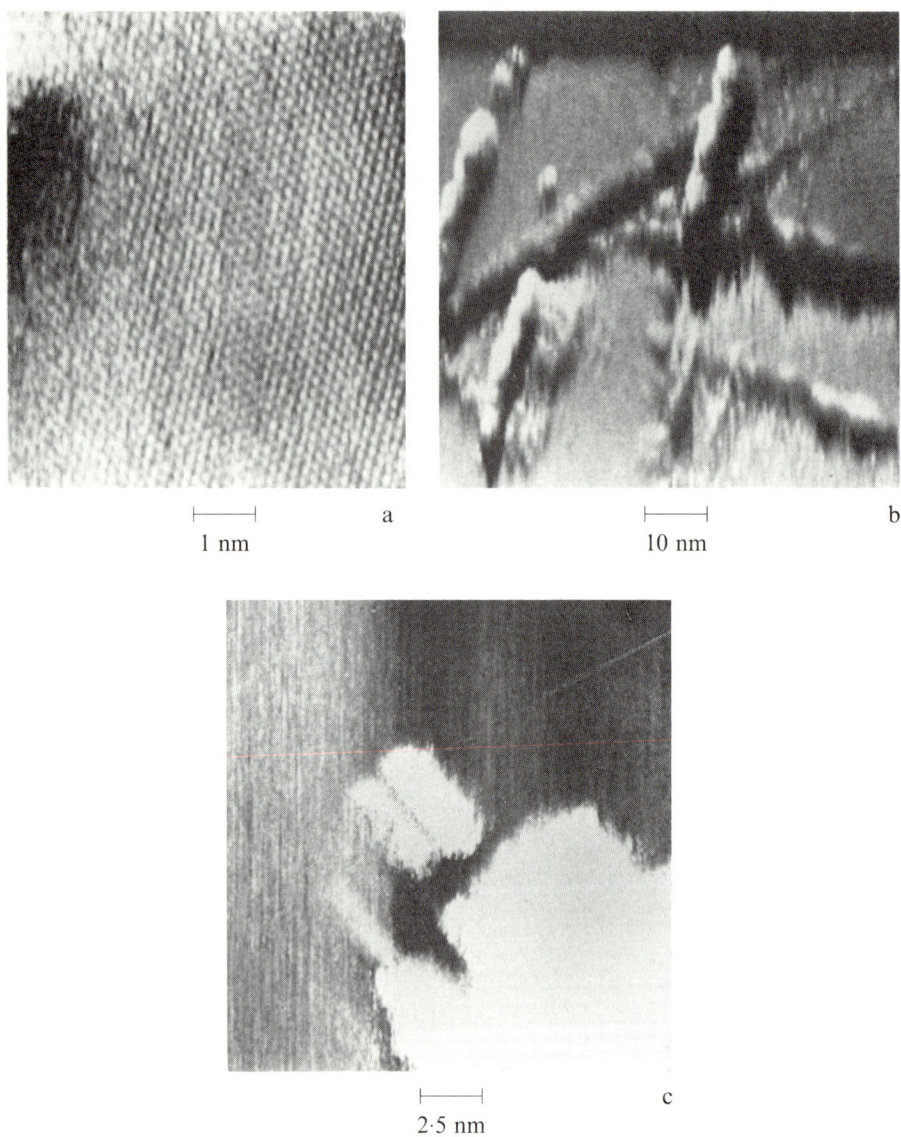

Fig. 2.17 Scanning tunnelling microscopy images of (*a*) graphite, scale bar = 1 nm, (*b*) immobilized bovine serum albumin, scale bar = 10 nm, (*c*) as (*b*), scale bar = 2·5 nm.

1975). The fact that a stable picture could be obtained indicates strong binding of the molecule to the surface.

Scanning tunnelling microscopy showed that protein deposition was successful, and that an image of a single protein can be made and the molecular dimensions estimated. The binding density and arrangement of bovine serum albumin on poorly oxidized graphite is not expected to be indicative of albumin coating on highly oxidized surfaces such as silicon and quartz.

2.4.4 Protein Immobilization – Comparison of Binding Densities Using Radioactivity and Fluorescence

2.4.4.1 Experiments Using Radiolabelled Bovine Serum Albumin
Comparisons of the immobilization methods (tresyl and glutaraldehyde activation, as well as adsorption) were made using quartz or silicon pieces that were cleaned and coated with 3-aminopropyl triethoxy silane in the same reaction set, and performing the protein incubation, washing and radiation counting steps in parallel. The results show that the protein coverage was reproducible for quartz or silicon in the same set, but varied more between sets. Relative comparisons between sets were made by standardizing the results with respect to one sample.

Activated surfaces were expected to immobilize the most protein, but APTES and octadecyltrichlorosilane derivatized quartz incubated in radiolabelled protein were often more radioactive than activated surfaces incubated in the same solutions (e.g., Tables 2.12 and 2.13, no Tween 20), even though parallel studies with fluorescently labelled albumin demonstrated that APTES immobilized much less albumin than tresyl-activated surfaces (Fig. 2.18).

Table 2.12
Coverage of Radiolabelled Bovine Serum Albumin on Quartz – With and Without Protein Incubation in 0.1% Tween 20

Surface chemistry	Tween	Protein density ($\mu g/cm^2$)		% decrease with Tween 20	
		trial # 1	trial # 2	trial # 1	trial # 2
plain quartz	no	0.077	0.021	36	67
	yes	0.049	0 007		
APTES	no	0.168	0.200	90	99
	yes	0.017	0.003		
tresyl	no	0.187	0.248	34	54
	yes	0.124	0.114		
glutaraldehyde	no	0.149	0.199	69	77
	yes	0.046	0.046		

APTES = 3-aminopropyl triethoxy silane

Table 2.13
Coverage of Radiolabelled Bovine Serum Albumin on Quartz – With and Without Protein
Preincubation in 0·1% Tween 20

Surface chemistry	Tween	Protein density ($\mu g/cm$)		% decrease with Tween 20	
		trial # 1	trial # 2	trial # 1	trial # 2
plain	no	0·0079			
	yes	0·0123			
APTES	no	0·2560	0·1560	62	26
	yes	0·0963	0·1148		
tresyl	no	0·1450	0·1600	34	37
	yes	0·0962	0·1005		
glutaraldehyde	no	0·0988	0·1056	44	42
	yes	0·0552	0·0616		

APTES = 3-aminopropyl triethoxy silane

To test if $^{125}I^-$ was adsorbing to the surfaces, each surface was incubated with free radiolabel, washed, and the radioactivity measured. Considerable $^{125}I^-$ was adsorbed to APTES and glutaraldehyde-activated surfaces, and less $^{125}I^-$ was adsorbed to plain quartz and tresyl-activated substrates (Table 2.14). After incubation in sodium iodide, all surfaces decreased in

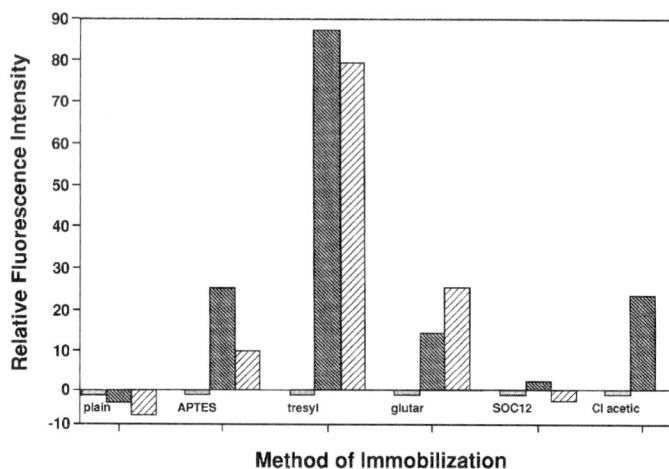

Fig. 2.18 FITC-Bovine serum albumin. FITC = fluorescein isothiocyanate; plain = plain quartz; APTES = 3-aminopropyl triethoxy silane; glutar = glutaraldehyde-activated; SOC12 = acid chloride-activated; cl acetic = chloroacetic-activated. Results from two experiments are shown.

Table 2.14

Effect of Sodium Iodide Wash on Adsorption of $^{125}I^-$ to Derivatized Quartz Substrates

Surface chemistry	Normalized counts		% of original counts after NaI wash
	original	after 0·5 M NaI wash	
plain quartz	6	2	5
APTES	100	100	13
tresyl	77	89	15
glutaraldehyde	142	138	12

APTES = 3-aminopropyl triethoxy silane

counts by 81% or more, demonstrating that $^{125}I^-$ does not react covalently, or reacts reversibly. Since the relative proportions of $^{125}I^-$ adsorption to the various surfaces were different from the relative proportions of radioactivity observed for surfaces incubated in radiolabelled protein, $^{125}I^-$ was not directly adsorbing to the surfaces to produce the unusually high counts on certain samples.

To test if $^{125}I^-$ was adsorbing to the bovine serum albumin and partitioning into the APTES, three lots of samples with immobilized radiolabelled bovine serum albumin were incubated in 0·5 M unlabelled sodium iodide overnight. As shown in Table 2.15, counts decreased about 20% for the covalently bound protein samples, and about 40–60% for adsorbed protein samples. If most of the APTES-quartz counts were

Table 2.15

Effect of Sodium Iodide Wash on Adsorption of Radiolabelled Bovine Serum to Derivatized Quartz

Surface chemistry	Incubation with Tween 20	Average % decrease in counts after NaI wash
plain quartz	no	54 ± 10
	yes	
APTES	no	38 ± 14
	yes	
tresyl	no	27 ± 10
	yes	
glutaraldehyde	no	17 ± 6
	yes	

APTES = 3-aminopropyl triethoxy silane
average ± standard deviation

due to adsorbed [125]I⁻, greater decreases should have been observed. A further test was made by performing a 'mock' iodination of bovine serum albumin where the activating reagent, chloramine T, was omitted. Upon separation of bound and unbound iodide, no significant counts were seen in the protein peak. This correlates with the results from a microtitration for iodine and iodide that was performed on fractions of the chromatographed iodination mixture – none was seen in the albumin fraction. Thus, [125]I⁻ is not likely the main interferent. However, Table 2.14 illustrated some interesting chemistry of the activated substrates, since it served as an iodide microtitration, and is therefore discussed below.

Plain quartz is negatively charged in water (Iler, 1979) and thus repels [125]I⁻, accounting for the small amount of radioactivity observed. On the other hand, APTES-substrates have a net positive charge (Chiang *et al.*, 1982) at neutral pH since the pK_A of the amine group is 10·3. The [125]I⁻ can electrostatically bind to protonated amines, thus large amounts of radiolabel are associated with APTES surfaces.

Tresyl-activated surfaces were 23% less radioactive than APTES substrates. The tresyl-activated surface has few available protonated amines for interaction with the [125]I⁻ in comparison to the APTES surface. The tresyl group itself provides only partial positive charges (on the carbons and sulfur). This results in less [125]I⁻ interaction with tresyl-activated surfaces than with APTES surfaces.

It is difficult to anticipate how iodide and glutaraldehyde might interact since the polymerization and condensation reactions of glutaraldehyde are not fully characterized. One notable difference in chemistry between APTES and glutaraldehyde-activated surfaces is the presence of double bonds on the glutaraldehyde-activated surface. Iodide, along with a cation, can undergo 1,4-addition to conjugated dienes. Iodine free radicals produced by decomposition product of [125]I⁻, may also add to double bonds. Iodide is very basic and polarizable, and may adsorb to partially positive centres. A combination of the above interactions results in the high counts associated with glutaraldehyde-activated surfaces.

Protein was radioiodinated by the chloramine T method, which adds one or two ^{125}I atoms to tyrosine residues, and may also react with sulfhydryl groups ('Radioiodination Techniques', Review 18, Amersham, Oakville, ON, Canada). One byproduct of this reaction is [125]I₂ (Hunter & Greenwood, 1962; 'Radioiodination Techniques', Review 18, Amersham, Oakville, ON, Canada). To test if [125]I₂ was adsorbing to bovine serum albumin, a quantity of albumin was iodinated, then separated into two portions. Unlabelled iodine was added to one portion until a yellow colour persisted, indicating saturation of the solution with iodine. Chromatography was then used to separate free and protein-bound radiolabel. With the untreated albumin sample, 95% of the counts exiting the column were associated with the protein. In contrast, where iodine displacement was performed, only 3% of the eluting counts were in the bovine serum albumin peak. Assuming that the unlabelled iodine does not displace any covalently bound radiolabel, $^{125}I_2$ and/or $^{125}I_3^-$ (formed in iodine and water solutions) may have been adsorbed to bovine serum albumin. If $^{125}I_3^-$ contributes to the adsorbed counts on APTES substrates, it is not primarily due to the negative charge since the pattern of immobilized counts when albumin is bound to the activated supports (Tables 2.12 and 2.13) is in different proportions than when $^{125}I^-$ was incubated with the supports (Table 2.14).

To test if a difference in APTES coverage or conformation made a difference in the amount of radioactivity that was immobilized, radiolabelled bovine serum albumin was incubated with surfaces prepared under different conditions of APTES deposition, and various activation methods. APTES surfaces were prepared using distilled and dried toluene or reagent grade toluene solvent, with and without base catalysis. For this experiment, free $^{125}I^-$ was separated from protein by dialysis rather than chromatography. After the radiolabelled albumin was incubated with various quartz surfaces, it was found that APTES surfaces were more radioactive than glutaraldehyde-activated surfaces (Table 2.16). All samples demonstrated high variability in counts. Dialysis takes hours to remove the iodination reagents from the protein, and may have allowed more adsorption of radioactive species to albumin and substrates than when the reagents were quickly removed by chromatography. The slides were washed in scintillation fluid overnight, then recounted. Between 5 and

Table 2.16

Radioactive Counts Adsorbed to Derivatized Quartz after Incubation with Dialysed Radiolabelled Bovine Serum Albumin

Surface chemistry	*Original counts per minute*	*% decrease in radioactivity*	
		after scintillation fluid wash	*after NaI wash*
APTES	$800\,000 \pm 200\,000$	20 ± 4	28 ± 6
tresyl	$800\,000 \pm 200\,000$	29 ± 5	32 ± 5
glutaraldehyde	$200\,000 \pm 30\,000$	5 ± 2	13 ± 3

APTES = 3-aminopropyl triethoxy silane
value ± standard deviation

29% of the adsorbed counts were removed, substrates with the higher counts yielding the higher proportions, showing that the displaced counts were from an adsorbed species. Incubation in sodium iodide decreased the radioactivity by an additional 12–32%, again more significantly in the substrates with higher initial counts. Thus, the adsorbing radioactive species binds both by hydrophobic and electrostatic interactions to the surface, and is not entirely displaceable as seen in Table 2.16. This experiment also demonstrates the superiority of the quick chromatographic separation of the free label from the protein, and the insensitivity of the radiolabel adsorption to APTES surface morphology.

To determine if the adsorbed radiolabel was affecting protein binding calculations, identically prepared APTES surfaces were incubated with the two types of bovine serum albumin, untreated and treated with unlabelled iodine. The two sources of albumin had very different specific activities, 65 570 cpm/μg for untreated bovine serum albumin, and 94 700 cpm/μg for iodine-incubated albumin, due to varying amounts of adsorbed radiolabel. Significantly, both types of bovine serum albumin gave the same quantitative measure of protein surface coverage, 0.9 ± 0.1 μg/cm^2. However, the binding density of albumin on APTES was very high compared to the amount of albumin immobilized on tresyl-activated surfaces, which bound more albumin than APTES, as seen using fluorescently labelled albumin. Therefore, the amount of radioactive interferent adsorbed to APTES is proportional to the amount of radioactive interferent adsorbed to bovine serum albumin.

These observations of radiolabel adsorbing to protein are not unique. Bovine serum albumin that was radioiodinated using the lactoperoxidase method was demonstrated to adsorb $^{125}I^-$, which could be removed from albumin by column chromatography (Hansma & Tersoff, 1987). However, after incubation of the protein at 37°C for 200 minutes, 7.4 % of the total radioactivity was free (Crandall et al., 1981). It was suggested that diffusion of loosely bound iodine, oxidation of sulfide bonds and radioactive degradation of the carbon-iodine bond of tyrosine may all be involved in the process of radiolabel release (Crandall et al., 1981).

It is possible that the retained activity could be due to radiolabelled tyrosine adsorbing to the albumin. Recall that tyrosine is added after incubation of albumin with radiolabel to help quench the labelling reaction. Albumin may bind tyrosine in the same manner that Tween 20 was bound. This may also indicate why the presence of Tween 20 during radiolabelled protein–surface incubation decreases the counts bound. The total radioactivity of surfaces incubated in radiolabelled bovine serum albumin is then probably due to a combination of sources: radioactive iodide, iodine, tyrosine and protein.

2.4.4.2 *Use of Detergent for Prevention of Adsorption of Radiolabel*

The usefulness of a detergent in the prevention of non-specific adsorption of radioactive molecules or of protein onto substrates was investigated. Detergents increase the solubility of a wide range of substances, thereby decreasing the non-specific adsorption of proteins at interfaces. The detergent Tween 20 was chosen because it is non-ionic, has a gentle solubilization action that tends to keep proteins in their native conformation, and has been used with success by other researchers (Elwing *et al.*, 1987, 1989). There are dramatic differences between surface radioactivity with and without Tween 20 present during the protein incubation, as seen in Table 2.12. With no detergent present, APTES and glutaraldehyde- and tresyl-activated surfaces had high apparent protein coverages. When Tween 20 was present during protein incubation, the tresyl activation method clearly immobilized the most protein. These results are in accord with fluorescence observed for surfaces incubated with fluorescein-labelled bovine serum albumin (Fig. 2.18). When surfaces were pre-incubated with Tween 20 for 30 minutes prior to protein incubation and incubated in protein without Tween 20, the protein coverage decrease in the samples with Tween 20 was not so pronounced (Table 2.13). APTES surfaces immobilized as much protein by adsorption as tresyl-activated surfaces did by covalent attachment. Therefore, co-incubation with Tween 20 decreased nonspecific adsorption of radioactive interferent(s). Detergent does not inhibit covalent binding in general, since tresyl-activated substrates bound albumin, and tresyl- and glutaraldehyde-activated surfaces bound immunoglobulin G in good quantity.

Since Tween 20 helped to solubilize radioactive species during protein incubation, thereby decreasing the amount of adsorbed radiolabel, Tween 20 was included in all subsequent protein incubation solutions.

2.4.4.3 *Desorption of Adsorbed Protein and Radiolabel by Detergent and Non-Labelled Protein*

Not all non-covalent adsorption of protein to substrates was prevented by Tween 20 co-incubation since some radiolabel was observed on plain substrates. To desorb these counts, various supports with immobilized antibody-^{125}I-antigen were incubated in a 1 mg/ml bovine serum albumin solution for 24 hours, followed by incubation in 0.1% (v/v) Triton X-100 for 24 hours. Table 2.17 shows that both agents were effective, as bound radioactivity decreased by 10–20% for each incubation. After this experi-

Table 2.17

The Effect of Washing Antibody-Radiolabelled Antigen Surfaces with Bovine Serum Albumin, Sodium Chloride, Urea and Triton X-100

Surface chemistry	% decrease in antigen after washing ± standard deviation	
	no Tween 20	with Tween 20
plain	24 ± 14	44 ± 26
APTES	15 ± 8	13 ± 9
tresyl	14 ± 8	14 ± 9
glutaraldehyde	10 ± 8	10 ± 7

APTES = 3-aminopropyl triethoxy silane

ment, substrates were always incubated in albumin and Triton solutions to desorb adsorbed protein and/or radiolabel. However, even with these post-incubation treatments, adsorbed protein sometimes persisted on plain and APTES substrates, substantiating observations that proteins can irreversibly adsorb to surfaces (Andrade & Hlady, 1987; Andrade *et al.*, 1984; 1987; Dillman & Miller, 1973; Norde, 1986). The results show that when surfaces are incubated in protein and Tween 20, followed by washing in albumin and Triton X-100, the binding densities of radio-labelled and fluorescently labelled protein qualitatively agree. Therefore, when this procedure is used, more confidence may be placed in the results.

Proteins adsorbed to a solid support have been found to be partially or fully displaced by the same (Bohnert & Horbett, 1986) or different proteins (Andrade *et al.*, 1987; Brash & Samak, 1978; Brash *et al.*, 1974; Chan & Brash, 1974; Norde, 1986) in solution. Proteins exchange by sequential detachment of the adsorbed protein, and attachment of a segment of the new protein (Norde, 1986). The new protein must first adsorb to the surface at the correct proximity and orientation (Bohnert & Horbett, 1986) to the first protein. Long incubation times, such as the 24 hours used in this experiment for albumin displacement, are necessary for maximum displacement of originally adsorbed protein.

The effectiveness of detergent in displacing adsorbed molecules is due to the solvation ability. Non-ionic detergents interrupt contact sites between adsorbates and the solid surface (Mittal & Lindman, 1983), and show orientation phenomena when adsorbed to solid surfaces (Crandall *et al.*, 1981). Solubilization ability may be intimately associated with the way the detergent molecules are oriented (Elwing *et al.*, 1989). Triton X-100 was chosen as the washing detergent because: (1) it is non-ionic, thus not likely to denature protein, (2) it is structurally different from Tween

20, so may have different orientation effects that will solvate molecules that Tween 20 cannot, and (3) it interacts strongly with bovine serum albumin (Jennissen, 1986). For protein displacement, the effectiveness of detergent action is a function of the time protein has incubated with the surface (Nemat-Gorgani & Karimiar, 1984). Longer residence times of protein with substrate lead to stronger and more extensive protein-surface interactions due to slow conformational changes in the protein (Andrade & Hlady, 1987; Andrade *et al.*, 1987; Norde, 1986; Soderquist & Walton, 1980). Supports were incubated with detergent for 24 hours to increase the amount of protein desorbed as much as possible.

2.4.4.4 Fluorescence Spectroscopy

In order to confirm the relative efficiencies of binding by the various immobilization methods, as measured using radiolabelled protein, fluorescently labelled proteins were similarly bound, and the fluorescence intensities compared. To compare fluorescence spectra from different surfaces, the area under a wavelength versus intensity plot was integrated. In some cases, especially where the signal was weak, a sloping baseline prevented accurate peak area measurement. For simplicity of data analysis, the assumption was made that the baseline slope was linear, but in some cases this may not hold true. Therefore, there is a relatively large amount of uncertainty in the peak integration. Since instrument realignment was necessary before each set of experiments, absolute intensities cannot be directly compared between experimental sets, and only relative intensities are used.

Monolayers of fluorescent materials require high quantum yield fluorophores to produce enough fluorescence intensity for detection. Fluorescein isothiocyanate was used to label proteins as it attaches easily to the protein, can be excited by a nitrogen laser, and has a high quantum yield. However, the free nucleophilic groups in the protein needed for the fluorescent labelling are also required for covalent binding to glutaraldehyde and tresyl chloride. The fluorescence labelling process was therefore a compromise between leaving enough unreacted amine groups for covalent binding to substrates, while labelling the protein in sufficient yield to give good sensitivity by fluorescence detection. Figure 2.19 provides an indication of the data obtained for comparison of immobilization methods.

Spectra of bovine serum albumin immobilized on various substrates were recorded, the baseline subtracted, and the areas under the peaks from 500–640 nm were compared. Immobilized and soluble albumin had the same wavelength maximum, indicating that the chemical environment of the fluorophore did not change dramatically on immobilization. Fluorescently labelled concanavalin A also had the same wavelength maxima

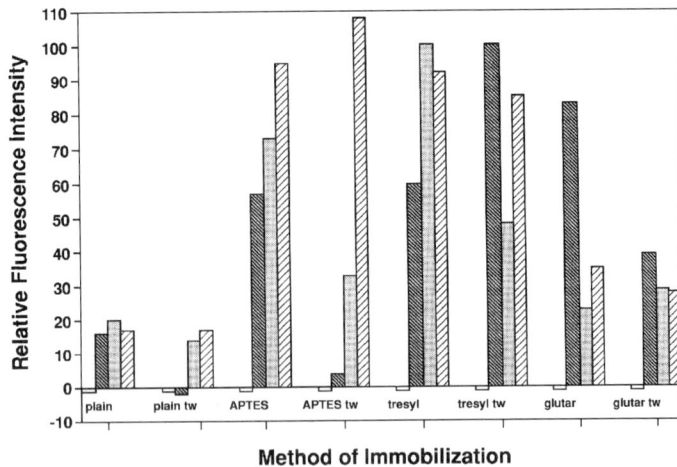

Fig. 2.19 FITC-Acetylcholinesterase. FITC = fluorescein isothiocyanate; plain = plain quartz; APTES = 3-aminopropyl triethoxy silane; glutar = glutaraldehyde; tw = protein incubation in Tween 20. Results from three experiments are shown.

when immobilized and in solution. However, since fluorescein is not very sensitive to the local environment, it is still possible that changes in conformation occur on protein immobilization.

In summary, the radiolabelled protein may non-specifically adsorb radiolabel, which may partition into APTES and interfere with quantitation of protein immobilization. Fluorescently labelled proteins may show decreased densities of binding for those reactions which require free amine groups. Therefore, radiolabelled and fluorescently labelled protein binding studies should be compared, and more confidence can be placed in the results that are in agreement.

2.4.4.5 Protein Immobilization Results

X-ray photoelectron spectroscopy and scanning electron microscopy results demonstrated that APTES-derivatized surfaces were covered by APTES to different extents depending on the reaction conditions used. To see if the differences in APTES coverage affected the amount of protein that was immobilized, APTES was deposited under different conditions (distilled vs undistilled toluene, base-catalysed vs uncatalysed). The surfaces were activated, incubated in radiolabelled bovine serum albumin, and the protein binding densities were compared. There was no Tween 20 present in the protein incubation and the supports were not washed in albumin or Triton solutions, therefore the apparent protein

Table 2.18
Comparison of Radiolabelled Bovine Serum Albumin Binding Density for Several APTES Deposition Conditions

APTES deposition conditions		*Protein binding density (μg/cm^2) $\pm$ 10% activation method*	
distilled toluene	*base catalysed*	*glutaraldehyde*	*tresyl*
yes	no	0·129	0·144
no	no	0·145	0·133
yes	yes	0·111	0·150
no	yes	0·145	0·157

APTES = 3-aminopropyl triethoxy silane

binding densities include adsorbed radiolabel and protein. Table 2.18 indicates that apparent protein binding density is the same within error for all APTES types within a particular activation method. Therefore, adsorption and binding of protein and radiolabel to activated surfaces are invariant with the conditions of APTES deposition.

To test further for interactions of radiolabel with surfaces of various surface free energies, a variety of substrates were incubated with radio-labelled bovine serum albumin. On quartz surfaces, more radiolabel was seen on APTES surfaces than on plain substrates (hydrophilic), or with substrates coated with octadecyl trichloro silane (hydrophobic) (Table 2.19). This correlates well with the binding results of fluorescently-

Table 2.19
Adsorption of Radiolabelled and Fluorescently Labelled Bovine Serum Albumin to Control Surfaces

Surface chmistry	*Protein density (μg/cm^2) $\pm$ standard deviation*		*Fluorescence (relative) $\pm$ 2%*	
	quartz	*silicon*	*quartz*	*silicon*
uncleaned	0·0768 ± 4%	0·526 ± 5%	6	35
methanol-cleaned		0·422 ± 3%		9
hydrophilic	0·0074 ± 7%	0·604 ± 25%	−1	75
OTS		0·0217 ± 9%		31
methanol/APTES	0·3027 ± 9%	0·854 ± 7%	79	96
hydrophilic/APTES	0·2926 ± 14%	0·948 ± 0·9%	99	58
hydrophilic/RT-APTES		1·070 ± 6%		84
hydrophilic/TEA-APTES		1·202 ± 13%		84

APTES = 3-aminopropyl triethoxy silane; RT-APTES = room temperature-deposited APTES; TEA-APTES = base catalysed APTES; OTS = octadecyl trichloro silane labelled bovine serum albumin, indicating better chemisorption of albumin to the intermediate free energy surface of APTES, rather than low energy (quartz) or high energy (octadecyl trichloro silane) surfaces. On silicon surfaces, more fluorescently labelled and radiolabelled bovine serum albumin bound to plain surfaces, perhaps because they have a higher surface free energy than plain quartz surfaces, as shown by contact angle measurements. The amount of adsorbed protein varied, indicating non-specific adsorption of protein to plain silicon.

Quartz shows consistency in immobilization of radio- and fluorescently-labelled bovine serum albumin, whereas silicon does not. Quartz, being microscopically rough, also has more real surface area than molecularly smooth silicon of the same dimensions. Therefore, most work was done on quartz substrates to maximize binding and thus minimize relative error. Fluorescence experiments were performed to confirm qualitatively the results of radiolabel experiments with respect to which methods of protein immobilization gave the highest coverages. It is interesting that there are negligible differences between the binding densities observed for substrates incubated in protein with and without Tween 20 present. This indicates that it is not primarily non-specific protein adsorption to quartz that Tween 20 prevents, but rather it is non-specific adsorption of a radio-labelled interferent that is hindered by the detergent. No fluorescence was seen on the plain quartz substrates. Glutaraldehyde-activated and APTES surfaces bound moderate amounts of fluorescent protein, but tresyl-activated surfaces clearly bound the greatest amount of fluorescently-labelled bovine serum albumin.

Derivatized quartz and silicon substrates were incubated with radio-labelled bovine serum albumin, washed in albumin and Triton X-100, and compared with respect to the density of the protein immobilized. Table 2.20 shows that only 6–17% as much albumin is immobilized on silicon as compared to quartz. When Tween 20 was present during protein incubation, the total amount of immobilized protein on APTES and activated surfaces decreased by 44–71% on silicon, and by 11–79% on quartz, demonstrating the effectiveness of Tween 20 in preventing non-specific adsorption. The effect of Tween 20 was greater with silicon substrates, presumably because they adsorbed more radiolabelled species. When Tween 20 was used, all quartz supports bound more protein than did silicon supports, and tresyl-activated supports bound the most. Detergent presence decreased the binding density of bovine serum albumin on tresyl-activated quartz by very little, and the binding densities of albumin immobilized in the presence of Tween 20 qualitatively agree with the fluorescently-labelled albumin binding densities.

Table 2.20

Comparison of Radiolabelled Bovine Serum Albumin Binding Density to Derivatized Quartz and Silicon

Surface chemistry	Tween	Protein binding density (μg/cm^2) ± standard deviation	
		quartz	silicon
plain	no	0·026	0·05 ± 0·02
	yes	0·001	0·0016 ± 0·0003
APTES	no	0·9 ± 0·2	0·11 ± 0·04
	yes	0·3 ± 0·4	0·05 ± 0·04
tresyl	no	0·9 ± 0·2	0·09 ± 0·03
	yes	0·8 ± 0·7	0·05 ± 0·01
glutaraldehyde	no	0·71 ± 0·02	0·07 ± 0·02
	yes	0·2 ± 0·1	0·02 ± 0·01

It can be calculated that 23% of the surface is covered with bovine serum albumin, assuming that all radioactive counts are associated with bovine serum albumin, that albumin binds to the surface side-on, as seen by scanning tunnelling microscopy, and that the shape can be approximated by a rectangle of 140 × 40 Å. This calculated coverage may not be accurate due to uncertainty associated with the source of the counts, since albumin may not adsorb in the shape assumed, and since quartz has a greater true surface area than its dimensions indicate, due to surface roughness. Nevertheless, it is useful as a first approximation.

The previous experiments were repeated using another radiolabelled protein, immunoglobulin G, to see if the same trends held. When Tween 20 was used in the protein incubation, the binding density of immunoglobulin G on quartz supports was roughly equal for APTES, and tresyl- and glutaraldehyde-activated surfaces (Table 2.21). All plain and derivatized silicon supports bound a low amount of immunoglobulin G, except tresyl-activated silicon which bound the same amount of albumin as tresyl-activated quartz. When fluorescently-labelled immunoglobulin G was covalently bound to the derivatized quartz in the presence of Tween 20, the binding density was about equal for APTES, tresyl and glutaraldehyde surfaces, but less for plain substrates. This agrees with the radio-labelled immunoglobulin G results, except that the plain surface was of similar radioactivity to the activated surfaces, again indicating adsorption of a radioactive interferent, or variable protein adsorption. Immunoglobulin G is known to adsorb strongly to a variety of surfaces. This latter property is used in immunosorbent assays. Kits are available

Table 2.21

Comparison of Radiolabelled Immunoglobulin G Binding Density to Derivatized Quartz and Silicon

Surface chemistry	*Tween*	*Protein binding density* ($\mu g/cm^2$) $\pm$ *standard deviation*	
		quartz	*silicon*
plain	no	0·112	0·119
	yes	0·314	0·016
APTES	no	0·26 ± 0·01	0·10 ± 0·03
	yes	0·32 ± 0·06	0·028 ± 0·006
tresyl	no	0·193 ± 0·007	0·10 ± 0·03
	yes	0·32 ± 0·04	0·36 ± 0·03
glutaraldehyde	no	0·2 ± 0 .1	0·10 ± 0·05
	yes	0·300	0·05 ± 0·02

commercially (for example from Cyberfluor and Becton-Dickinson) which employ immunoglobulin G adsorption to plastic surfaces, and use detergent to wash out non-antigenic proteins.

The protein binding density was approximately equal for bovine serum albumin and immunoglobulin G on all substrates. The protein binding densities reported here compare favourably with those reported in the literature. Covalent binding of immunoglobulins to glutaraldehyde-activated surfaces has been reported at 0·2–0·5 $\mu g/cm^2$ (Sutherland *et al.*, 1984; Williamson *et al.*, 1989), whereas densities of 0·2–0·3 $\mu g/cm^2$ were observed here. Adsorption of a different protein, calmodulin, to APTES was measured at approximately 0·09 $\mu g/cm^2$ (Williamson *et al.*, 1989) and was of the same order of magnitude as observed for immunoglobulin G and bovine serum albumin to APTES in this work. Bovine serum albumin has been adsorbed in very small amounts to hydrophilic surfaces at about 0.2 ng/cm^2 (Golander & Kiss, 1988; Hlady *et al.*, 1987; Morrisey *et al.*, 1976), Adsorption of albumin to quartz was variable in this study, but adsorbed amounts of up to 50 ng/cm^2 were measured.

Because of the unknown extent to which the radiolabel may be attributed to labelled protein or to free label, the protein binding results must be interpreted with caution. Protein incubation in the presence of Tween 20, followed by washing in albumin and Triton X-100 solutions, seems to be highly effective in removing loosely associated counts from the substrates. However, since adsorbed radiolabel may be present, the protein binding densities should be used for comparison purposes only, and not be considered correct in the absolute sense.

It is interesting that the APTES surface adsorbs protein in quantities comparable to covalently bound protein, as in the case of immunoglobulin G. APTES has both hydrophilic sites (amines) and hydrophobic sites (methylene groups), therefore it can adsorb a wide variety of amino acid residues. Electrostatic interactions are also possible between the protonated amines of APTES and negatively charged amino acids (glutamate, aspartate). Due to the length of protein incubation, adsorbed protein has sufficient time to reorient and change conformation to maximize binding to the surface (Andrade & Hlady, 1986). APTES may also change configuration to maximize interaction with the protein (Andrade *et al.*, 1984), which leads to essentially irreversible adsorption (Soderquist & Walton, 1980). Vertical polymerization of APTES can create cavities where proteins can be partially surrounded with the APTES. If islands of APTES on silicon exist, proteins may maximize electrostatic interactions at the APTES/silicon interface since the silicon surface is negatively charged and APTES is partially positively charged. Literature confirms that hydrophobic and electrostatic interactions are important for immunoglobulin adsorption to APTES (Lin *et al.*, 1988). The extent of strong adsorption of protein to activated substrates is not distinguishable from covalent binding for immunoglobulin G (Table 2.21).

It can be concluded that the preferred surface for protein immobilization is partially protein specific, but tresyl-activated surfaces bind larger amounts of all the proteins studied. Optimization of reaction times for the activated surfaces are described in the next section.

2.4.4.6 *Optimization of Protein Immobilization*

The incubation times of APTES surface with activating agent and of protein with activated substrate were varied to see the effect on amount of protein subsequently bound. Table 2.22 and Fig. 2.20 demonstrate that tresyl chloride/APTES-quartz incubation time between one and 60 minutes is irrelevant to the amount of protein ultimately bound. Glutaraldehyde/APTES-quartz incubation time, in contrast, made a marked difference in the amount of protein bound, and at least two hours was needed for maximum activation (Table 2.23 and Fig. 2.21). The effect of detergent solubilization of bound protein is seen in the last columns of Tables 2.22 and 2.23. The amount of protein released by the detergent was independent of incubation time for tresyl-activated quartz because all surfaces were equally reactive. Glutaraldehyde-activated surfaces desorbed slightly less protein after detergent incubation than tresyl-activated surfaces. The proportion of covalently attached versus adsorbed proteins

Table 2.22
Effect of Incubation Time of Tresyl Chloride with APTES-Quartz on the Binding Density
of Radiolabelled Immunoglobulin G

Incubation time	Tween 20	Protein density (10^{-2} μg/cm^2) after Triton		% decrease after Triton
		original	X-100 wash	X-100 wash
1 min	no	1·062	0·832	22
	yes	1·047	1·001	4
5 min	no	1·015	0·828	18
	yes	0·896	0·716	20
10 min	no	0·952	0·754	21
	yes	0·938	0·745	21
20 min	no	0·955	0·745	22
	yes	1·153	0·907	21
40 min	no	1·068	0·809	24
	yes	1·097	0·853	22
1 hour	no	1·178	0·551	53
	yes	0·802	0·672	16

APTES = 3-aminopropyl triethoxy silane

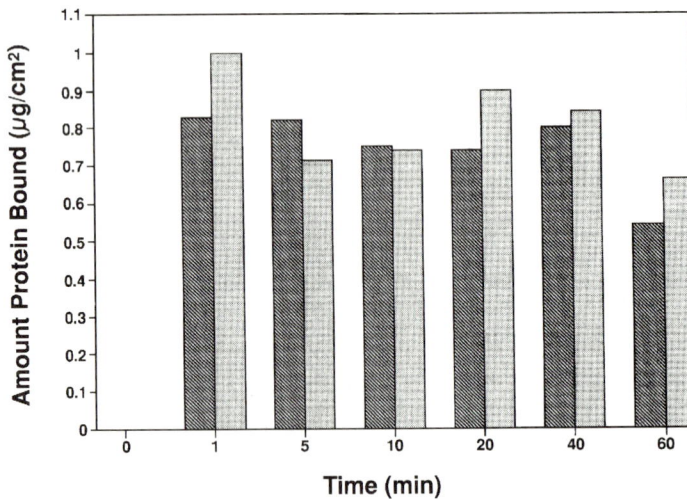

Fig. 2.20 Tresyl Chloride activated quartz showing effect of Tween 20. Dark areas – no
Tween; light areas – with Tween

Table 2.23

Effect of Incubation Time of Glutaraldehyde with APTES-Quartz on the Binding Density
of Radiolabelled Immunoglobulin G

Incubation time	*Tween 20*	*Protein density* (10^{-2} µg/cm^2) *after Triton*		*% decrease after Triton*
		original	*X-100 wash*	*X-100 wash*
10 min	no	1·116	1·016	10
	yes	1·083	1·030	6
21 min	no	0·568	0·518	10
	yes	1·289	1·214	7
1 hour	no	1·432	1·389	4
	yes	1·308	1·237	6
2 hours	no	1·620	1·602	2
	yes	1·752	1·691	5
3 hours	no	1·651	1·610	4
	yes	1·460	1·439	2

APTES = 3-aminopropyl triethoxy silane

may be greater on glutaraldehyde-activated surfaces, or else adsorbed protein is more tightly bound to these surfaces.

When protein was incubated with tresyl-activated substrates for various periods of time, it was discovered that half the total amount of protein immobilized did so in the first 10 minutes (Table 2.24 and Fig. 2.22).

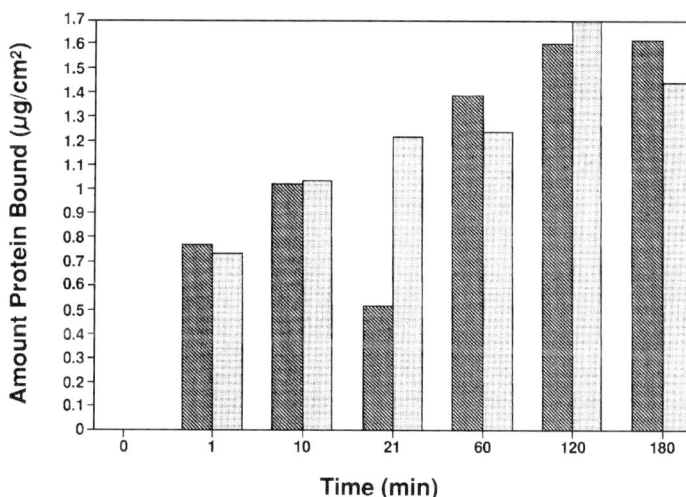

Fig. 2.21 Glutaraldehyde activated quartz showing effect of Tween 20. Dark areas – no Tween; light areas – with Tween

E. T. Vandenberg, R. S. Brown & U. J. Krull

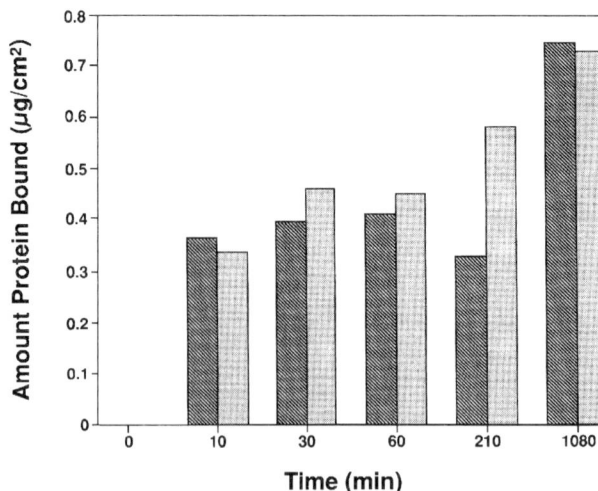

Fig. 2.22 Protein incubation time. Tresyl-activated quartz surfaces were incubated with radiolabelled bovine serum albumin and Tween 20 for the times indicated. Dark areas – no Tween; light areas – with Tween

In the next 3·6 hours, 30% more was bound, and the remaining 20% was bound after 18 hours of incubation. The first 10 minutes may represent the period required for reversible adsorption as described in Saderquist & Walton (1980), and the half-maximum coverage may be due to washing away of loosely adsorbed protein. With time, the initially reversibly adsorbed protein may reorient until the protein and surface are in the correct orientation for covalent binding, increasing the amount of immobilized protein. Alternatively, time may simply be required for diffusion of protein to the surface (Norde, 1986).

Table 2.24
Effect of Incubation Time of Radiolabelled Immunoglobulin G with Tresyl-Activated Quartz on the Binding Density of Protein

Incubation time	Protein density $(10^{-2}\ \mu g/cm^2)$		% decrease after Triton X-100 wash	
	– Tween 20	+ Tween 20	– Tween 20	+ Tween 20
10 min	0·364	0·336	50	47
30 min	0·393	0·457	54	64
1 hour	0·407	0·446	55	62
3·6 hours	0·321	0·577	44	80
18 hours	0·734	0·718	100	100

2.4.4.7 Summary

During the radiolabelling of bovine serum albumin, unidentified radio-active species adsorbed to the protein, and apparently desorbed from albumin and adsorbed to APTES during protein incubation. This process could be prevented or lessened by the addition of 0·1% Tween 20 to the protein incubation solution. Reversibly adsorbed protein could be removed from surfaces by washing in bovine serum albumin and Triton X-100 solutions. The protein binding density results were correlated qualitatively with fluorescence intensities of surfaces having the same protein type which was fluorescently labelled. Where agreement was observed, the results were considered valid. Non-agreement indicated that interferences or artifacts existed in the experiment. When protein incubation with surfaces was performed in the presence of Tween 20, the radiolabel and fluorescence experiments always qualitatively agreed. Plain quartz surfaces, as well as all silicon supports, immobilized varying amounts of protein. Tresyl-activated quartz bound the most bovine serum albumin (about $0·2 \, \mu g/cm^2$), and tresyl- and glutaraldehyde-activated quartz, as well as APTES quartz, bound about $0·3 \, \mu g/cm^2$ immunoglobulin G. Tresyl chloride reacted almost immediately with APTES, whereas glutaraldehyde required two hours for maximum activation. Protein immobilization was more effective at long incubation times. For most applications, a high surface density of protein is desirable, but high physiological activity of the bound protein is even more important. The high density of coverage was established in this section. The next section will evaluate the retention of activity of the bound protein.

2.5 PROTEIN IMMOBILIZATION – COMPARISON OF PROTEIN ACTIVITY WITH VARIOUS IMMOBILIZATION METHODS

2.5.1 Antibody–Antigen Binding

A high density of protein bound to a biosensor surface will give a better response than a low density of protein only when the protein also retains its selective binding function. When comparing the binding function of proteins immobilized by different methods, it is important not to confuse binding density with activity. The binding function of immunoglobulin G was tested since it was immobilized in similar amounts on all derivatized surfaces, and can be used as a receptor (antibody) or as a ligand (antigen). By using an immunoglobulin G–anti-immunoglobulin G system,

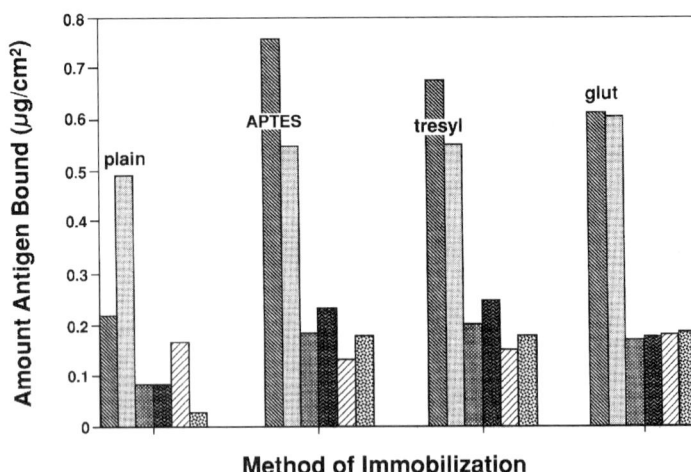

Fig. 2.23 Antibody-antigen binding to quartz. plain = plain quartz; APTES = 3-amino-propyl triethoxy silane; glut = glutaraldehyde; ab = antibody; IgG = immunoglobulin G (antigen). Shading scheme from left to right indicating order of immobilization procedure quartz-IgG-ab (no Tween); quartz-IgG-ab (with Tween); quartz-ab-IgG (no Tween); quartz-ab-IgG (with Tween); quartz-ab-IgG (no Tween); quartz-ab-IgG (with Tween)

the experiment was free from the effects of size and shape differences between antibody and antigen. Antibody was fluorescently labelled and antigen was radiolabelled so that relative amounts of receptor and ligand immobilization could be distinguished.

Fluorescein-labelled anti-rabbit immunoglobulin G (antibody) and radio-labelled rabbit immunoglobulin G (antigen) were investigated for the retention of biological activity when immobilized on various surfaces. As seen in Fig. 2.23, the amount of antigen immobilized was about the same (0.6–0.8 μg/cm^2) for APTES surfaces and for tresyl- and glutaraldehyde-activated surfaces, and somewhat less for plain quartz. Upon binding of antibody to the antigen-substrate, the surface density of antigen decreased by about 0.1 μg/cm^2, likely due to displacement of adsorbed antigen by non-radiolabelled antibody. This displacement occurred even though the immobilized antigen samples were washed with albumin to displace ad-sorbed protein. The antibody is physically more similar to the antigen than albumin is, thus is better able to displace antigen (Jennissen, 1986). The relative binding efficiencies of the methods remained the same after antibody incubation.

The reverse situation was also investigated, where antigen was incu-bated with immobilized antibody surfaces. The density of antigen bound to the antibody-immobilized surfaces was about 0.2 μg/cm^2, corresponding

Table 2.25

Comparison of Antibody and Antigen Immobilization on Derivatized Quartz with Respect to Order of Binding

Surface chemistry	Tween 20	Relative amounts of protein bound (%)			
		substrate-ag-ab		substrate-ab-ag	
		ag	ab	ag	ab
plain	no	29	20	95	91
	yes	65	44	19	15
APTES	no	100	80	80	72
	yes	72	100	100	97
tresyl	no	89	56	85	81
	yes	73	71	98	96
glutaraldehyde	no	80	33	93	97
	yes	79	38	95	100

APTES = 3-aminopropyl triethoxy silane; *ab* = fluorescein-isothiocyanate-labelled anti-immunoglobulin G (antibody); *ag* = radiolabelled immunoglobulin G (antigen); 100% binding was defined as the most intense fluorescence for antibody, or the highest radio-activity for antigen.

to only 22–45% of the amount bound when antigen was immobilized on the surface directly. Thus, only a fraction of antibody immobilized on the surface retained physiological activity. Since protein immobilization is random, a sizeable proportion of antibodies will have their binding site facing the surface or facing adjacent proteins where antigen cannot access the binding site. Therefore, an apparent activity of 22–45% is reasonable, and does not necessarily indicate a denaturation of antibody. APTES surfaces bound comparable amounts of active antibody to tresyl- and glutaraldehyde-activated surfaces. Immunoglobulin G is known to adsorb tightly to hydrophobic surfaces (Brash & Lyman, 1969), and the high adsorption to APTES could be due to hydrophobic interaction of the protein with exposed methylene groups on the APTES.

The antibody–antigen surfaces of Fig. 2.23 were also examined for fluorescence, relative fluorescence intensities and radioactivity, which are listed in Table 2.25. It was not possible to relate fluorescence intensity to concentration with the instrumentation used, so the surface with the highest fluorescence intensity (for antibody) or radioactivity (antigen) was set to 100% for comparison purposes. The results show that approximately equal amounts of antibody and of antigen were immobilized on APTES and activated surfaces when the antibody bound first. When antigen was immobilized first, the APTES surface bound most antibody, tresyl-activated surfaces bound somewhat less, and glutaraldehyde-activated surfaces bound only 33–38% of the antibody immobilized on APTES. It is unlikely that the antigen changed conformation and lost antigenicity upon binding to activated surfaces, since antibody suffered no similar loss of activity when immobilized to the same surfaces. Therefore, the preferred binding orientation of antigen on different surfaces resulted in the exposure of different proportions of the antigenic determinants. This contrasts with a report that immunoglobulin G changed conformation when immobilized on APTES (Lin *et al.*, 1988). However, the orientation effects are expected to be system-specific.

To ensure that the radioactive counts observed in these antibody-antigen binding experiments were only due to selectively adsorbed antigen, quartz supports incubated with antibody then antigen were washed with several biochemical agents to see if antigen was removable (Table 2.17). No significant loss of radioactivity occurred after 1 M sodium chloride and 6 M urea treatment, in contrast to the radioactivity displaced when radiolabelled bovine serum albumin-incubated surfaces were washed in salt solution. Therefore, radiolabelled antibody seems to be free of adsorbed radioactive interferent(s), unlike bovine serum albumin which adsorbed large amounts of interferent(s). Antigen losses of about 10% were observed after washing in 1 mg/ml bovine serum albumin and in 0·1% Triton X-100. Since urea, a known denaturing agent (Stryer, 1981), did not cause antigen release, it is unlikely that albumin or Triton X-100 treatment denatured the antibody, resulting in the antigen displacement. Instead, these solutions must have desorbed reversibly adsorbed antigen. In general, 80% of the antigen originally immobilized on the surface was tightly bound.

Protein–protein interactions are known to occur (e.g. protein A–immunoglobulin G, the aggregation of concanavalin A) and may result in a false positive response if antigen non-specifically adsorbs to antibody. To further ensure that antibody–antigen binding was specific, antigen was incubated with substrates having several types of protein immobilized on their surfaces. Bovine serum albumin-immobilized surfaces bound 0·5% and non-specific gamma-globulin immobilized surfaces bound 2·4% of the amount of antigen bound to antibody-immobilized surfaces, giving strong evidence that the antibody–antigen reaction is selective.

Activity as a function of time was tested for antibody-derivatized quartz substrates, which were stored for various lengths of time in a buffered solution at 4°C before incubation with antigen. A sizeable

Table 2.26
Time-Denaturation of Antibody-Bound Quartz as Detected by Binding of Radiolabelled
Antigen

Storage time (days)	Protein binding density (μg protein/cm²) ± standard deviation	% of original activity after 66 days at 25°C
1	0·1473 ± 2%	22
2	0·1176 ± 5%	34
11	0·0815 ± 0·8%	58
19	0·0784 ± 7%	41
31	0·0721 ± 2%	72
42	0·0542 ± 8%	71

decrease in antibody activity (determined by antigen binding) was seen after one day of storage, possibly due to desorption of non-covalently bound antibody as well as time-dependent denaturation (Table 2.26 and Fig. 2.24). As storage time increased, the activity of the immobilized antibody decreased. Note that loss of activity was indistinguishable from physical loss of protein.

Loss of activity with time has been documented for antibodies (Darst *et al.*, 1988; Lin *et al.*, 1988) as well as other proteins (Kiba *et al.*, 1986; Moody *et al.*, 1986*a*,*b*). The antibody–antigen-immobilized samples were incubated at room temperature from the day that antigen was bound

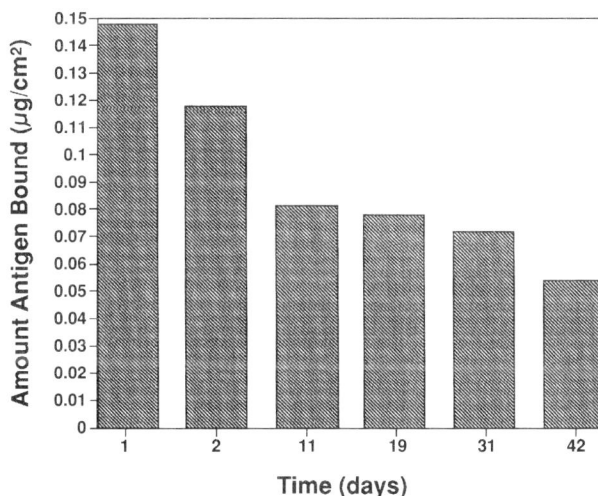

Fig. 2.24 Antigen binding to antibody on quartz. Tresyl-activated quartz was incubated with antibody, then stored at 4°C for the times indicated. Radiolabelled antigen was then incubated with the antibody-quartz.

until day 66. Then the samples were washed and recounted. After adjusting for the decay of radioisotope, the % antigen remaining on the surface after room temperature incubation was calculated. Samples stored for 65 days at room temperature retained only 22% of the original amount of antigen, whereas samples stored 24 days retained 71% of the antigen. This indicated either desorption of the antibody–antigen complex, or release of antigen due to denaturation of the antibody.

In summary, active antibody is immobilized to APTES and tresyl- and glutaraldehyde-activated surfaces in similar amounts. The antigen studied (rabbit immunoglobulin G) lost varying amounts of antigenicity, perhaps due to binding orientation, when immobilized on activated substrates. The antibody–antigen reaction was selective, and the complex was not dissociated by mild biochemical agents. Immobilized antibody lost 30% activity over 42 days, and immobilized antigen was lost with time.

2.5.2 Enzyme Binding and Activity

To confirm that proteins other than immunoglobulins can retain activity after immobilization, an enzyme-substrate system was studied. The enzyme β-galactosidase was immobilized, since an assay exists for quantitation of as little as 1 ng of active enzyme. β-galactosidase was immobilized on quartz, APTES, tresyl-activated, and two types of glutaraldehyde-activated substrates. The sources of glutaraldehyde were an old supply from BDH, and a fresh supply from Sigma. An activity assay was performed on all β-galactosidase immobilized surfaces, utilizing the ability of the enzyme to hydrolyse the substrate fluorescein-di-β-D-galactopyranoside, which forms the fluorescent products fluorescein-mono-β-D-galactopyranoside and fluorescein. The fluorescence of the system was quantitatively measured after five or eight minutes' incubation. The enzyme was not fluorescently labelled so that its label would not interfere with the assay. With this system, adequate sensitivity to quantitate the amount of enzyme that was immobilized was achieved.

Activated and APTES surfaces immobilized roughly equal amounts of active enzyme (5·2–0·3 μg/cm^2), and plain quartz immobilized slightly less (4·9 μg/cm^2). New glutaraldehyde-activated surfaces were slightly more effective in immobilizing protein than old glutaraldehyde-activated surfaces. This may be due to increasing glutaraldehyde polymerization with age, resulting in less aldehyde groups per volume of glutaraldehyde in the older sample. After 15 days, β-galactosidase activity decreased by 34–41% for all surfaces. The decrease in enzyme activity is similar to that reported by others (Kiba *et al.*, 1986; Moody *et al.*, 1986*a,b*). Due to the difficulty in quantitating the small amount of β-galactosidase immobilized, the % active enzyme was not calculated.

2.6   CONCLUSIONS

The objective of this work was to characterize protein immobilized onto planar surfaces by a variety of techniques, with respect to the binding density and retention of physiological activity of the protein. Ellipsometry and contact angle measurements were successfully used to monitor substrate cleaning and reaction with APTES. Scanning electron microscopy and scanning tunnelling microscopy were informative for qualitative observations of 3-aminopropyl triethoxy silane (APTES) and bovine serum albumin surfaces respectively. Fluorescence spectroscopy and radiation counting were used to compare relative amounts of immobilized, appropriately labelled protein. By correlating the results from the latter two techniques, more confidence could be placed in the results. X-ray photoelectron spectroscopy gave useful information about the chemistry of all the surfaces.

Protein was covalently attached and/or physically adsorbed to surfaces with a variety of chemistries, with varying degrees of success. Tresyl-activated surfaces consistently bound the most protein. Glutaraldehyde-activated and APTES surfaces also immobilized moderate to large quantities of protein, but the protein binding density varied depending on the type of protein and conditions used.

This work has shown strengths and limitations of ellipsometry, contact angle measurements and scanning tunnelling microscopy for analysis of derivatized surfaces. Of major importance is the successful X-ray photo-electron spectroscopy analysis of protein-deposited surfaces, since few applications of this instrumentation to biological macromolecules have been made. Some interesting chemistry was observed in the course of this work. It was discovered that radioactive species were adsorbing to radio-labelled bovine serum albumin. If Tween 20 was present during bovine serum albumin incubation with derivatized planar surfaces, the adsorbed radiolabel did not seem to interfere with protein binding density calculations. Non-specifically, reversibly bound protein was removed by washing the surfaces for 24 hours each in bovine serum albumin and Triton X-100 solutions.

This chapter provides detailed studies comparing different methods of protein immobilization in the same experiment. No studies have been published to date comparing covalent binding of protein to glutaraldehyde- and tresyl-activated surfaces, or comparing the immobilization of protein by covalent deposition and by adsorption to APTES. Glutaraldehyde-activated and APTES surfaces were suitable for protein immobilization in certain instances, in accord with their use in the literature. However, tresyl-activated surfaces were found to be superior for immobilization of the proteins studied here due to the higher consistency of protein binding observed for all proteins studied here.

Direct confirmation is needed that tresyl chloride binds to APTES in the manner depicted in Fig. 2.2. There is evidence from X-ray photoelectron and infra-red spectroscopy that tresyl chloride sublimes at low vacuum, or possibly even on exposure to atmosphere. A measure of the bond strength of tresyl to APTES would be of interest in understanding the mechanism by which protein functional groups, but not water or buffer salts, displace the tresyl group.

No method of analysis was available to differentiate between covalent binding and strong adsorption of proteins to the surfaces. The activated surfaces may well immobilize protein by a combination of adsorption and covalent binding. In order to distinguish between the two, mass spectrometry with high molecular weight analysis, or fluorescence recovery after photobleaching could be used to analyse the surfaces.

This work has demonstrated that chloramine T radioiodination of bovine serum albumin may result in the adsorption of radioactive species to, and desorption from, this protein. As demonstrated here, adsorption of radioactive species to proteins may give misleading results if free radiolabel exists in the system. Identification of the radioactive species should be made by those interested in working with radiolabelled protein.

To discern between thickness and refractive index changes in systems involving complex macromolecules, *in situ* ellipsometry has been successfully used (Jonsson *et al.*, 1985). This technique could be used to follow the deposition of APTES, activating agent, and protein, and gain reliable estimates of the coating thickness and refractive index of each layer.

The results of this work have applications in fields where the study of proteins at solid surfaces is important. Immobilization of protein is necessary for the selective layer in biosensors, and the methodology developed in this study may be applied directly to fibre optic sensors and piezoelectric devices, as they contain quartz surfaces. The method also may be applied to electrochemical sensors with an oxide layer, since surface hydroxyls have been used for attachment of protein to electrodes (Watanabe *et al.*, 1988; Narasimban & Wingard, 1986). Immobilization of protein is also desirable in the preparation of affinity chromatography supports. Many chromatographic packing materials have available hydroxyl groups, so the chemistry described herein is directly applicable. Permanent binding of physiologically active proteins to the surface may increase the efficiency of routine solid-phase assays. Living cells and bacteria may be immobilized by the tresyl- (Mandenius *et al.*, 1984) and glutaraldehyde-activation techniques, if there are available proteins on the cell surface. The biocompatibility of *in vivo* objects may be increased by binding proteins to the surface, and *in vitro* studies of biological materials may require quantitation of protein adsorption to surfaces.

The decrease in protein activity with time and desorption of protein are potentially serious complications for the applications discussed above. Further studies into the prevention of protein desorption (or prevention of the initial adsorption), and the retention of biological activity of immobilized proteins, are required. These problems are complex due to the varying properties of each protein, and any new methodology developed is expected to be protein-specific. For example, different detergents and preserving solutions may need to be used for different proteins, although the method of immobilization could be the same for all proteins.

2.7   REFERENCES

Afshar, T., Bailey, A. I., Luckham, P. F., MacNaughton, W. and Chapman, D. (1987) *Biochim. Biophys. Acta*, **915**, 101.

Alarie, J. P., Bowyer, J. R., Sepaniak, M. J., Hoyt, A. M. and Vo-Dinh, Y. (1990) *Anal. Chem. Acta*, **236**, 237.

Andrade, J. D., Hlady, V. L. and Van Wagenen, R. A. (1984) *Pure Appl. Chem.*, **56**, 1345.

Andrade, J. D., Smith, L. M. and Gregonis, D. E. (1985) ed., *Surface and Interfacial Aspects of Biomedical Polymers*, Plenum Press, New York, p. 249.

Andrade, J. D. and Hlady, V. (1986) *Adv. Polymer Sci.*, **79**, 1.

Andrade, J. D. and Hlady, V. (1987) *Ann. N.Y. Acad. Sci.*, **516**, 158.

Andrade, J. D., Herron, J. Hlady, V. and Horsley, D. (1987) *Croatica Chemica Acta*, **60**, 495.

Arnold, M. A. (1985) *Anal. Chem.*, **57**, 565.

Azzam, R. M. A. and Basharah, N. M. (1977) *Ellipsometry and Polarized Light*, North-Holland Publishing Co., New York.

Azzam, R. M. A., Rigby, P. G. and Krueger, J. A. (1977) *Phys. Med. Biol.*, **22**, 422.

Baier, R. E. and Manley, R. S. (1975) ed., *Adhesion in Biological Systems*, Academic Press, New York, p. 15.

Baier, R. E. and Glantz, P. O. (1978) *Acta Odontol. Scand.*, **36**, 289.

Baker, A. D. and Brundle, C. R. (1977/78) *Electron Spectroscopy*, Academic Press, New York, Vols 1 & 2, 1978.

Ballantine, D. S. and Wohltjen, H. (1989) *Anal. Chem.*, **61**, 704A.

Behrens, P. Q., Spiekerman, A. M. and Brown, J. R. (1975) *Fed. Proc. Fed. Am. Soc. Exp. Biol.*, **34**, 591.

Boerio, F. J. Armogan, L. and Cheng, S. Y. (1980) *J. Colloid Interface Sci.*, **73**, 416.

Bohnert, J. L. and Horbett, T. A. (1986) *J. Colloid Interface Sci.*, **111**, 363.

Boutelle, M. G., Stanford, C., Fillenz, M., Albery, W. J., and Bartlett, P. N. (1986) *Neurosci. Lett.*, **72**, 283.

Brash, J. L. and Lyman, D. J. (1969) *J. Biomed. Mater. Res.*, **3**, 175.

Brash, J. L., Uniyal S., and Samak, Q. (1974) *Trans. Amer. Soc. Artif. Int. Organs*, **20**, 69.

Brash, J. L. and Samak, Q. M. (1978) *J. Colloid Interface Sci.*, **65**, 495.

Brown, J. R. (1975) *Fed. Proc. Fed. Am. Soc. Exp. Biol.*, **34**, 591.

Caras, S. and Janata, J. (1980) *Anal. Chem.*, **52**, 1935.

Caras, S. D., Janata, J., Saupe D. and Schmitt, K. (1985) *Anal. Chem.*, **57**, 1917.

Caravajal, G. S., Leyden, D. E., Quinting G. R. and Maciel, G. E. (1988) *Anal. Chem.*, **60**, 1776.

Chan, B. M. C. and Brash, J. L. (1974) *J. Colloid Interface Sci.*, **48**, 345.

Chan, B. M. C. and Brash, J. L. (1981) *J. Colloid Interface Sci.*, **84**, 263.

Chiang, C-H., Liu, N-I. and Koenig, J. L. (1982) *J. Colloid Interface Sci.*, **86**, 26, 1982.

Comfort, A. R., Mullon, C. J-P. and Langer, R. (1988) *Biotechnol. Bioeng.*, **32**, 554.

Crandall, R. E., Janatova, J. and Andrade, J. D. (1981) *Prep. Biochem.*, **11**, 111.

Cuypers, P. A., Hermens, W. Th. and Hemker, H. C. (1978) *Anal. Biochem.*, **84**, 56.

Cuypers, P. A., Hermens, W. Th. and Hemker, H. C. (1988) *Ann. N.Y. Acad. Sci.*, **283**, 77.

Czaban, J. D. (1984) *Anal. Chem.*, **57**, 345A.

Damodaran, S. and Song, K. B. (1988) *Biochim. Biophys. Acta*, **954**, 253.

Darst, S. A., Robertson, C. R. and Berzofsky, J. A. (1988) *Biophys. J.*, **53**, 533.

Dexter, S. C. (1979) *J. Colloid Interface Sci.*, **70**, 346.

Dezelic, G., Dezelic, N. and Telisman, Z. (1971) *Eur. J. Biochem.*, **23**, 575.

Dillman, W. J. Jr and Miller, I. F. (1973) *J. Colloid Interface Sci.*, **44**, 221.

Dulm, P. van, Norde, W. and Lyklema, J. (1981) *J. Colloid Interface Sci.*, **82**, 77.

Eldefrawi, M. E. (1989) Biosensors Symposium, Chapel Hill, NC, Sept. 7–9.

Elwing, H., Askendal, A. and Lundstrom, I. (1989) *J. Colloid Interface Sci.*, **128**, 296.

Elwing, H., Welin, S., Askendal, A., Nilsson U. and Lundstrom, I. (1987) *J. Colloid Interface Sci.*, **119**, 203.

Fusek, M., Capka, M. and Turkova, J. (1987) *Biotechnol. Lett.*, **9**, 561.

Garcia, A. III, Oh, S. and Enger, C. R. (1989) *Biotechnol. Bioeng.*, **33**, 321.

Genshaw, M. A. (1988) *Clin. Chem.*, **34**, 1717.

Gillett, R. and Gull, K. (1972) *Histochemie*, **30**, 162.

Girotti, S., Roda, S., Ghini, S., Piacentini, A. L., Carrea, G. and Bovara, R. (1986) *Anal. Chim. Acta*, **183**, 187.

Golander, C-G. and Kiss, E. (1988) *J. Colloid Interface Sci.*, **121**, 240.

Guilbault, G. G. (1984) *Analytical Uses Of Immobilized Enzymes*, Marcel Dekker, New York, p. 112.

Hansma, P. K. and Tersoff, J. (1987) *J. Appl. Phys.*, **61**, R1.

Hayashi, Y., Zaitsu, K. and Ohkura, Y. (1986) *Anal. Chim. Acta*, **186**, 131.

Hendrickson, W. A. *et al.* (1975) *Proc. Natl. Acad. Sci.*, U.S.A., **72**, 2160.

Hlady, V., Golander, C. and Andrade, J. (1987) *Colloids and Surfaces*, **25**, 185.

Hunter, W. M. and Greenwood, F. C. (1962) *Nature*, **194**, 495.

Ianello, R.M. and Tacynych, A. M. (1981) *Anal. Chem.*, **53**, 2090.

Ikariyama, Y., Furuki, M. and Aizawa, M. (1985) *Anal. Chem.*, **57**, 496.

Ikeda, T., Miki, K., Fushimi, F. and Senda, M. (1987) *Agric. Biol. Chem.*, **51**, 747.

Iler, K. (1979) *The Chemistry of Silica*, John Wiley and Sons Inc., Toronto, 1979, p. 628.

Ishida, H. and Koenig, J. L. (1980) *J. Polymer Sci.: Polymer Phys. Ed.*, **18**, 1931.

Jennissen, H. P. (1986) *J. Colloid Interface Sci.*, **111**, 570.

Jonsson, U. (1986) *Immobilization and Interaction of Biomolecules at the Liquid–Solid Interface. Applications to Solid-Phase Analytical Techniques*, Linkoping Studies in Science and Technology, Dissertations No. 145, Linkoping Institute of Technology, S-581 83 Linkoping, Sweden.

Jonsson, U. Ivarsson, B., Lundstrom, I. and Berghem, L. (1982) *J. Colloid Interface Sci.*, **90**, 148.

Jonsson, U., Malmqvist, M. and Ronnberg, I. (1985*a*) *Biochem. J.*, **227**, 363,

Jonsson, U., Malmqvist, M. and Ronnberg, I. (1985*b*) *Biochem. J.*, **227**, 373.

Kallury, K. M. R., Krull, U. J. and Thompson, M. (1988) *Anal. Chem.*, **60**, 169.

Karube, I., Tamija, E., Dicks, J. M. and Gotoh, M. (1986) *Anal. Chim. Acta*, **185**, 195.

Kern, W. and Puotinen, D. A. (1970) *RCA Review*, June, 187.

Kiba, N., Goto K. and Furusawa, M. (1986) *Anal. Chim. Acta*, **185**, 287.

Kinkel, J. N. and Unger, K. K. (1984) *J. Chromatog.*, **316**, 193.

Kitamura, K., Matsuda, K., Ide, M., Tokunaga, T. and Honda, M. (1989) *J. Immunol. Meth.*, **121**, 281.

Kobos, R. K. (1980) in *Ion Selective Electrodes in Analytical Chemistry*, Plenum Press, New York, Vol. 2, 1980.

Kobos, R. K. (1987) *Trends Anal. Chem.*, **6**, 6.

Krull, U. J. and Thompson, M. (1985) *Trends Anal. Chem.*, **4**, 90.

Krull, U. J., Brown, R. S., Koilpillai, R. N. and Nespolo, R. (1989) *Analyst*, **114**, 33.

Kvitek, R. J., Watson, M. W., Evans, J. F. and Carr, P. W. (1981) *Anal. Chim. Acta*, **129**, 269.

Lin, J-N., Herron, J., Andrade, J. D. and Brizgys, M. (1988) *IEEE Trans. Biomed. Eng.*, **35**, 466.

Lindstrom, J., Anholt, R., Einarson, B., Engel, A., Osame, M. and Montal, M. (1980) *J. Biol. Chem.*, **255**, 8340.

Loeb, G. E., McHardy, J., Kelliher, E. M. and Brummer S. M. (1982) In *CRC Biocompatibility in Clinical Practice*, Vol. 2, Williams, D. F. ed., Boca Raton: CRC Press Inc.

Losche, M. and Mohwald, H. (1984) *Rev. Sci. Instrum.*, **55**, 1968.

McCrackin, F. L. (1969) Technical Note 479, Washington D.C., 20234.

McLaren, A. D. (1954) *J. Phys. Chem.*, **58**, 129.

MacRitchie, F. (1972) *J. Colloid Interface Sci.*, **38**, 484.

Makino, S., Reynolds, J. A. and Tanford, C. (1973) *J. Biol. Chem.*, **248**, 4926.

Mandenius, C. F., Welin, S., Danielsson, B., Lundstrom, I. and Mosbach, K. (1984) *Anal. Biochem.*, **137**, 106.

Mannens, G. Slegers, G. Lambrecht R. and Claeys, A. (1988) *Biochim. Biophys. Acta*, **959**, 214.

Masoom, M. and Townshend, A. (1984) *Anal. Chim. Acta*, **166**, 111.

Masoom, M. and Townshend, A. (1986) *Anal. Chim. Acta*, **18**, 49.

Mathot, C. and Rothen, A. (1969) *Surf. Sci.*, **16**, 428.

Matsumoto, K., Hamada, O., Ukeda, H. and Osajima, Y. (1986) *Anal. Chem.*, **58**, 2732.

Mattiasson, B. and Mosbach, K. (1976) *Methods Enzymol.*, **44**, 335.

Messing, R. A. (1987) *Methods Enzymol.*, **44**, 148.

Mierendorf, R. C. Jr. and Dimond, R. L. (1983) *Anal. Biochem.*, **135**, 221.

Miller, A. W. and Robyt, J. F. (1983) *Biotechnol. Bioeng.*, **25**, 2795.

Miller, K. F., Bolt, D. J. and Goldsby, R. A. (1983) *J. Immunol. Meth.*, **59**, 277.

Mittal, K. L. and Lindman, B. (1983) *Surfactants in Solution*, Plenum Press, New York.

Monsan, P. and Combes, D. (1988) *Methods Enzymol.*, **137**, 584.

Monsan, P. (1977/78) *J. Molec. Catalysis*, **3**, 371.

Monsan, P., Puzo, G. and Mazarguil, H. (1975) *Biochimie*, **57**, 1281.

Moody, G. J., Sanghera, G. S. and Thomas, J. D. R. (1986*a*) *Analyst*, **111**, 1235.

Moody, G. J., Sanghera, G. S. and Thomas, J. D. R. (1986*b*) *Analyst*, **111**, 605.

Morrisey, B. W. and Stromberg, R. R. (1974) *J. Colloid Interface Sci.*, **46**, 152.

Morrisey, B. W., Smith, L. E., Stromberg, R. R. and Fenstermaker, C. A. (1976) *J. Colloid Interface Sci.*, **56**, 557.

Morrisey, B. W. and Han, C. C. (1978) *J. Colloid Interface Sci.*, **65**, 423.

Mosbach, K. (1984) *Anal. Biochem.*, **137**, 106.

Mosbach, K. (1976) *Methods Enzymol.*, **44**, 3.

Mullen, W. H. and Vadgama, P. M. (1986) *J. Appl. Bacteriology*, **61**, 181.

Narasimban, K. and Wingard, L. B. Jr. (1986) *Anal. Chem.*, **58**, 2984.

Naviroj, S., Culler, S. R., Koenig, J. L. and Ishida, H. (1984) *J. Colloid Interface Sci.*, **97**, 308.

Nelson, J. M. and Griffin, E. G. (1916) *J. Am. Chem. Soc.*, **38**, 1109.

Nemat-Gorgani, M. and Karimiar, K. (1984) *Biotechnol. Bioeng.*, **26**, 565.

Nilsson, K. and Mosbach, K. (1981) *Biophys. Biochem. Res Commun.*, **102**, 449.

Norde, W. and Lyklema, J. (1978*a*) *J. Colloid Interface Sci.*, **66**, 295.

Norde, W. and Lyklema, J. (1978*b*) *J. Colloid Interface Sci.*, **66**, 266.

Norde, W. and Lyklema, J. (1978*c*) *J. Colloid Interface Sci.*, **66**, 257.

Norde, W. (1986) *Adv. Colloid Interface Sci.*, **25**, 267

Norde, W., MacRitchie, F., Nowicka, G. and Lyklema, J. (1986) *J. Colloid Interface Sci.*, **112**(2), 447.

Nygren, H. (1988) *J. Immunol. Methods*, **114**, 107.

Pace, C. N. (1975) *CRC Crit. Rev. Biochem*, **3**, 1.

Pace, J. F., Sutherland, R. M. and Dahne, C. (1985) *Biosensors*, **1**, 321.

Pavari, R., Pecht, I. and Soreq, H. (1983*a*) *Anal. Biochem.*, **133**, 450.

Pavari, R., Pecht, I. and Soreq, H. (1983*b*) *Anal. Biochem.*, **131**, 180.

Pinkerton, T. C. and Lawson, B. L. (1982) *Clin. Chem.*, **28/9**, 1946.

Porter, S. and Froehner, S. C. (1983) *J. Biol. Chem.*, **258**, 10034.

Poste, G. and Moss, C. (1972) *Prog. Surf. Sci.*, **2**, 67.

Powell, C. J. (1974) *Surface Science*, **44**, 29.

Raconczay, Z. (1986) *Neuromethods. Series 1: Neurochemistry. 5. Neurotransmitter Enzymes*, eds Boulton, A. A., Baker, G. B. and Yu, P. H. Humana Press, Clifton, NJ, p. 335.

Ratner, B. D. (1983) *Ann. Biomed. Eng.*, **11**, 313.

Reekes, G., Becker, J. and Eldeman, G. (1975) *J. Biol. Chem.*, **250**, 1525.

Richards, F. M. and Knowles, J. R. (1968) *J. Mol. Biol.*, **37**, 231.

Rinderknecht, H. (1962) *Nature*, **193**, 167.

Schultz, P. G. (1988) *Science*, **240**, 426.

Scouten, W. H. (1987) *Methods Enzymol.*, **135**, 30.

Scouten, W. H., Van Den Tweel, W., Delhaes, D., Kranenberg, H. and Dekker, M. (1986) *J. Chromatog.*, **376**, 289.

Soderquist, M. E. and Walton, A. G. (1980) *J. Colloid Interface Sci.*, **75**, 386.

Stone, I. (1955) U.S. Patent 2717852.

Stryer, L. (1981*a*) *Biochemistry*, 2nd edition, W. H. Freeman and Sons, San Francisco, p. 33.

Stryer, L. (1981*b*) *Biochemistry*, 2nd edition, W. H. Freeman and Co., San Francisco, p. 112.

Sudholter, E. J. R., Huis, R., Hays, G. R. and Alma, N. C. M. (1985) *J. Colloid Interface Sci.*, **103**, 554.

Sutherland, R. M., Dahne, C., Pace, J. F. and Ringose, A. R. (1984) *J. Immunol. Meth.*, **74**, 253.

Thompson, M. and Vandenberg, E. T. (1986) *Clin. Biochem.*, **19**, 255.

Thompson, M., Dhaliwal, G. K., Arthur, C. L. and Calabrese, G. S. (1987). *IEEE Trans. Ultrason. Ferroelect. Freq. Control*, **34**, 127.

Tramontano, A., Ammann, A. A., and Lerner, R. A. (1988) *J. Am. Chem. Soc.*, **110**, 2282.

Tscharner, V. von & McConnell, H. M. (1981) *Biophys. J.*, **36**, 421.

Turner, A. P. F. (1987) in *Biosensors, Fundamentals and Applications*, eds Turner, A. P. F., Karube, I. and Wilson, G. S. Oxford Science Publications, Oxford, p. v.

Valdes, J. J., Wall Jr., J. G., Chambers, J. P. and Eldefrawi, M. E. (1988) *The Johns Hopkins APL Technical Digest*, **9**, 4.

Vroman L. (1967) In *Blood Clotting Enzymology*, ed. Seegers, W H. Academic Press, New York, p. 279.

Watanabe, T., Okawa, Y., Tsuzuki, H., Yoshida, S. and Nihei, Y. (1988) *Chem. Lett.*, 1183.

Wienhausen, G. and DeLuca, M. (1982) *Anal. Biochem.*, **127**, 380.

Williamson, M. L., Atha, D. H., Reeder, D. J. and Sundaram, P. V. (1989) *Anal. Lett.*, **22**, 803.

Worsfold, P. J. and Nabi, A. (1986) *Anal. Chim. Acta*, **179**, 307.

Zaitsu, K., Nakayama, M. and Ohkura, Y. (1987) *Anal. Chim. Acta.*, **201**, 351.

Chapter 3

IMMOBILIZED ENZYMES AND CELLS IN BIOCHEMICAL REACTIONS

Staffan Birnbaum

Process Analysis, Research and Development, Kabi Pharmacia AB, Stockholm, Sweden

CONTENTS

3.1   INTRODUCTION

The primary use of immobilized enzymes and cells is to exploit their catalytic properties. Thus, we have chosen to focus our attention on this particular aspect, that is the employment of immobilized biocatalysts for specific biochemical reactions. Examples have been taken for applications in energy production and analysis. The examples and areas are those which we feel are of significant interest to those involved with immobilized enzymes and cells.

Immobilized biocatalysts may be defined as enzymes or cells (or organelles though not covered here) which are physically confined to a defined spatial region. In general, this has meant that the biocatalyst is associated to a support. In the case of enzymes, immobilization is often accomplished by covalently coupling the enzyme to a support, such as beaded agarose. For cells, entrapment or encapsulation is often the method of choice. Adsorption, primarily through either ionic or hydrophobic interactions is another popular method for both enzyme and cell immobilization. Additional, and more in depth, information on immobilization methodology may be found either in the other chapters in this book or from other sources (Scouten, 1983; Mosbach 1987a,b; Mosbach 1988; Taylor, 1991; Birnbaum & Mosbach, 1991; Birnbaum, 1992).

The primary impetus for employing immobilized enzymes or cells are the advantages which are attained. With respect to using enzymes or the enzymic activity in cells as catalysts, what are the characteristics that make them the catalysts of choice? One highly appreciated property of enzymes is their high selectivity. This is of course a major advantage when using a crude substrate solution. This selectivity is usually also a stereoselectivity which is of great interest since stereoselective synthesis is a major problem in classical organic chemistry. Another advantage of enzymes versus other catalysts is that it can be used at moderate temperature and pressure and in water based systems.

With respect to immobilization, several advantages are obtained in comparison to the free system. Foremost among the benefits is that separation and recovery of the biocatalyst is facilitated. Thus, the immobilized preparation can be reused either in batch or employed in a continuous process, hence diminishing the cost of the process as the amount of biocatalyst used can often be decreased. Immobilization also often improves the stability or longevity of the enzyme or cell, for example, through multipoint fixation or physical protection of the enzyme or cell from adverse conditions. Enhanced stability is often a prerequisite for using an immobilized preparation as immobilization *per se* often involves an additional cost to the process.

Often the support material itself will also impart favorable conditions for the reaction under study. For example, the optimal pH or temperature of a reaction will often be altered and/or broadened. Partitioning effects between the bulk and immobilized phase may occur for various compounds which either effect or are involved in the particular reaction, i.e. the substrate or product. Mass transfer limitation may also take place which can be advantageous, for example when substrate inhibition occurs.

Immobilization may invoke limitations as well. Not only does it involve an additional cost in catalyst preparation, but many of the characteristics described above such as diffusion limitations may have negative effects on the process studied. The particular immobilization method used may destroy or significantly alter the enzyme activity or cell functionality. In short, every particular catalytic process and immobilization method used must be individually evaluated in order to determine whether it should be employed.

3.2 ENERGY PRODUCTION WITH IMMOBILIZED BIOCATALYSTS

The desire to minimize the dependence on foreign petroleum imports as well as to decrease atmospheric carbon dioxide levels (and thereby global warming) has revitalized an interest in alternative energy production methods, in particular from renewable domestic resources. Ethanol, produced by microbial fermentation, is one of the fuel alternatives considered. Brazil, for example, annually uses approximately three billion gallons of ethanol and in the United States 7% of all gasoline sold contains ethanol derived from fermentation (Lynd *et al.*, 1991).

Ethanol production by yeast is among the oldest manufacturing processes known to man dating back to 6000 B.C. (Bailey & Ollis, 1986). The production of industrial ethanol by fermentation developed significantly early in the twentieth century but declined after World War II as a result of the lower price for petroleum produced ethanol (Villet, 1981). The petroleum price rise reinitiated investigations during the mid-1970s to optimize classical fermentation procedures. The simultaneous developments in immobilization technology spurred the convergence of this technique with the application of energy production.

The forum of energy production with immobilized biocatalyst is not solely limited to ethanol production from glucose with yeast cells as catalysts. Other substrates, biocatalysts and energy products have been investigated. In addition, a number of refinements in bioreactor design as well as process development have been studied to optimize the yield and productivity of the system and some of these studies are discussed further below. But first, let us return to the subject of ethanol production by immobilized yeast. Margaritis and Merchant wrote an excellent review on the topic of ethanol production using immobilized cell systems (Margaritis & Merchant, 1984).

3.2.1 Ethanol Production

The microaerobic synthesis of ethanol from glucose by *Saccharomyces cerevisiae* entails twelve enzymic steps including two coenzyme regeneration steps and produces two moles ATP per mole glucose as illustrated in Fig. 3.1. The conversion is inhibited both at high substrate concentrations and at high product concentrations. Presently, the majority of fermentative ethanol is produced by classical batch method with an average ethanol productivity of 2–3 g/l/hr (Rose, 1976). The use of cell recycle in a continuous reactor has been investigated (Cysewski & Wilke, 1977). When operated under vacuum, an enhanced ethanol productivity up to

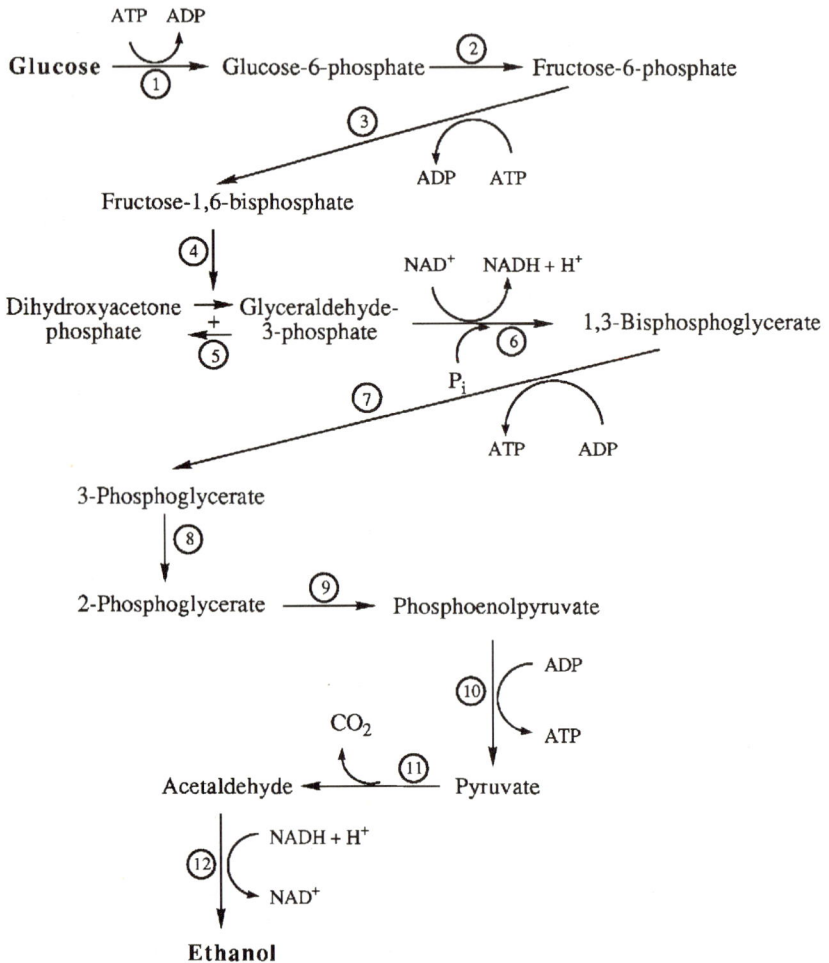

Fig. 3.1 Metabolic pathway from glucose to ethanol by *Saccharomyces cerevisiae*.

82 g/l/hr at a sugar feed concentration of 334 g/l was obtained. The cell density in this case was 124 g/l.

The first publication on ethanol production with immobilized yeast cells was probably that reported by Delbruck in 1892 where he described a porous matrix for yeast retention (Delbruck, 1892). Several methods have been described since then for immobilizing biocatalysts. Due to the limited scope of this chapter we do not intend to give a detailed description for each of the immobilization methods which have been applied to energy production systems and we therefore refer the reader to other excellent reviews which describe these methods in more detail (Birnbaum *et al.*, 1983; Woodward, 1985; Mosbach, 1987*a*). We will however mention some of these methods in which high ethanol productivity has been achieved.

One of the earlier examples of immobilized cells for ethanol production was the use of dialysis bioreactors by Gerhardt and Gallup (Gerhardt & Gallup, 1963). The problems of slow diffusion and membrane fouling in this early membrane type reactor were subsequently overcome by the introduction of the Rotorfermenter (Margaritis & Wilke, 1972). This early membrane reactor achieved productivities of 27·3 g/l/hr. The next step in development of membrane reactors was the use of hollow-fiber reactors. With these systems maximum productivities of 170 and 133 g/l/hr have been reported (Cheryan & Mehaia, 1984; Inloes *et al.*, 1985). However, the problems with membrane fouling and gas venting remain.

Cells may also be immobilized by adsorption or covalent attachment to a support surface. In general, the former method has been used with more success as it is considerably gentler. Among the various carriers studied, sawdust and anion exchanger resins gave the best results (Dougalis *et al.*, 1981; Michaud *et al.*, 1982). Ethanol productivity rates for adsorbed *S. cerevisiae* of 60 and 62 g/l/hr have been achieved. Adsorption methods are however limited since the maximum cell loading is determined by the surface area of the carrier and maximum flow rate possible is constrained to minimize shearing and cell detachment. The possible use of macroporous carriers may alleviate this to some degree (Nilsson *et al.*, 1986).

The third method of cell immobilization which we will consider is entrapment. Calcium alginate and κ-carrageenan are the two most popular supports though others have also been used. With calcium alginate, an ethanol productivity of 46·0 g/l/hr for entrapped *S. cerevisiae* has been achieved in a horizontal-packed bed reactor (Shiotano & Yamane, 1981). the horizontal design was employed to minimize gas hold-up and subsequent back-mixing. An ethanol productivity of 38 g/l/hr for a vertical

Table 3.1

Maximum Ethanol Productivity for Various Immobilized Cell Systems

Retention method	Cell	Substrate	Comments	Productivity (g/l/h)	References
Recycle	*S. cerevisiae*	Glucose	Vacuum	82	Cysewski & Wilke, 1977
Hollow-Fiber	*S. cerevisiae*	Glucose		170	Cheryan & Mehaia, 1984
Adsorption	*S. cerevisiae*	Glucose		62	Dougalis *et al.*, 1981
Entrapment					
Calcium alginate	*S. cerevisiae*	Glucose	Horizontal	46	Shiotani & Yamane, 1981
Calcium alginate	*Z. mobilis*	Glucose	Reactor series	108	Klein & Kressdorf, 1986
Calcium alginate	*K. marxianus*	Inulin		118	Bajpai & Margaritis, 1983
Ceramic microsphere	*S. cerevisiae*	Glucose		46	Salter *et al.*, 1990

packed-bed reactor with *S. cerevisiae* cells entrapped and grown in alginate beads has been reported (Lee *et al.*, 1983). For κ-carrageenan, ethanol productivities of 50 g/l/hr (based on bed volume) have been achieved (Wada *et al.*, 1980). A new entrapment method, termed hydrodynamic deposition in ceramic microspheres has been used to obtain ethanol productivities of 46·1 g/l/h (Salter *et al.*, 1990). A comparison of the maximum ethanol, productivities obtained with various immobilized cell systems is given in Table 3.1.

As discussed in the introduction, biocatalyst immobilization imparts a number of advantages in comparison to free biocatalyst processes. Among the advantages mentioned above we should note that it has been shown that both the immobilizing matrix and immobilization *per se* can have positive (and even negative) effects on ethanol production. These effects can be due both to partitioning effects and/or alterations in cell metabolism. For example, it has been found that the carrier used for cell adsorption could enhance productivity (Navarro & Durand, 1977). It has also been shown that the components found in the the immobilizing matrix can enhance ethanol yields significantly. For example, it was found that the ethanol yield for fermentation of 320 g/l glucose by *S. bayanus* increased from 10·5% (v/v) for free cells to 14·5% and 13·1% for celite adsorbed and κ-carrageenan entrapped cells respectively (Viera *et al.*, 1989). The ethanol yield could be increased for the free cells if the media were supplemented with the compounds present in the immobilization matrix. The authors hypothesized that the increase in calcium and zinc concentration were responsible for this increase. Alterations in cell metabolism upon immobilization were clearly shown by Bailey and coworkers. For example, the specific rate of ethanol production for *S. cerevisiae* cells adsorbed to gelatin was 40 to 50% higher than the comparable suspended cells while growth rate was reduced by 45% (Doran & Bailey, 1986). In addition, it was shown that the DNA content of the cells was dramatically increased while RNA levels were significantly decreased. The immobilized cells showed a one-third reduction in size and protein content, a reduction in the proportion of budding cells by one half, and a significantly higher concentration of reserve carbohydrates and structural polysaccharide.

In the case of calcium alginate immobilized cells, Galazzo and Bailey found the rates for glucose uptake and ethanol production to be considerably higher than that for suspended cells (Galazzo & Bailey, 1989, 1990). *In vivo* NMR measurements of intracellular metabolite and effector concentrations with *in vitro* enzyme activity measurements showed phosphofructokinase, ATPase and polysaccharide synthesis maximum activities two- to nine-fold higher than their suspension counterparts. In addition, polyphosphate concentrations were also observed to be increased. One would expect that increases in the polyphosphate as well as the polysaccharide concentration in the cells would result in lower water activity in the cells which has been postulated to alter metabolism in the cells (Mattiasson & Hahn-Hägerdahl, 1982).

Metabolic control theory analysis of flux control coefficients for alginate grown and immobilized cells indicated that the glucose uptake step is primarily responsible for determining the ethanol production rate for the immobilized cells under the conditions studied. This is similar to the results for suspension-grown cells when studied in suspension while in contrast to suspension-grown cells studied when immobilized where flux control was determined to occur at the phosphofructokinase step (Galazzo & Bailey, 1989, 1990). Several conclusions can be drawn from these in-depth studies. First, the enhanced ethanol production rate for yeast cells immobilized and grown in calcium alginate is a result of enhanced glucose transport into the cells presumably as a result of alterations in membrane permeability. Recent findings indicating alteration in the fatty acid composition of immobilized *S. cerevisiae* support these conclusions (Hilge-Rotmann & Rehm, 1991). Secondly, further improvements in ethanol production rate would occur if the glucose uptake rate could be increased further, for example through molecular cloning. One should soon expect detailed kinetic analysis of the pyruvate decarboxylase step to complete the picture of metabolic control of ethanol production with immobilized *S. cerevisiae* and thereby reveal targets for further metabolic engineering endeavors. Attempts in this line have already been made (Diaz-Ricci *et al.*, 1992).

3.2.2   Carrier, Bioreactor, and Process Development

One of the detrimental characteristics of ethanol production with immobilized yeast is the significant CO_2 gas evolution which occurs during fermentation. Extensive gas production may cause bead rupture for entrapped cells and cause channelling, pressure build-up and will significantly decrease the active catalytic volume in the reactor thereby decreasing ethanol productivity. In addition, for a reaction such as ethanol production which exhibits product inhibition, optimal operation is obtained in an ideal plug-flow reactor. The vigorous upward flow of CO_2 gas in a vertical reactor will impart back-mixing characteristics to the plug-flow reactor. The difficulty with gas build-up has been addressed in numerous reports as discussed below.

Bead disruption by internal gas evolution will not occur when the cells are surface bound. Alternatively, when the cells are entrapped the carrier can be chosen or manufactured with increased mechanical stability. For example, we and others have reported various methods for stabilizing alginate gels containing baker's yeast without significantly affecting their fermentative capacity (Birnbaum *et al.*, 1981; Veliky & Williams, 1981; Smidsrød & Skjåk-Braek, 1990). In addition, methods have been described for preparing immobilized cells in open-pore carriers of gelatin, agar and polyacrylamide (Sivaraman *et al.*, 1982; Rao *et al.*, 1986; Pundle *et al.*, 1988). In the latter case, the preparation performed superior to a comparable calcium alginate preparation. The authors reasoned that the improved productivity was due to the lower diffusion hindrance of the porous carrier.

The problem of gas hold-up within the packed-bed reactor has led to the investigation of alternative reactor design. To this end, the horizontal packed-bed (Shiotani & Yamane, 1981; Navarro *et al.*, 1984; Toda *et al.*, 1986), the cross-flow (Chotani & Constantinides, 1984) the ceramic monolith (Ariga *et al.*, 1984) and the compartmentalized column reactors (Quereshi *et al.*, 1987) have been investigated and approach an idealized plug-flow mode. Reactors which alleviate the problems with gas hold-up but resemble a CSTR in operation include the rotating (Del Borghi *et al.*, 1985; Parekh & Wayman, 1987; Amin & Doelle, 1990), the packed drum (Lee & Chang, 1985), and the stirred basket reactors (Gamarra *et al.*, 1986). Fluidization of the bed will also alleviate gas hold-up but again resemble a CSTR (Cho *et al.*, 1981).

Swaine and Daugulis have examined the degree of back-mixing which occurs in packed beds of *S. cerevisiae* adsorbed on an ion-exhange of 0·5 mm diameter (Swaine & Daugulis, 1989). They found that the 2·0 liter packed-bed reactor was equivalent to between 15 and 37 tanks-in-series depending on the dilution rate. In comparison an abiotic bed resembled 75 to 175 tanks-in-series. As would be expected, back-mixing increases as the dilution rate increases. Furthermore, the results for the ion exchange adsorbed cells were considerably better than data obtained for cells entrapped in larger particles. For example, a vertical bed with *S. cerevisiae* entrapped in calcium alginate particles of 3·7 mm diameter resemble 4–6 tanks-in-series. Thus, it was concluded that better plug-flow operation is obtained with smaller, more rigid particles.

Alternative methods have been used to improve productivity by enhancing ethanol separation and/or recovery in conjunction with immobilized biocatalysts. Apart from the use of microfiltration membranes for cell retention, membranes may also be employed for the selective separation of ethanol from the fermentation broth (Groot *et al.*, 1991; Jeong *et al.*, 1991). The ethanol may then be recovered by pervaporation or solvent extraction (Kang *et al.*, 1990; Groot *et al.*, 1991). Alternatively, ethanol may be recovered directly from the gas phase of the reactor by passing a stripping gas through the reactor (Lin *et al.*, 1990).

3.2.3 *Zymomonas mobilis*

An alternative approach to increasing ethanol productivity is to employ the gram negative bacterium *Z. mobilis*. Compared to *S. cerevisiae, Z. mobilis* produces one mole ATP/glucose via the Entner–Duodoroff pathway instead of two moles of ATP/glucose via the Embden–Meyerhoff pathway. Thus, less energy is available for growth and consequently less substrate is diverted to biomass production with higher ethanol yield as a result. Specific rate of glucose uptake and ethanol production are higher for *Z. mobilis* than for *S. cerevisiae* (Rogers *et al.*, 1982). Futhermore, unlike *S. cerevisiae, Z. mobilis* is capable of growth in the complete absence of oxygen. Hence, strict control of microaerobic conditions is obviated. *Z. mobilis* does have the disadvantage that the range of sugars which can be metabolized is more limited than for yeast. One can however extend the range of possible sugar substrates by the addition of auxiliary enzymes for example by co-immobilization.

Klein and Kressdorf have demonstrated the outstanding performance of immobilized *Z. mobilis* for ethanol production (Klein & Kressdorf, 1986). The authors employed a three stage reactor systems incorporating initially two fluidized bed reactors followed by a single plug-flow reactor. At a dilution rate of 2·97 (residence time 20 min), complete conversion of 150 g/l glucose was obtained with an ethanol productivity of 108 g/l/hr. Higher productivity was obtained at higher dilution rates though per cent conversion (as well as ethanol concentration) decreased. At the high dilution rate employed, non sterile conditions were used without contamination.

3.2.4 Alternative Substrate Sources for Ethanol Production

A number of alternative substrates for ethanol production other than glucose have been examined. Among these are lactose (from whey), inulin (from Jerusalem artichokes) as well as starch or cellulose derived hexoses and pentoses. Whey, a major by-product in cheese manufacturing, is a significant waste problem. Whey contains approximately 5% lactose and thus is a potential substrate for ethanol production. However, lactose is not readily fermented by *S. cerevisiae*. Recent advances in cloning the lactose operon into *S. cerevisiae* hold promise in exploiting this organism for the direct conversion of lactose to ethanol (Jeong *et al.*, 1991; Kumar *et al.*, 1992, Guimaraes *et al.*, 1992).

The organisms which have been studied in connection with lactose fermentation are *Kluyveromyces fragilis, Kluyveromyces marxianus* and *Saccharomyces fragi* (Gianetto *et al.*, 1986; Marwaha & Kennedy, 1984; King & Zall, 1983). Gianetto *et al.* reported a productivity of 17·2 g/l/hr and a maximum ethanol concentration of 49·5 g/l with immobilized *K. fragilis* while a higher productivity of 44 g/l/hr was reported with immobilized *K. marxianus* (Marwaha & Kennedy, 1984).

Another approach was used by Hahn-Hägerdahl. Lactose was converted to ethanol with *S. cerevisiae* immobilized in calcium alginate in which β-galactosidase was covalently attached to the alginate polymers (Hahn-Hägerdahl, 1985). When this co-immobilized preparation was employed in series with galactose-adapted *S. cerevisiae*, an ethanol concentration of 71 g/l was obtained from a 15% lactose solution. The co-immobilized preparation performed superior to a comparable system which employed immobilized *K. fragilis*.

Inulin derived from Jerusalem artichokes is another emerging carbohydrate source for ethanol production (Bajpai & Bajpai, 1989). Jerusalem artichoke has one of the highest carbohydrate yields reported per acre, approximately equal to sugar beets. It grows well on poor soil and hence can be cultivated on marginal land without competing with food crops. Furthermore, it exhibits robust resistance to both frost and disease. It thus represents an attractive alternative to the conventional substrate resources utilized for ethanol production.

Bajpai and Margaritis have examined ethanol production by immobilized cells from Jerusalem artichoke extract (Bajpai & Bajpai, 1989; Bajpai & Margaritis, 1983, 1987). A number of yeast strains were examined. *K. marxianus*, immobilized in calcium alginate, was found to give the best results with a maximum ethanol productivity of 118 g/l/hr (Bajpai & Margaritis, 1983). The productivity where the final ethanol concentration of 75 g/l could be obtained was 80 and 60 g/l/hr at 80% and 90% substrate utilization (Bajpai & Bajpai, 1989). The maximum ethanol concentration achievable was 110 g/l at an initial sugar concentration of 250 g/l (Bajpai & Bajpai, 1989). The maximum productivity of the immobilized cells was approximately 15 times higher than that which could be obtained in an ordinary CSTR with free cells of *K. marxianus* (Margaritis & Bajpai, 1982). The half life of the continuous immobilized cell reactors assuming exponential decay was estimated to range from 72 to 105 days. Bajpai & Bajpai have estimated the potential ethanol yield to be approximately up to 800 tons/year/km^2 (Bajpai & Bajpai, 1989).

3.2.5 Cellulose and Starch

Cellulosic biomass represents an enormous renewable resource potential for energy production. However, biomass residues are generally considered as waste, are not utilized and are either processed with a net energy input or allowed to decompose naturally. Agricultural and forestry residues consist primarily of lignocellulosic material, the major components of which are lignin, α-cellulose and hemicellulose. Lignin is the prominent obstacle in crop and forest residue utilization as it is totally recalcitrant and requires extremely oxidative conditions for biological oxidation. Furthermore, it physically surrounds cellulose and hemicellulose in biomass and thereby sterically hinders cellulases and hemicellulases from their substrate. Thus, the biomass must first be pretreated to remove lignin prior to fermentation (Zeikus, 1980). Recently, interest has also been shown in microbial and enzymatic lignin degradation which should improve the potential of using cellulosic biomass for ethanol production (Kirk & Farrell, 1990; Neidleman, 1991). In this respect, even immobilized biocatalysts have been studied (Kirkpatrick *et al.*, 1990; Chen *et al.* 1991).

Cellulose is converted to glucose enzymatically by the combined action of exo- and endo-cellulases as well as by β-glucosidase (Enari & Niki-

Paavola, 1987). The glucose produced can then be converted to ethanol as described previously, for example with immobilized S. cerevisiae. Cellulose is insoluble and thus requires the exo- and endocellulases to be extracellular enzymes. These enzymes cannot be employed in the form of entrapped biocatalyst for initial cellulose hydrolysis. Cellobiose, on the other hand, is soluble and therefore β-glucosidase can be employed in the entrapped form. As cellobiose and glucose are inhibitory to the cellulytic enzymes, it is desirable to convert the cellobiose produced directly to ethanol. This has been achieved by co-immobilizing β-glucosidase with entrapped S. cerevisae (Hahn-Hägerdahl, 1984).

The crystalline nature of cellulose dictates the use of extracellular and non-entrapped enzymes for its hydrolysis. In this light, Hahn-Hägerdahl and co-workers have investigated the use of two-phase systems for enzymatic hydrolysis of cellulose to glucose (Tjerneld et al., 1985a,b,c). When used in conjunction with an ultrafiltration membrane an average reducing sugar concentration of 74 g/l in the permeate was achieved in a continuously operated system from an 80 g/l cellulose concentration (Tjerneld et al., 1985c). The glucose produced can subsequently be converted to ethanol by the methods described earlier.

Wayman and co-workers have examined the production of ethanol from hardwood cellulytic biomass with immobilized cells as well as co-immobilized preparations (Parekh et al., 1989; Chen & Wayman, 1989). In the first case, aspen or pine chips were prehydrolyzed with SO_2 to remove hemicellulose and subsequently subjected to enzymatic saccharification to obtain 90–95% glucose yield. This enzyme derived hydrolysate, without filtration, was then fermented continuously with a productivity of 7 g/l/h at a residence time of 0·33 h with 96% conversion using alginate immobilized Z. mobilis in a rotating (fiber disc) reactor. Subsequently, Chen and Wayman described a continuous simultaneous saccharification and fermentation process of prehydrolyzed and delignified aspen cellulose using alginate entrapped S. cerevisiae co-immobilized with cellulose and β-glucosidase in the same type of reactor. Over 90% of the theoretical ethanol yield was achieved over a period of eight days.

The third major component in wood is hemicellulose of which xylose is the major constituent. Previously, xylose has been considered non-fermentable. Recently however, Pachysolen tannophilus, Pichia stipitios and Candida shehatae were shown to be able to ferment xylose, though generally with very low yields (Gong et al., 1981; Slininger et al., 1982; Bruinenberg et al., 1983). Skoog and Hahn-Hägerdahl have recently reviewed xylose fermentation (Skoog & Hahn-Hägerdahl, 1988). Good results have been obtained with immobilized growing P. tannophilus cells

(Amin *et al.*, 1988). By strictly controlling aeration and nitrogen concentration in a continuous system, 26·9 g/l ethanol could be produced with 81% conversion with a productivity of 1·1 g/l/hr which was twice that compared to results obtained earlier for immobilized *Pichia stipitis*. Higher productivity (3·8 g/l/hr) could be achieved in a continuous plug-flow reactor at the expense of decreased product concentration (Nunez *et al.*, 1991).

Starch is yet another substrate which has been used for ethanol production. Ethanol produced by immobilized cells using starch derived from sago, cassava, potato and banana have been reported (Rosario & Pamatong, 1985; Lee *et al.*, 1986; Tanaka *et al.*, 1986; Ramakrishna *et al.*, 1988). Ethical questions will undoubtedly arise which concern the appropriateness of utilization of food starch for energy production.

Starch can be hydrolyzed either chemically (cooking under pressure with acid) or enzymatically (with amylase and amyloglucosidase) or by a combination of both. *Aspergillus awamori* an amylolytic mold, and *Z. mobilis*, co-immobilized in alginate were examined for the direct conversion of starch to ethanol (Tanaka *et al.*, 1986). A natural habitat segregation occurred with the aerobic mycelia present at the oxygen-rich surface of the alginate bead and the anaerobic bacteria present mainly in the oxygen deficient center of the bead. The authors later employed *S. cerevisiae* in place of *Z. mobilis* for ethanol production (Kurosawa *et al.*, 1989). By adding Vantocil (a biocidal component) the yeasts were segregated primarily to the center of the gel bead.

A co-immobilized alginate preparation consisting of the enzyme amyloglucosidase and *S. cerevisiae* has also been investigated. In this case a productivity of 41·4 g/l/hr (based on gel volume) was achieved when the preparation was fed with starch hydrolysate (Chithra & Baradarajan, 1989). A number of yeast strains have also been examined for the direct conversion of starch to ethanol (Amin *et al.*, 1985). A maximum ethanol concentration of 70·4 g/l with an ethanol productivity of 12·7 g/l/hr was obtained in a two-step immobilized cell reactor with *Saccharomyces diastaticus* and *Endomycopses fibuligera*.

3.2.6 Non-ethanolic Energy Products

The production of fuels other than alcohol by immobilized biocatalysts can be found in the literature. For example, both hydrogen and methane production have been reported. Hydrogen in particular is attractive as it represents a completely clean fuel alternative. Hydrogen is evolved photosynthetically by various algae and photosynthetic bacteria. The purple non-sulfur bacteria *Rhodospirillum rubrum* and *Rhodopseudomonas capsulata* produce hydrogen via the nitrogenase enzyme which acts as a hydrogenase on substrates such as lactate, glutamate, and acetate, as well as other organic acids. The bacteria utilize only a single photosystem and thus do not evolve oxygen which would otherwise inactivate the nitrogenase. Thus, both *R. rubrum* and *R. capsulata* have been immobilized in either agar, alginate or carrageenan and utilized to produce hydrogen gas up to 57 m/l/hr (Francou & Vignais, 1984; von Felten *et al.*, 1985). The hydrogen gas produced by alginate immobilized *R. rubrum* can be used to drive a phosphoric acid fuel cell and thereby generate electricity (Karube & Suzuki, 1988). Thus, two twenty liter reactors could generate an electrical current of 0·5 A with an electric power output of 0·16 W (0·3 V).

The generation of methane from organic wastes by microorganisms retained to solid supports is well documented in the literature (Nishio *et al.*, 1985; Scharer *et al.*, 1988). The conversion consists of two phases in which the organic substrate is initially converted anaerobically to organic acids such as formic, acetic, propionic and butyric acid. In the case where the organic waste consists of digestible solids, this initial step can take place in a separate reactor operating with free cells. After separating the solids from the effluent stream the product is pumped into the second reactor where the organic acids are further converted anaerobically by the immobilized methanogens to methane and carbon dioxide. In the case of acetate conversion, which is generally the most prominent acid present in the substrate stream, the conversion is usually close to theoretical. Acetate is known, however, to partially inhibit the conversion of the other acids present in the substrate, i.e. propionate and butyrate. As hydrogen production is also based on the conversion of organic acids, Beneman *et al.* assessed the cost of both hydrogen and methane production and concluded that the former would be approximately four times as expensive as the latter (Benemann *et al.*, 1982).

Methane can also be produced from synthesic gas (syngas) with immobilized microorganisms.This is an attractive alternative for energy production since synthesic gas can be obtained from coal which in many countries (such as the US) is amply available. Syngas, which consists primarily of H_2, CO and CO_2, is produced from coal by gasification, quenching and purification. Methane has been produced employing trickle-bed reactor containing the triculture *R. rubrum, M. bareri* and *M. formicicum* with a productivity of 2–3 mmol/l/hr (Kimmel *et al.*, 1991). Alternatively, methane can be produced with immobilized *Methanothrix* sp. from acetate produced from syngas by *P. productus* (Klasson *et al.*, 1991).

The continuous production of methanol from methane has been demonstrated (Mehta *et al.*, 1991). The methanotroph *Methlosinus trichosporium* was covalently immobilized to DEAE-cellulose and utilized in an ultrafiltration reactor. A maximum productivity of 5·3 mmol/l/hr was attained with an operational half-life of 108 hours.

3.3 IMMOBILIZED ENZYMES AND CELLS IN ANALYSIS

Among the various applications of immobilized biocatalysts, those that involve their use in analysis are among the most prevalent. One journal, *Biosensors and Bioelectronics*, is dedicated to publishing articles which involve novel developments and applications of immobilized proteins and cells for analytical purposes. Other analytical journals, such as *Analytical Biochemistry, Analytical Chemistry, Analytica Chimica Acta, Analyst* and *Analytical Letters* among others, regularly include articles which involve immobilized preparations. In addition, several books and reviews have appeared on this subject and the reader is referred to these for further insight (Guilbault, 1984; Turner *et al.*, 1987; Mosbach, 1988; Coughlan *et al.*, 1988; Wise, 1989; Cass, 1990; Buck *et al.*, 1990; Blum & Coulet, 1991; Birnbaum *et al.*, 1991, 1992).

The fundamental conceptual flow of events as it occurs in the analytical regime with immobilized biocatalysts is shown schematically in Fig. 3.2. The physical form of the bioanalytical unit varies considerably and will depend, in part, on the intended use, requirements and characteristics of the method and its application. For example, the immobilized entity can be in the form of a membrane positioned at the tip of an electrode. This form would be suitable for measurements in discrete samples and would exhibit rapid response time due to the small diffusion distance. Alternatively, the biological transducer may be immobilized to a carrier and placed in a reactor spatially separated from the physiochemical transducer. This form may be preferred for continuous monitoring in a flow injection analysis type construction. The latter form would have the advantage that the larger capacity of the reactor in comparison to the membrane would provide increased operational stability. Combinations of these forms, for example where an immobilized enzyme reactor is in direct contact to the physiochemical transducer, do also exist.

The term biosensor has generally implied the very close proximity and preferably the direct attachment of the biological transducer to the physiochemical transducer. The distance between the two varies considerably for the various immobilized bioanalytical systems found in the literature

ANALYTE

↓

**BIOLOGICAL
TRANSDUCER**

↓

**PHYSIOCHEMICAL
PRODUCT**

↓

**PHYSIOCHEMICAL
TRANSDUCER**

↓

**ELECTRICAL
SIGNAL**

Fig. 3.2 Flow of events for analysis with immobilized biomolecules

and we do not intend to define biosensor with a particle distance. Biosensors may also be termed direct or indirect where the former implies that the analyte of interest sequesters the physiochemical response in itself whereas in an indirect biosensor a secondary component, such as an enzyme-labelled antibody, is responsible for the physiochemical product or invokes the physiochemical response.

3.3.1 The Biochemical Transducer

In terms of the biological component, it is often the case that an enzyme is used. An enzyme is chosen which has the particular specificity desired and which produces a product which can be detected by the physiochemical transducer. Alternatively, a second enzyme system can convert the initial product which can then be detected by the secondary transducer. Though the numbers of enzymes with known specificities are great, their number and availability are limited. In addition, the binding constants of an enzyme for its substrate are usually on the order of $10^6 \, M^{-1}$ or less and thus the sensitivities are in the order of micromolar.

Alternatively, the analytical system can be constructed with immunological detection of the analyte using either polyclonal or monoclonal antibodies. The range of analytes thus seems to be limitless as long as antibodies against the analyte can be raised. As the binding constants are generally of the order of 10^9–10^{12} M^{-1} we see that the sensitivity in the range of nano- to picomolar is possible. Other biorecognition systems are also possible for example by the use of lectins (for sugar specificity), nucleic acid (hybridization) receptors or even whole cells or tissues.

As most antibodies are not catalytic, the analytical system which employs immunological recognition may require the use of enzyme-labelled antibody conjugates to produce a compound that is measurable by the physiochemical transducer employed thus constituting an indirect analytical embodiment. There are, however, other constructions, in particular when considering the physiochemical transducer (as discussed below) that allow a direct immunosensor form.

There are, as yet, relatively few catalytic antibodies described. Yet, it was recently reported that catalytic antibodies could be used in the immobilized form (Blackburn *et al.*, 1990). Thus, the catalytic antibody 20G9 was immobilized onto a pH electrode and used to detect phenyl-acetate. The sensor was selective and reversible and thus allowed serial analysis. Repeated use of an immunosensor is generally limited as regeneration usually requires harsh treatment of the sensor. Though this example of catalytic antibodies is of great interest, it remains to be shown that catalytic antibodies of comparable binding constants (10^9–10^{12} M^{-1}) in conjunction with catalytic rates comparable to the enzymes normally used in analytical systems can be made.

3.3.2 The Physiochemical Transducer

The particular physiochemical transducer used has varied considerably. Table 3.2 describes and categorizes these. The earliest report on the use of electrochemical transducer in conjunction with a soluble enzyme for a specific analyte determination, glucose, was reported by Clark and Lyons in 1962 (Clark & Lyons, 1962). They employed a potentiometric pH electrode to detect the change in proton concentration as produced by glucose oxidase.

3.3.2.1 Potentiometry
The potentiometric transducer measures the potential difference generated at the sensing electrode against an accurate reference electrode. The potential developed is proportional to the logarithm of the ion concentration. The pH electrode is the potentiometric electrode which has found the most application in conjunction with immobilized enzymes and cells. This is due to its long life time, high selectivity for protons and

Table 3.2
Physiochemical transducers used in immobilized bioanalytical systems

Electrochemical	Potentiometric	pH
		Ion selective
		Field effect transistors
		Streaming potential
	Amperometric	Polarographic
	Conductivity	
Calorimetric		
Piezoelectric		
Optical	Photometric	LED-photodiode
		Optical wave guide
	Fluorescent	
	Luminescent	
	Surface-oriented	Reflectometry
		Ellipsometry
		Surface-plasmon resonance

wide working range. It does, however, have the disadvantage of suffering from variations in the buffering capacity of the sample.

The ion-selective electrode (ISE) incorporates a permeability-selective membrane. A typical example, the carbon dioxide electrode, is shown schematically in Fig. 3.3. Several examples of the use of ion-selective electrodes have been described in the literature including gas sensing electrodes which contain hydrophobic gas permeable membranes (Guilbault & Shu, 1972). In this case, even microbes could be used as the biological transducer for example to produce carbon dioxide via microbial metabolism of glutamic acid (Hikuma *et al.*, 1980).

An interesting example of the ISE was described by Rechnitz and co-workers (Keating & Rechnitz, 1984). They reported a potentiometric ionophore modulation immunoassay (PIMIA) which can be categorized

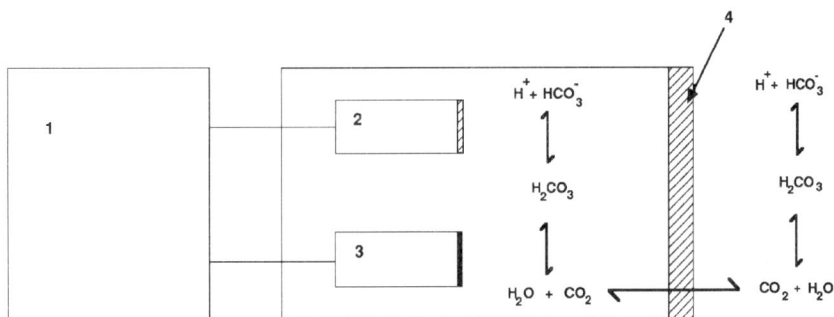

Fig. 3.3 The carbon dioxide electrode. 1 = voltmeter, 2 = silver/silver chloride electrode, 3 = reference electrode, 4 = gas permeable membrane.

as a direct immunosensor. This system utilizes an antigen-ionophore conjugate which is incorporated into the support membrane and used as a conventional ISE. When antibody is present in the sample together with a constant supply of a specific ion, for example potassium, the conjugate will bind to the antibody and consequently affect the ability of the membrane to transport ions. The reading from the ISE will thus be proportional to the amount of conjugate bound to the antibody which in turn will depend on the antibody concentration. With a constant and limiting antibody concentration this system can be employed to determine antigen concentration in a competitive assay.

The main drawbacks with ISEs are related to limited selectivity of the membrane. Thus, interference from other ions in the sample can be significant. In addition, the stability of the membrane is affected by agents in the sample. For example, antifoaming agents present in the fermentation media cause irreversible damage to certain types of ISEs (Clarke *et al.*, 1982). Membrane fouling and protein deposition are other factors which limit the lifetime. Miniaturization is another problem which has partially been addressed by the development of coated wire ISEs (Solsky, 1982).

Another potentiometric device is the field effect transistor (FET). The use of FETs marks the introduction of solid-state semiconductor technology to the field of analytical biochemistry. These analytical devices allow the possibility of mass production at low cost and miniaturization. A schematic diagram of an FET is shown in Fig. 3.4. In short, the FET consists of a current source and drain embedded in silicon that has been doped with, for example, boron (e.g. p-substrate). A silicon dioxide insulator

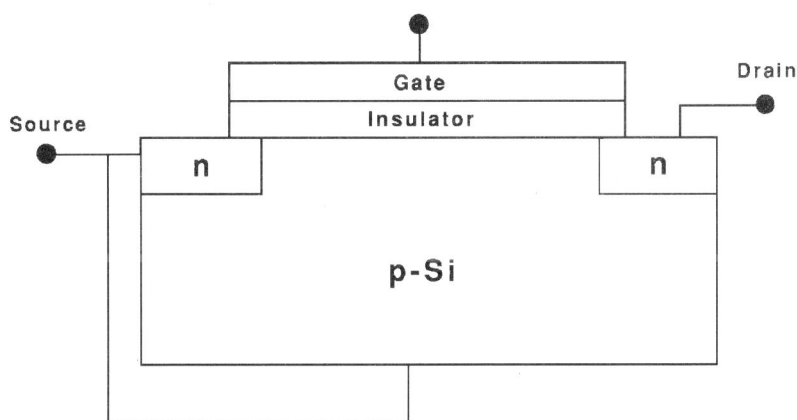

Fig. 3.4 Schematic diagram of a p-type field effect transistor.

and metal gate are deposited on the surface of the semiconductor between the current source and drain. As the voltage of the gate is increased (positive potential) the mobile electrons position themselves on the surface of the semiconductor and a conductance path is formed between source and drain thus producing an increase in the drain current registered.

Various FET devices have been described. The metal oxide semiconductor FET (MOSFET) can be used to detect hydrogen and ammonia gas down to *circa* 1 ppm (Winquist *et al.*, 1989). In conjunction with immobilized enzyme reactors which produce ammonia or hydrogen, a number of substrates have been analyzed (Winquist *et al.*, 1988). Ion-selective FETs (ISFET) which utilize ion-selective membrane on the gate instead of metal oxide have also been described. For example, various pH-ISFETs have been constructed for glucose as well as penicillin analysis (Caras *et al.*, 1985; Caras & Janata, 1980). ImmunoFETs have also been reported though these still show some difficulty (Janata & Blackburn, 1984).

Potentiometric measurements can also be made directly over a packed bed. This method has been termed as 'streaming potential' and has been used, for example, to quantitate albumin as well as monoclonal antibodies (Mattiasson & Miyabayashi, 1988; Miyabayashi & Mattiasson, 1990).

3.3.2.2 *Amperometry*
Amperometric electrodes measure the current which is produced as the analyte undergoes a redox reaction at the sensing electrode as a constant

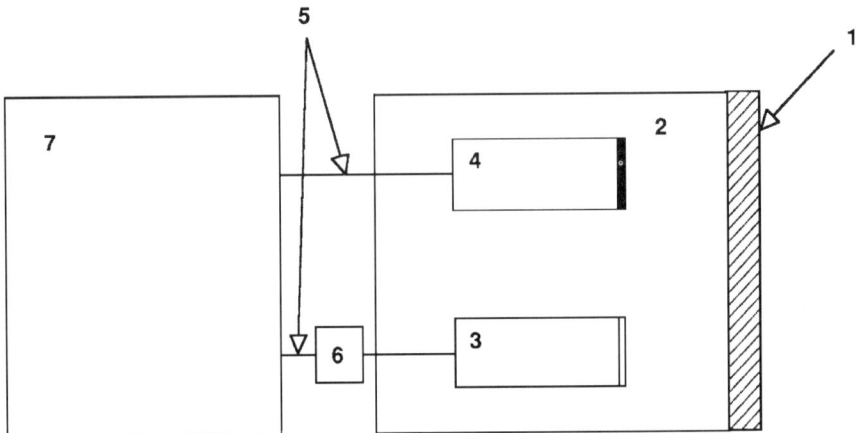

Fig. 3.5 The Clark oxygen electrode. 1 = semi-permeable membrane, 2 = electrolyte (sat. KCl), 3 = silver/silver chloride anode, 4 = platinum cathode, 5 = coaxial cables, 6 = external polarizing potential supply and 7 = amperometer.

potential is applied. As opposed to the potentiometric electrode, the analyte (at the electrode) is consumed and thus makes amperometric measurement dependent on the rate of mass transfer to the electrode surface. The typical example of the amperometric electrode is the Clark oxygen electrode which is shown diagrammatically in Fig. 3.5.

The polarographic electrode contains a platinum cathode (poised at *circa* −700 mV) and a silver/silver chloride anode which is protected by an oxygen-permeable membrane. Oxygen is reduced at the cathode which produces a current proportional to the oxygen concentration. Hydrogen peroxide can also be monitored where the platinum anode is poised at *circa* + 800 mV.

A number of oxidases have been used in conjunction with the amperometric electrode (Updike & Hicks, 1967; Karube & Suzuki, 1983; Clark, 1987). Thus, for example, oxygen consumed as cosubstrate is proportional to the analyte concentration These type of electrodes do suffer some drawbacks such as the requirement for oxygen. This has led to the search for artificial electron acceptors which can replace oxygen and which can mediate electron transfer to the electrode surface (Turner, 1988). Among others, ferricyanide and various quinones may be employed. The use of mediators has had the result that a lower applied potential is required and a wider range of oxidoreductases may be employed such as dehydrogenases.

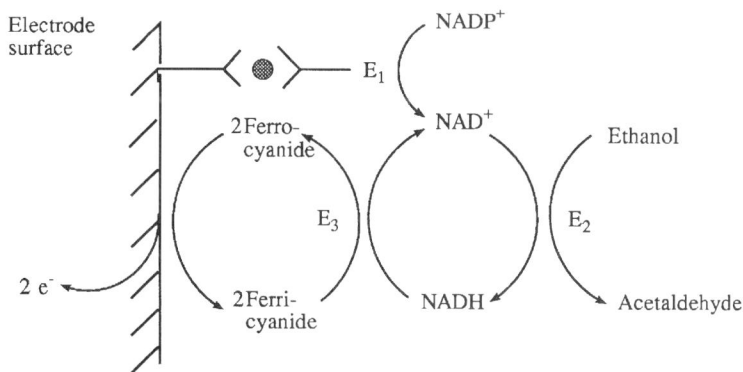

E_1 = Alkaline phosphatase

E_2 = Alcohol dehydrogenase

E_3 = Lipoamide dehydrogenase

Fig. 3.6 Signal amplification by mediator recycling.

Another development in this area is the ability to amplify the signal by recycling the mediator. Thus, a TSH amperometric assay with a sensitivity of 0·2 mIU/l has been described (Cardosi *et al.*, 1986). TSH is detected with an alkaline phosphatase labelled antibody. The enzyme initiates the sequence of events as shown in Fig. 3·6. Another example of amplification via recycling is the use of GOD and GDH as described by Scheller and co-workers (Scheller *et al.*, 1985*a*) .

3.3.2.3   *Calorimetry*

Most enzymic reactions are characterized by a relatively large enthalpy change. Thus, sensitive thermometric measurements represent a nearly universal means for determining substrate concentration. This principle has been utilized for measurement of numerous analytes (Danielsson & Mosbach, 1987).

Enzyme labelled antibodies may also be used and a thermometric ELISA procedure can be employed (Mattiasson *et al.*, 1977*a*). As an example, human proinsulin produced by recombinant *E. coli* was determined with this method (Birnbaum *et al.*, 1986). The sensitivity in this case was approximately 10 picomole. The sensitivity of the enzyme thermistor can also be improved considerably by substrate or coenzyme recycling in close analogy with the recycling methods described above for amperometric measurements (Scheller *et al.*, 1985*b*). Other applications of the thermistor unit include the use of microbial transducer for BOD measurements and, more recently, the measurement of analytes in organic solvents which give amplified signals due to the increased heat capacity of organic solvents (Mattiasson *et al.*, 1977*b*; Stasinska *et al.*, 1989).

3.3.2.4   *Piezoelectric crystals*

The resonant frequency of a piezoelectric crystal changes with the mass of the crystal. A biospecific microgravimetric transducer of this type has been described for formaldehyde (Guilbault, 1983). Generally, piezoelectric crystals require measurements to be made in the gas phase. However, by employing surface acoustic wave devices, analytes can be monitored while the transducer is still in contact with the sample. This has been demonstrated for the analysis of human IgG (Roederer & Bastiaans, 1983).

3.3.2.5   *Optical transducers*

Probably the most common analytical use of an immobilized protein or cell entails its conjunction with an optical measuring cell such as a spectrophotometer, fluorometer or luminometer. High sensitivity has been reported, in particular for fluorometric or luminometric determinations (Hornby *et al.*, 1977; Jablonski & DeLuca, 1979). However, as the immobilized biological transducer is usually displaced significantly from the measuring cell, these constructions do not fulfill the prerequisites for biosensors. Therefore, we do not intend to further describe these and refer the reader to other references for further review (DeLuca, 1984; Carr & Bowers, 1980). However, recent developments in optoelectronic construction, such as fiber optics and guided wave systems, are described below.

Goldfinch and Lowe describe a solid-phase albumin optoelectronic sensor (Lowe & Goldfinch, 1988). Triphenylmethane dyes were covalently attached to transparent membranes sandwiched between an LED and photodiode. When albumin was adsorbed a yellow to blue-green color change occurred which caused a voltage output drop. Similar constructions with pH sensitive dyes could be used to measure penicillin (in conjunction with penicillinase) or other proton forming systems.

Schultz *et al.* have described the use of fiber optics to determine glucose concentration (Schultz & Mansouri, 1988). The principle relies on the displacement of FITC-labelled dextran by glucose from Con A immobilized on the interior surface of a hollow-fiber. The hollow-fiber, which is sealed at one end, is fixed onto an optic fiber. Light at 490 nm is used for excitation. Light emission is maximum at 520 nm. Other examples of the use of optic fibers have been reported, many of which utilize the evanescent wave produced with a wave-guide coated internally with the sensing layer.

Another optical principle which has been utilized for bioanalytical measurements relies on the reflection of polarized light (Arwin & Lundström, 1988). Thus, reflectometry, ellipsometry and surface plasmon resonance have been used to monitor protein presence at a surface. Among these, ellipsometry has shown the highest sensitivity. Whereas reflectometry measures the change in intensity of reflection, ellipsometry measures the complex reflectance ratio which shows the exchange in state of polarization of the reflection. Phase measurements with ellipsometry are used to quantitate 'average' film thickness at the surface which, in turn, is correlated to surface coverage or mass per unit area. Applications of ellipsometry include quantitation of antibody, cells and even nucleic acid (Jonsson *et al.*, 1988; Mandenius *et al.*, 1988, 1989).

Vibrational electron motion at the surface of a metal conductor is termed surface plasmon resonance and occurs at a particular angle of light incidence. It depends not only on the medium but also on absorbates at the surface as well as wave length. It is manifested as a sharp decrease in reflectance. A shift in angle of incidence at the reflectance minimum occurs as protein is absorbed onto the surface. This method has been employed for kinetic studies of antigen–antibody interactions at a surface (Arwin & Lundström, 1988).

3.4 REFERENCES

Amin, G. & Doelle, H. (1990). *Enzyme Microb. Technol.*, **12**, 443.

Amin, G., DeMot, R., Van Dijck, K. & Verachtert, H. (1985). *Appl. Microbiol. Biotechnol.*, **22**, 237.

Amin, G., Khallafalla, G. & Doelle, H. W. (1988). *Appl. Microbiol. Biotechnol.*, **27**, 325

Ariga, O., Kimura, M., Taya, M. & Kobayashi, T. (1986). *J. Ferment. Technol.*, **64**, 327.

Arwin, H. & Lundström, I. (1988). *Meth. Enzymol.*, **137**, 366–381.

Bailey, J. E. & Ollis, D. F. (1986). *Biochemical Engineering Fundamentals*, McGraw-Hill, New York.

Bajpai, P. & Bajpai, P. K. (1989). *Biotechnol. Appl. Biochem.*, **11**, 155.

Bajpai, P. & Margaritas, A. (1983). *Ann. N.Y. Acad. Sci.*, **413**. 474.

Bajpai, P. & Margaritas, A. (1987). *Appl. Microbiol. Biotechnol.*, **26**, 447

Benemann, J. R., Augenstein, D. C., Weissman, J. C. & Goebel, R. P. (1982). Final report to Argonne National Laboratory.

Birnbaum, S. (1992). In *Immobilised Macromolecules: Application Potentials*, eds Sleytr, U. B. & White, J. P. N., Springer, Heidelberg.

Birnbaum, S. & Mosbach, K. (1991). *Curr. Opin. Biotechnol.*, **2**, 44.

Birnbaum, S., Pendelton, R., Larsson, P. O. & Mosbach, K. (1981). *Biotechnol. Lett.*, **3**, 391.

Birnbaum, S., Larsson, P. O. & Mosbach, K. (1983). In *Solid Phase Biochemistry. Analytical and Synthetic Aspects*, ed. W. H. Scouten, J. Wiley & Sons, New York; pp. 679–762.

Birnbaum, S., Bulow, L., Hardy, K., Danielsson, B. & Mosbach, K. (1986). *Anal. Biochem.*, **158**, 12.

Birnbaum, S., Mosbach, K. & Wallace R. B. (1991). *Curr. Opin. Biotechnol.*, **2**.

Birnbaum, S., Mosbach, K. & Wallace R. B. (1992). *Curr. Opin. Biotechnol.*, **3**.

Blackburn, G., Talley, D., Booth, P., Durfor, C., Martin, M., Napper, A., & Rees, A. (1990). *Anal. Chem.*, **62**, 2211.

Blum, L. J. & Coulet, P. R. (1991). *Biosensor Principles and Applications*, Marcel Dekker, New York.

Breddam, K., Widmer, F. & Johansen, J. T. (1980). *Carlsberg Res. Commun.*, **45**, 361.

Bruinenberg, P. M., DeBot, H. M., Van Dijken, P. & Scheffers, W. A. (1983). *Eur. J. Appl. Microbiol. Biotechnol.*, **18**, 287.

Buck, R. P., Hatfield, W. E., Umana, M. & Bowden, E. F. (1990). *Biosensor Technology*, Marcel Dekker, New York.

Caras, S. D., Petelenz, D. & Janata, J. (1985). *Anal. Chem.*, **57**, 1920.

Caras, S. & Janata, J. (1980). *Anal. Chem.*, **52**, 1935.

Cardosi, M. F., Stanley, C. J. & Turner, A. P. F. (1986). In *Chemical Sensors*, eds Acouturir, J. L. *et al.* Imprimerie Biscaye, Bordeaux, p. 634.

Carr, P. W. & Bowers, L. D. (1980). *Immobilized Enzymes in Analytical and Clinical Chemistry*, Wiley, New York.

Cass, A. E. G. (1990). *Biosensors, A Practical Approach.*, IRL Press, Oxford.

Chen, H. C. & Zall R. R. (1982). *Process Biochem.*, **17**, 20.

Chen, A. H. C., Dosoretz, C. G. & Grethlein, H. E. (1991). *Enzyme Microb. Technol.*, **13**, 404.

Chen, S. & Wayman, M. (1989). *Process. Biochem.*, **24**, 204.

Cheryan, M. & Mehaia, M. A. (1984). *Process. Biochem.*, **19**, 204.

Chithra, N. & Baradarajan, A. (1989). *Process. Biochem.*, **24**, 208.

Cho, G. H., Choi, C. V., Choi, Y. D. & Han, M. H. (1981). *Biotechnol. Lett.*, **31**, 667.

Chotani, G. K. & Constantinides, A. (1984). *Biotechnol. Bioeng.*, **26**, 217.

Clark, L. C. (1987). In *Biosensors: Fundamentals and Applications*, eds Turner, A. P. F., Karube, I. & Wilson, G. S., Oxford University Press, p. 153.

Clark, L. & Lyons, C. (1962). *Ann. N. Y. Acad. Sci.*, **102**, 29.

Clarke, D. J., Kell, D. B., Morris, J. G. & Burns, A. (1982). *Ion Select. Elect. Rev.*, **4**, 75.

Coughlan, M. P., Kierstan, M. P. J., Border, P. M. & Turner, A. P. (1988). *J. Microbiol. Methods*, **8**, 1.

Cysewski, G. R. & Wilke, C. R. (1977). *Biotechnol. Bioeng.*, **19**, 1125.

Danielsson, B. & Mosbach, K. (1987). In *Biosensors. Fundamentals and Applications*, eds Turner, A. P. F., Karube, I. & Wilson, G. S., Oxford University Press, p. 575.

Davies, P. & Mosbach, K. (1974). *Biochim. Biophys. Acta.*, **370**, 329.

Deetz, J. S. & Rozzell, J. D. (1988). *Enz. Eng.*, **9**, *Ann. N. Y. Acad. Sci.*, **542**, 230.

Del Borghi, M., Converti, A., Parisi, F. & Ferraiolo, G. (1985). *Biotechnol. Bioeng.*, **27**, 761.

Delbruck, M. (1892). *Wochenschr. Brau.*, **9**, 645.

DeLuca, M. (1984). In *Analytical Applications of Bioluminescence and Chemiluminescence*, eds Kricka, L. J., Stanley, P. E., Thorpe, G. H. G. & Whitehead, T. P., Academic Press, London, p. 111.

Diaz-Ricci, J. C., Tsu, M. & Bailey, J. E. (1992). *Biotechnol. Bioeng.*, **39**, 59.

Doran, P. M. & Bailey, J. E. (1986). *Biotechnol. Bioeng.*, **28**, 73.

Dougalis, A. J., Brown, N. M., Cloett, W. R. & Dunlop, D. B. (1981). *Biotechnol. Lett.*, **3**, 651.

Effenberger, F., Ziegler, T. & Forster, S. (1987). *Angew. Chem.*, **99**, 491.

Enari, T. M. & Niku-Paavola, M.J . (1987). *Critical Reviews Biotechnol.*, **5**, 67.

Francou, N. & Vignais, P. M. (1984). *Biotechnol. Lett.*, **6**, 639.

Galazzo, J. & Bailey, J. E. (1989). *Biotechnol. Bioeng.*, **33**, 1283.

Galazzo, J. & Bailey, J. E. (1990). *Biotechnol. Bioeng.*, **36**, 417–426.

Gamarra, J. A., Cuevas, C. M. & Lescano, G. (1986). *J. Ferment. Technol.*, **64**, 25.

Gerhardt, P. & Gallup, D. M. (1963). *J.Bacteriol.*, **96**, 919.

Gianetto, A., Berruti, F., Glick, B. R. & Kempton, A. G. (1986). *Appl. Microbiol. Biotechnol.*, **24**, 277.

Gong, C. S., Chen, L. F., Flickinger, M. C. & Tsao, G. T. (1981). *Adv. Biochem. Engineer.*, **20**, 93.

Groot, W. J., van der Lans, R. G. J. M. & Luyben, K. Ch. A. M. (1991). *Appl. Biochem. Biotechnol.*, **28/29**, 539.

Guilbault, G. G. (1983). *Anal. Chem.*, **55**, 1682.

Guilbault, G. G. (1984). *Analytical Uses of Immobilized Enzymes*, Marcel Dekker, New York.

Guilbault, G. G. & Shu, F. R. (1972). *Anal. Chem.*, **44**, 2161.

Guimaraes, W. V., Dudley, G. L. & Ingram, L. O. (1992). *Biotechnol. Bioeng.*, **40**, 41.

Hahn-Hägerdahl, B. (1984). *Biotechnol. Bioeng.*, **26**, 771.

Hahn-Hägerdahl, B. (1985). *Biotechnol. Bioeng.*, **27**, 914.

Hikuma, T., Obana, H., Yasuda, T., Karube, I. & Suzuki, S. (1980). *Anal. Chim. Acta,* **116**, 61.

Hilge-Rotmann, B. & Rehm, H.-J. (1991). *Appl. Microbiol. Biotechnol.*, **3**, 502.

Hornby, W. E., Noy, G. A. & Bin Salleh (1977). In *Biotechnological Applications of Proteins and Enzymes*, eds Bohak, Z. & Sharon N. Academic Press, New York.

Inloes, D. S., Michaels, A. S., Robertson, C. R. & Matin, A. (1985). *Appl. Microbiol. Biotechnol.*, **23**, 85.

Jablonski, E. & DeLuca, M. (1979). *Clin. Chem.*, **25**, 1622.

Janata, J. & Blackburn, G. F. (1984). *Ann. N.Y. Acad. Sci.*, **428**, 286.

Jeong, Y. S., Vieth, W. R. & Matsuura, T. (1991). *Biotechnol. Bioeng.*, **37**, 587–590.

Jönsson, U., Malmquist, M., Olofsson G. & Rönnberg, I. (1988). *Meth. Enzymol.*, **137**, 381.

Kang, W., Shukla, R. & Sirkar, K. K. (1990). *Biotechnol. Bioeng.*, **36**, 826.

Karube, I. & Suzuki, S. (1983). *Ion Sel. Electrode Rev.*, **6**, 15

Karube, I. & Suzuki, S. (1988). *Meth. Enzymol.*, **137**, 668

Keating, M. Y. & Rechnitz, G. A. (1984). *Anal. Chem.*, **56**, 801.

Kimmel, D. E., Klasson, K. T., Clausen E. C. & Gaddy, J. L. (1991). *Appl. Biochem. Biotechnol.*, **28/29**, 457.

King, V. A-E. & Zall, R. R. (1983). *Process. Biochem.*, **18**, 17.

Kirk, T. K. & Farrell, R. L. (1990). *Ann. Rev. Microbiol.*, **41**, 465.

Kirkpatrick, N., Reid, I. D., Ziomek, E. & Paice M. G. (1990). **33**, 105.

Klasson, K. T., Ackerson, M. D., Clausen, E. C. & Gaddy, J. L. (1991). *Fuel*, **70**, 605.

Klein, J. & Kressdorf, B. (1986). *Biotechnol. Lett.*, **8**, 739.

Kumar, V., Ramakrishnan, S., Teeri, T. T., Knowles, J. K. C. & Hartley, B. S. (1992). *Biotechnology*, **10**, 82.

Kurosawa, H., Namura, N. & Tanaka, H. (1989). *Biotechnol. Bioeng.*, **33**, 716.

Lee, C. W. & Chang, H. N. (1985). *Enzyme Microb. Technol.*, **7**, 561.

Lee, G. M., Kim, C. H., Lee, K. J., Yousef, Z. A. M., Han, M. H. & Rhee, S. K. (1986). *J. Ferment. Technol.*, **64**, 293.

Lee, T. H., Ahn, J. C. & Ryu, D. D. Y. (1983). *Enzyme Microb. Technol.*, **5**, 41.

Lin, J. J., Dale, M. C. & Okos, M. R. (1990). *Process Biochem.*, **25**, 61.

Lowe, C. R. & Goldfinch, M. J. (1988). *Meth. Enzymol.*, **137**, 338–348.

Lynd, L. R., Cushman J. H., Nichols, R. J. & Wyman, C. E. (1991). *Science*, **251**, 1318–1323.

Macrae, A. R. (1983). *J. Am. Oil. Chem.*, **60**, 291.

Mandenius, C-F., Chollet, A., Mecklenburg, M., Lundström, I. & Mosbach, K. (1989). *Anal. Lett.*, **22**, 2961.

Mandenius, C-F., Welin, S., Lundstrom, I. & Mosbach, K. (1988). *Meth. Enzymol.*, **137**, 388.

Margaritis, A. & Wilke, C. R. (1972). *Dev. Ind. Microbiol.*, **13**, 159.

Margaritis, A. & Bajpai, P. (1982). *Biotechnol. Bioeng.*, **23**, 1483.

Margaritis, A. & Merchant, F. J. A. (1984). *Critical Reviews Biotechnol.*, **1**, 339.

Margolin, A. L., Tai, D. F. & Klibanov, A. M. (1987). *J. Am. Chem. Soc.*, **109**, 7885.

Marwaha, S. S. & Kennedy, J. F. (1984). *Process Biochem.*, **19**, 79.

Mattiasson, B. & Hahn-Hägerdal, B. (1982). *Europ. J. Appl. Microbiol. Biotechnol.*, **16**, 52.

Mattiasson B. & Miyabayashi, A. (1988). *Analytica Chimica Acta*, **213**, 79.

Mattiasson, B., Borrebaeck, C., Sanfridson, B. & Mosbach, K. (1977a). *Biochim. Biophys. Acta.*, **483**, 221.

Mattiasson, B., Larsson, P-O. & Mosbach, K. (1977b). *Nature*, **268**, 519.

Mehta, P. K., Mishra, S. & Ghose, T.K. (1991). *Biotechnol. Bioeng.*, **37**, 551.

Michaud, M., Paguot, M., Baijot, B. & Thonart, Ph. (1982). *Biotechnol. Bioeng. Symp.*, **12**, 475.

Miyabayashi, A. & Mattiasson, B. (1990). *Anal. Biochem.*, **184**, 165–171.

Mosbach, K. (ed.) (1987a). *Meth. Enzymol.*, **135**.

Mosbach, K. (ed.) (1987b). *Meth. Enzymol.*, **136**.

Mosbach, K. (1988). *Meth. Enzymol., Vol. 137*, Academic Press, San Diego.

Navarro, A., Marangoni, H., Plaza, I. M. & Callieri, D. (1984). *Biotechnol. Lett.*, **6**, 465.

Navarro, J. M. & Durand, G. (1977). *Eur. J. Appl. Microbiol.*, **4**, 243.

Neidleman, S. L. (1991). *Curr. Opin. Biotechnol.*, **2**, 390.

Nilsson, K., Buzasky, F. & Mosbach, K. (1986). *Bio/Technology*, **4**, 989.

Nishio, N., Kayawake, E. & Nagai, S. (1985). *J. Ferment. Technol.*, **63**, 205.

Nishio, T., Kamimura, M., Murata, M., Terao, Y. & Achiwak, K. (1988). *J. Biochem.*, **104**, 681.

Nunez, M. J., Chamy, R., Dominguez, H., Sanroman, A. & Lema, J. M. (1991). *Appl. Biochem. Biotechnol.*, **28/29**, 731.

Parekh, S. R. & Wayman, M. (1987). *Enzyme Microb. Technol.*, **9**, 406.

Parekh, S. R., Parekh, R. S. & Wayman, M. (1989). *Process. Biochem.*, **24**, 88.

Pundle, A., Prabhune, A. & Sivaraman, H. (1988). *Appl. Microbiol. Biotecnnol.*, **29**, 426.

Quereshi, N., Pai, J. S. & Tamhane, D. V. (1987). *J. Chem. Tech. Biotechnol.*, **39**, 75.

Ramakrishna, S. V., Sreedharan, V. P. & Prema, P. (1988). In *Bioreactor Immobilized Enzymes and Cells; Fundamentals and Applications*, ed. M. Moo-Young, Elsevier, London.

Ramos-Tombo, G. M., Schaer, H. P., Busquets, F. I. & Ghisalba, O. (1986). *Tetrahedron Lett.*, **27**, 5707.

Rao, B. S., Pundle, A. V., Prabhune, A. A., Shankar, V. & Sivaraman, H. (1986). *Appl. Biochem. Biotechnol.*, **12**, 17.

Roederer, Z. E. & Bastiaans, G. J. (1983). *Anal. Chem.*, **55**, 2333.

Rogers, P. L., Lee, K. J., Skotnicki, M. L. & Trube, D. E. (1982). *Adv. Biochem. Eng.*, **23**, 37.

Rosario, E. J. & Pamatong, F. V. (1985). *Biotechnol. Lett.*, **7**, 819.

Rose, D. (1976). *Process Biochem.*, **11**, 10.

Salter, G. J., Kell, D. B., Ash, L. A., Adams J. M., Brown A. J. & James, R. (1990). *Enzyme Microb. Technol.*, **12**, 419–30.

Scharer, J. M., Bhadra, A. & Moo-Young, M. (1988). In *Bioreactor Immobilized Enzymes and Cells; Fundamentals and Applications*, ed. M. Moo-Young, Elsevier, London.

Scheller, F. W., Rennenberg, R., Mueller, H.-G., Jaenchen, M. & Weise, H. (1985a). *Biosensors*, **1**, 135.

Scheller, F., Siegbahn, N., Danielsson, B. & Mosbach, K. (1985b) *Anal. Chem.*, **57**, 1740.

Schultz, J. S. & Mansouri, S. (1988). *Meth. Enzymol.*, **137**, 349–366.

Scouten, W. H. (1983). *Solid Phase Biochemistry; Analytical and Synthetic Aspects*, J. Wiley & Sons, New York.

Shiotani, T. & Yamane, T. (1981). *Eur. J. Appl. Microbiol. Biotechnol.*, **13**, 96.

Sivaraman, H., Rao, B. S., Pundle, A. V. & Sivaraman, C. (1982). *Biotechnol. Lett.*, **4**, 359.

Skoog, K. & Hahn-Hägerdahl, B. (1988). *Enzyme Microb. Technol.*, **10**, 66–80.

Slininger, P. J., Bothast, R. J., Van Cœvwenberg, J. E. & Kurtman, C. R. (1982). *Biotechnol. Bioeng.*, **24**, 371.

Smidsrød, O. & Skjåk-Braek, G. (1990). *Trends Biotechnol.*, **8**, 71.

Snijder-Lambers, A. M., Doddema, H. J., Graude, H. J. & van Lelyveld, P. H. (1986). *Biocatal. Org. Media, Proc. Int. Symp.*, **87**.

Solsky, R. L. (1982). *Anal. Chem.*, **14**, 1.

Stasinska, B., Danielsson, B. & Mosbach, K. (1989). *Biotechnol. Techniques*, **3**, 281.

Swaine, D. E. & Daugulis, A. J. (1989). *Biotechnol. Bioeng.*, 604.

Takahashi, K., Tamaura, Y., Kodera, Y., Mahama, T., Saito, Y. & Inada, Y. (1987). *Biochem. Biophys. Res. Commun.*, **142**, 291.

Tanaka, H., Kurosawa, H. & Murakami, H. (1986). *Biotechnol. Bioeng.*, **2**, 1761.

Taylor, R. F. (1991). *Protein Immobilization: Fundamentals and Applications*, Marcel Dekker, New York.

Tjerneld, F., Persson, I., Albertsson, P. Å. & Hahn-Hägerdahl, B. (1985a). *Biotechnol. Bioeng.*, **27**, 1036.

Tjerneld, F., Persson, I., Albertsson, P. Å. & Hahn-Hägerdahl, B. (1985b). *Biotechnol. Bioeng.*, **27**, 1044.

Tjerneld, F., Persson, I., Albertsson, P. Å & Hahn-Hägerdahl, B. (1985c). *Biotechnol. Bioeng. Symp.*, No. 15, 419.

Toda, K., Ohtake H. & Asakura T. (1986). *Appl. Microbiol. Biotechnol.*, **24**, 97.

Turner, A. P. F., Karube, I. & Wilson, G. S. (1987). *Biosensors: Fundamentals and Applications*, Oxford University Press.

Turner, A. P. F. (1988). *Meth. Enzymol.*, **137**, 90.

Updike, S. J. & Hicks, G. P. (1967). *Nature*, **214**, 986.

Veliky, I. A. & Williams, R. E. (1981). *Biotechnol. Lett.*, **3**, 275.

Viera, A. M., Sa-Correia, I., Novais, J. M. & Cabral, J. M. S. (1989). *Biotechnol. Lett.*, **11**, 137–40.

Villet, R. (1981). *Dev. Ind. Microbiol.*, **21**, 97.

von Felten, P., Zürrer, H. & Bachofen, R. (1985). *Appl. Microbiol. Biotechnol.*, **23**, 15.

Wada, M., Kato, J. & Chibata, I. (1980). *Eur. J. Appl. Microbiol. Biotechnol.*, **10**, 275.

Winquist, F., Dallielsson, B., Lundström, I. & Mosbach, K. (1988). *Meth. Enzymol.*, **137**, 232.

Winquist, F., Lundström, I. & Danielsson, B. (1989). In *Applied Biosensors*, ed. D. Wise, Butterworth, Stoneham, p. 291.

Wise, D. L. (1989). *Applied Biosensors*, Butterworth, Stoneham.

Woodward, J. (1985). *Immobilised cells and enzymes, a practical approach*, IRL Press, Oxford.

Zeikus, J. G. (1980). *Ann. Rev. Microbiol.*, **34**, 423.

Zhi-wei, G., Ngori, T. K., Scilimato, A., Fullig, G. & Sih, C. J. (1988). *Tetrahedron Lett.*, **29**, 5583.

Chapter 4

CELL IMMOBILIZATION BY ADSORPTION TO GLASS FIBRE MATS

FRANK DiCOSMO,[1,2,3] HIDEO TANAKA[4] AND A. WILHELM NEUMANN[2,3,5]

[1]*Centre for Plant Biotechnology, Department of Botany,*
[2]*Institute of Biomedical Engineering,* [3]*Department of Mechanical Engineering, University of Toronto, Canada*
[4]*Institute of Applied Biochemistry, University of Tsukuba, Japan*
[5]*Research Institute, Hospital for Sick Children, Toronto, Canada*

CONTENTS

4.1   INTRODUCTION

The immobilization of biocatalysts (e.g. cells, organelles and enzymes) onto, or within a support has become a powerful and well-accepted technology with biotechnological applications in biomedical, and industrial research and development. For example, various biosensors using immobilized enzymes are important diagnostic tools in medicine, and are useful probes in research and industry (Mosbach, 1988). Likewise, immobilized cells have broad applications in the production of useful biochemicals such as ethanol, amino acids, organic acids, antibiotics, enzymes and hormones, used by the commodity chemical, food and pharmaceutical industries (Mosbach, 1976, 1987*a,b*). The reasons for the utility of immobilized biosystems are that high biocatalyst densities with high catalytic activity are possible, within the bioreactor. The biocatalyst can be reused allowing for continuous or semi-continuous operation in which the system is maintained in the most active phase for extended periods, and immobilized systems allow for improved downstream processing. Immobilized biocatalysts include single enzymes or enzyme-systems comprising two or more enzymes coupled for sequential catalysis, and cells of plant, animal, fungal or bacterial origin that are anchored in some fashion to a support-matrix. The support materials include soluble or insoluble substances, such as hydrous transition metal oxides (Kennedy & Cabral, 1987; Messing, 1974), alginate (Bucke, 1987), agarose (Nilsson et al., 1987), polyacrylamide (O'Driscoll, 1976; Skryabin & Koshcheenko, 1987), celite (Gbewonyo et al., 1987; Keshavarz et al., 1989), cellulose (Kim et al., 1988), porous ceramics (Shimada & Shimahara, 1989; Shiraishi et al., 1989), controlled-pore glass (Messing, 1969, 1985), glass pads (Arcuri, 1982), sand (Dunn, 1987), stainless steel supports (Black et al., 1984), and insoluble polymeric materials such as polyurethane foam (Lindsey & Yeoman, 1987), non-woven fabrics (Sakurai et al., 1989), DEAE-Sephadex (Chibata et al., 1976), wood-chips (Lamptey & Moo Young, 1987), nylon fibres (Seiskari et al., 1985), and polyester pads (Black et al., 1984), among others.

Effective application of these materials in the process of immobilization requires that specific protocols are adopted that may include chemical activation and cross-linking of polymers and enzymes or cells, covalent attachment of enzymes or cells, or both to the support, entrapment, encapsulation, adsorption, or a combination of these. The literature provides many examples where such methods have been used to immobilize biocatalysts effectively and efficiently while maintaining viability and metabolic activity. However, some methods may result in adverse effects that include decreased activity because the biocatalyst is damaged during the immobilization procedure. For example, covalent attachment of α-chymotrypsin with Eupergit C for dipeptide synthesis was found to lead to enzyme deactivation because of exposure to n-propanol (Flörsheimer et al., 1989). None the less, the immobilization of cells and enzymes is a well established technology that will continue to provide viable solutions to complex medical and industrial applications. Excellent and detailed discussions, and principles of the various procedures of immobilization have been presented in the four volumes of Methods in Enzymology (Vol. XLIV, 135–137), edited by Mosbach. Recent reviews have also appeared that describe biocatalyst immobilization (Chibata et al., 1986; Hartmeir, 1988; Klein & Vorlop, 1985; Lewis et al., 1986; Messing, 1985; Vega et al., 1988; Webb et al., 1986).

In these reports the techniques and problems involved in selecting a support and achieving optimum biocatalyst loading are usually well presented. Most notably, the volumes of *Methods in Enzymology* (*loc. cit.*) are excellent in this regard. However, the technique of immobilization by adsorption of cells and enzymes to support-materials is the simplest method to apply, but perhaps the least well understood. Messing (1985) provided a synopsis of enzyme adsorption and Klein & Vorlop (1985) briefly discussed cell adsorption. Although this method of immobilization appears to be 'simple' (e.g. the particles in suspension are mixed with the support and adsorption ensues), the forces involved in mediating adsorption are complex, and conflicting opinions exist concerning the mechanism(s) of adhesion to various support materials. Archambault *et al.* (1989) observed a secretion from cells that appeared to act as a 'glueing agent'. Bornman & Zachrisson (1987) noted that protoplasts failed to adhere to non-lectin-treated Cytodex 1 or even to positively charged Cytodex 2 beads, and media with high levels of Ca^{2+} were found to promote attachment. Dunn (1987) found that the nature of the solid support was not influential in determining the adhesion of mixed cells to surfaces, including glass, metal and plastics; the formation of a polysaccharide slime could mediate attachment to such surfaces. Betinec & Mendelson (1988) noted the immobilization of *Bacillus subtilis* macrofibres onto several support materials including cellulose and steel and attributed immobilization to the morphogenesis of the macrofibres. Several reports indicate that hydrophobic interactions and electrostatic phenomena that occur when cells and solid supports are introduced into a liquid are paramount in determining efficient and rapid adsorption (Abbott *et al.*, 1983; Absolom *et al.*, 1985; DiCosmo *et al.*, 1988; Facchini *et al.*, 1988*b*, 1989*a*,*b*; Mozes *et al.*, 1987). The adhesion of cells in suspension to a solid surface immersed in a liquid is seen to include two independent processes. In order for cells to attach firmly to any solid support material the cells must first be transported to, and make contact with the surface; this involves the physical–chemical interactions prevalent in the system. The subsequent binding of cells to the surface and to adjacent cells will occur because of the time-dependent production of extracellular mucilaginous, adhesive secretions which act to agglutinate cells to the surface and other cells.

The process of cell adhesion will include at least van der Waals type of interactions and electrostatic phenomena. These will be influenced by the relative surface energies and surface charges of the interacting components, the pH and ionic strength of the suspending media (Absolom *et al.*, 1985; Facchini *et al.*, 1988*b*; Mozes *et al.*, 1987). The adhesion of bacteria to various surfaces has been evaluated from several perspectives, including specific interactions (Daly & Stevenson, 1987; Gibbons, 1980), electrical charge and hydrophobicity (Doyle *et al.*, 1982; Rosenberg *et al.*,

1980) and thermodynamic considerations (Absolom *et al.*, 1983, 1985; Busscher *et al.*, 1984; Pratt-Tepstra *et al.*, 1987, 1989). The physical-chemical properties of cell surfaces are modulated in a predictable pattern, throughout the growth-cycle of the cultured cells. For example, the relative hydrophobicities of *Catharanthus roseus* (Facchini *et al.*, 1988*a,b*) cells and *Corynebacterium glutamicum* (Buchs *et al.*, 1988) are altered as a function of age and nutrition. This inherent complexity has no doubt restricted the use of adsorption as a mode of immobilization because there appear to be no generally applicable techniques that serve to guide researchers toward achieving efficient biocatalyst loading on well-defined supports. It is generally believed that high biocatalyst densities adherent to the support are not achievable, and the strength of adhesion is insufficient for many practical applications.

However, the process of firm and irreversible adhesion of cells and microorganisms to surfaces is a well-documented phenomenon crucial to environmental, biomedical and technological areas. Adsorption and adhesion of particles, cells, proteins, and microorganisms is known to result in thrombus formation (Absolom *et al.*, 1985; Pratt-Tepstra 1989 and refs therein), dental plaque and caries formation (Abbott *et al.*, 1983 and references therein), and occurs in waste-water and sewage treatment (Ridgeway *et al.*, 1983), liquid-flow conduit-fouling (Duddridge *et al.*, 1985), and metal leaching by microorganisms (Ehrlich & Holmes, 1986; Keller, 1982) and demonstrates convincingly that particles and surfaces can be held together by attractive forces for extended periods. In all of these examples high populations of cells, usually greater than can be accommodated in free suspension, may occur in mixed consortia under natural and complex conditions. The strength of adhesion is sufficient to maintain firm attachment in altered environments and high fluid-flow velocities. Messing (1969, 1985) discussed the adsorption of proteins to controlled-pore glass and the evidence for ionic bonding between the NH^+_3 of the proteins and the dissociated SiO^- moieties of the glass support. The adsorption was so strong that strong acids, ammonium hydroxide, or various ionic strength buffers were unable to cause quantitative desorption of the proteins from the glass. The proteins, however, were eluted with acidic solutions of urea, but not by urea solutions alone. It was suggested that hydrogen bonding and amine–silicate bonds were involved in the strong adsorption of the proteins to the glass. The obvious conclusion from the above studies is that adsorption of proteins, and cells to surfaces is efficient and strong, and is irreversible under the proper conditions.

It appears that the use of adsorption techniques for immobilization of cells for biotechnological applications is perhaps restricted because of our limited understanding of the forces involved in spontaneous adsorption and adhesion of cells to surfaces and how to optimize these forces for adhesion under operational conditions. It is eminently clear that no single method of cell immobilization will fulfil the needs of every application. The most satisfactory method of immobilization is of course, dictated by the application; often it may be advantageous to couple one or more methods of immobilization to achieve the desired goal. It is our hope that this contribution will provide increased awareness and impetus for research and development of immobilization strategies using adsorption methodologies. The advantages may include chemical modification of the support-carrier to yield the most desirable surface properties (e.g. hydrophobicity, charge) while retaining the mechanical characteristics of the support material, coupled with the simplicity of application under physiological or operational conditions. The process seems however, to have been overlooked, or perhaps rejected in certain applications. Microorganisms and cells in culture, proteins and other particles in solution will become firmly and irreversibly attached or immobilized to solid surfaces. When the cells are viable, continued growth results in so-called 'biofilms'. This spontaneous adhesion of particles to surfaces can be used to advantage in many technological and biotechnological applications. However, conflicting opinions regarding the utility of the process and the methods that are useful in optimizing adhesion have led us to conclude that some clarity is required. Herein we will discuss the adhesion of cells in suspension to support-surfaces as predicted by a thermodynamic model and modified by electrostatic interactions prevalent in the system. We will show that the process of adsorption (adhesion) can be predicted, controlled and modified through an understanding of the forces mediating the response. This is not intended as an exhaustive review, but rather is to highlight some important principles that must be considered for efficient adsorption of cells to surfaces. We present evidence indicating the broad utility of the technique in establishing a plant cell immobilization system (Facchini & DiCosmo, 1990, 1991*a,b*).

4.2 CELL ADHESION AS PREDICTED BY A THERMODYNAMIC MODEL

The initial interaction between a particle, viz. a cell in suspension, and a solid surface, referred to as adsorption herein, is controlled by well-defined physical–chemical properties of the cell, the solid surface and the suspending liquid; the process can be studied, and appropriately

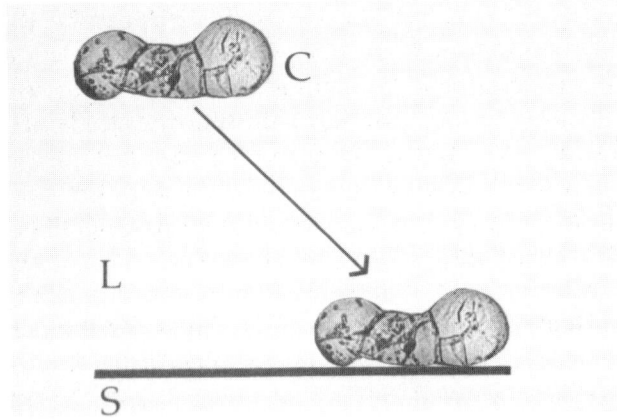

Fig. 4.1. A representation of the adhesion of a cultured plant cell (C) immersed in a liquid (L) to a solid substrate (S). The free energy of adhesion ΔF^{adh} is calculated using Eq. (1) as described in the text.

modelled with the principles of colloid and interface science (Adamson, 1982; Hiemenz, 1986; Neumann *et al.*, 1974). Consider the interacting components of the system in Fig. 1. The cell (C) is suspended in a defined liquid (L) in which a solid support-surface (S) is immersed.

We will define adsorption as the adhesion of the particle onto the surface of a support that has not been specifically altered for covalent or receptor-ligand-type bonding. In the system we must consider the surface tensions of the interacting components, pH, ionic strength of the liquid, the zeta potentials of the cell surface and the support, as well as the concentration of particles in the bulk fluid because these parameters have been shown to influence adsorption and the subsequent retention of the cells on the support (DiCosmo *et al.*, 1989). The initial impetus for cell adhesion to a surface is provided by a decrease in the free energy of adhesion (ΔF^{adh}) which is determined by the cell/liquid/solid interfacial energies, electrostatic attraction or repulsion induced by the pH and ionic strength of the suspending fluid (Buchs *et al.*, 1988; Champluvier *et al.*, 1988; Facchini *et al.*, 1989*a,b*; Pratt-Tepstra *et al.*, 1989; Thonart *et al.*, 1982).

The thermodynamic model of particle adhesion onto a surface considers van der Waals type interactions. If we assume constant electrostatic interactions, then the process of adhesion may be appropriately modelled by a simple free energy balance,

$$\Delta F^{adh} = \gamma_{cs} - \gamma_{cl} - \gamma_{sl} < O \qquad (1)$$

where γ_{cs} and γ_{cl} are the cell/solid and cell/liquid interfacial tensions respectively, and γ_{sl} is the solid/liquid interfacial tension.

The model predicts that the thermodynamic potential, ΔF^{adh}, defined as the free energy of adhesion per unit interface area will be minimized at equilibrium. If cell adsorption causes ΔF^{adh} to decrease adsorption is favoured. In contrast, if ΔF^{adh} becomes more positive adsorption should not occur. This model may be useful for the design of processes which would allow for the increased or decreased particle adhesion to surfaces as dictated by the application. The most important point in the use of the model is the determination of the values for γ_{cs}, γ_{cl}, and γ_{sl} of Eq. (1). These values can be generated using the equation-of-state approach enunciated by Neumann *et al.* (1974). This allows for the determination of the surface tension γ_{sv} of a solid (S) in contact with the vapour phase (V) from the measured liquid surface tension γ_{lv}, and the contact angle (θ), which a drop of the liquid makes with the solid. The technique considers Young's Equation,

$$\gamma_{sv} - \gamma_{sl} = \gamma_{lv} \cos \theta \qquad (2)$$

Only γ_{lv} and $\cos \theta$ can be measured experimentally, and γ_{sv} and γ_{sl} are required for application of the model. It was shown by Ward and Neumann (1974) using a two-component, liquid–vapour-ideal-solid system that an equation-of-state of the form

$$\gamma_{sl} = f(\gamma_{sv}, \gamma_{lv}) \qquad (3)$$

could be derived. Equation (3) was explicitly formulated (Neumann *et al.*, 1974) using experimental liquid–vapour interfacial tensions and contact angles to yield:

$$\gamma_{sl} = \frac{(\gamma_{sv}^{1/2} - \gamma_{lv}^{1/2})^2}{1 - 0 \cdot 0157 \, (\gamma_{sv} \, \gamma_{lv})^{1/2}} \qquad (4)$$

Equation (4) in conjunction with Eq. (2) can be arranged to yield:

$$\cos \theta = \frac{0 \cdot 015 \, \gamma_{sv} - 2 \cdot 00 \, (\gamma_{sv} \, \gamma_{lv})^{1/2} + \gamma_{lv}}{\gamma_{lv} \, [0 \cdot 0157 \, (\gamma_{sv} \, \gamma_{lv})^{1/2} - 1]} \qquad (5)$$

The mathematical limitations of Eqs (4) and (5) can be overcome by the use of appropriate tables and computer programs (Neumann *et al.*, 1980*a,b*); these are available upon request. Thus, the relevant interfacial tensions in Eq. (1) can be readily determined from the liquid surface tension and contact angles of the solids involved in the system. The data allow us to predict the direction of the thermodynamic process in question, i.e. will adsorption be favoured or not? Herein lies the practicality and utility of the equation-of-state approach for the purpose of cell immobilization. The equation-of-state approach has been considered and tested extensively, and compared with calculations arising from the unmodified Good equation, and Antonow's rule. Although the three relations can not be valid for a particular set of experimental circumstances, they do predict the behaviour of ΔF^{adh} which is used to interpret the experimental data. The independence in the qualitative features of ΔF^{adh} as noted by Smith *et al.* (1986) arises because the observed patterns in the ΔF^{adh} curves are ensured by the existence of an equation-of-state for interfacial tensions,

$$\gamma_{12} = f(\gamma_{1v}, \gamma_{2v}) \tag{6}$$

the subscripts 1 and 2 refer to different phases, and its symmetry:

$$\gamma_{12} = f(\gamma_{1v}, \gamma_{2v}) = f(\gamma_{2v}, \gamma_{1v}) \tag{7}$$

We are aware of no equation-of-state-type relation that does not satisfy relation (7). The existence of an equation-of-state (Eq. 3) has been verified using two independent methods. Initially, by a consideration of the Gibbs–Duhem equations for the interfaces in a two component, three-phase, solid/liquid/vapour system (Ward & Neumann, 1974) and recently, by analysis of the Gibbs phase rule for capillary systems (Li *et al.*, 1989). The conclusions that arise from ΔF^{adh} in concert with appropriate experimental data are indeed independent of any form of the equation-of-state for interfacial tensions (Smith *et al.*, 1986).

The equation-of-state approach predicts patterns in the ΔF^{adh} curves that are multifaceted. A theoretical calculation for ΔF^{adh} considering the adsorption of cells to solids as a function of increasing surface tensions of the solids (γ_{sv}) illustrates three circumstances (Fig. 4.2). When the liquid surface tension (γ_{1v}) is greater than the cell surface tension (γ_{cv}), ΔF^{adh} increases with increasing values of solid surface tension (γ_{sv}); this predicts decreasing cellular adsorption to the solid as a function of increasing solid support surface tension. In contrast, when γ_{1v} is less than γ_{cv}, ΔF^{adh} becomes more negative as a function of increasing γ_{sv}, predicting an increase in adsorption with increasing solid surface tension. Finally, when $\gamma_{1v} = \gamma_{cv}$, adsorption of cells to the solids is independent of the surface tension of the solid. For all practical purposes this means that in systems comprised of cells in liquid suspension adhesion to a surface immersed in the liquid will depend on the relative values of the surface tensions of the liquid, cell and solid. Thus, the question of whether the extent of cell adhesion to a support surface will increase or decrease as a function of increasing surface tension (increasing hydrophilicity) of the support will have a different answer depending on whether adhesion

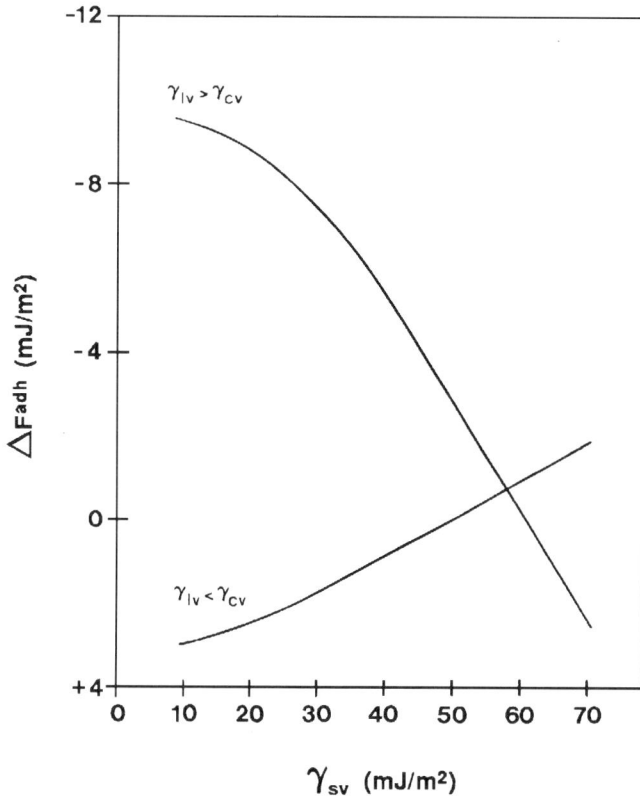

Fig. 4.2. Dependence of the free energy of adhesion (ΔF^{adh}) of plant cells in suspension as a function of substrate surface tension (γ_{sv}). The cultured plant cells have a surface tension, $\gamma_{cv} = 55$ mJ m^{-2}; for $\gamma_{lv} > \gamma_{cv}$, $\gamma_{lv} = 72.5$ mJ m^{-2} and for $\gamma_{lv} < \gamma_{cv}$, $\gamma_{lv} = 44.5$ mJ m^{-2}.

occurs out of solutions where $\gamma_{lv} > \gamma_{cv}$, $\gamma_{lv} < \gamma_{cv}$, or $\gamma_{lv} = \gamma_{cv}$. For example, Kolster & Beiderbeck (1987) studied the conditions for the promotion of attachment of *Calystegia sepium* cells to various solid supports. The conclusion was that attachment of cells required the presence of soluble carbohydrate(s) and a macroelement (i.e. KNO_3, $CaCl_2$, NH_4NO_3, $MgSO_4$, KH_2PO_4); the pH for maximum attachment was in the range of 5–7. Attachment was prevalent on hydrophillic supports under these conditions (i.e. surfaces with relatively high surface tensions). In contrast, Facchini *et al.* (1988*a,b*) observed an increase in the extent of adhesion of *Catharanthus roseus* cells as a function of decreasing hydrophilicity when the surface tension of the liquid was greater than that of the cells. The results of Kolster & Beiderbeck (1987) and Facchini *et al.* (1988*a,b*) are, however, in full agreement when the effects of γ_{lv}, γ_{cv}, and γ_{sv} are considered.

In the system of Facchini *et al.* (1988*a*) the plant cells were suspended in distilled water (γ_{lv} = 72·5 mJ m^{-2}) and the surface tension of the cells was considered to be approximately 56 mJ m^{-2}; thus the condition γ_{lv} > γ_{cv} is fulfilled. Under this set of conditions the thermodynamic model of cell adhesion predicts decreased adhesion as a function of decreasing hydrophobicity (increasing surface tension) and the results were in accord with the theoretical prediction. When the value of γ_{lv} was altered to below that of the cells (Facchini *et al.*, 1988*b*) the levels of adhesion increased in response to increasing hydrophilicity of the solid surface in agreement with model predictions. It would appear that these were the conditions under which *Calystegia sepium* cells attached to the solid surfaces because it is expected that the nutrient media will have a surface tension considerably lower than that of the plant cells due to the presence of organic nutrients. The surface tension of Murashige and Skoog growth medium for example, has a surface tension of approximately 47 mJ m^{-2}. The thermodynamic model has been used to predict the adhesion of a wide variety of microorganisms and cell types to solid supports, including granulocytes (Neumann *et al.*, 1979), bacteria (Absolom *et al.*, 1983), erythrocytes (Absolom *et al.*, 1985) and plant cells (Facchini *et al.*, 1988*a,b*).

4.3   ELECTROSTATIC CONSIDERATIONS

The thermodynamic model does not consider the contribution of electrostatic interactions to the overall process of cell adhesion. In biological systems the cells are usually bathed in a complex liquid that may be composed of proteins, amino acids, carbohydrates, vitamins, hormones, monovalent and plurivalent ions, and other constituents which define the physical-chemical parameters such as, the surface tension, the pH and ionic strength of the system. These interacting parameters must be considered in the process of cell adhesion because the magnitude of electrostatic interactions (either repulsion or attraction) between the cells and the solid support surface is expected to depend on the pH and ionic strength of the suspending medium and the magnitude of the surface charges of the cell and the support. This is because all cells have a negative surface potential and in aqueous media acquire an altered surface charge due to the adsorption of ions, or dissociation of functional groups present on the surfaces. This potential will cause the migration of negatively charged particles toward a positively charged surface, positively charged particles toward a negatively charged surface. The surface charge is seen to be balanced by a counter-potential of oppositely charged ions termed counter-ions. In this way an electrical diffuse double-layer is established. The van der Waals–London molecular forces of attraction, and electrostatic repulsion that result from overlapping diffuse ionic double-layers of the same sign of charge are considered by the DLVO theory. The overlapping ionic double-layers are considered to be in equilibrium. However, this equilibrium can be perturbed by increases in the density of the ionic electrical field in the area of the overlap, and in the physico-chemical composition of the suspending fluid. The addition of salts to a liquid in which cells are suspended will cause a compression of the diffuse double-layer and a reduction of the repulsion barrier; the greater the valency of the counter-ion, the less the amount of electrolyte required. Surface charges of unlike solid supports, or cells are not expected to be identical and when the charges are of opposite sign, the overlapping ionic double-layers will result in attraction between the cell and the support surface. The studies of Abbott *et al.*, 1983; Facchini *et al.*, 1989*a,b* and Mozes *et al.*, 1987 that considered the effect of ionic concentration and valency on adhesion of various cells to surfaces is entirely in accord with this. The reduction in repulsion increased the extent and strength of adhesion of cells to support surfaces; and an optimum ion concentration resulted in maximum adhesion. The DVLO theory explains the dependence of the repulsion interaction on the concentration of the counter-ions and their valency. Detailed considerations of the DLVO theory can be found in Adamson (1982), Iler (1973), Efremov (1976), Hiemenz (1986) and Visser (1976).

It is therefore essential to consider the effect of pH and ionic strength on the process of cell adhesion. Fe^{3+} at 0·05 M stimulated the adhesion of algae to glass (Nordin, J.S. 1965 cited in Visser 1976); however no significant alteration in the pattern of adhesion was observed with the addition of calcium or magnesium. Mozes *et al.*, (1987) also found Fe^{3+} to strongly influence the adsorption of yeast and bacteria to surfaces. Visser (1976) pointed out that the adhesion force for quartz and glass spheres adhering to a glass surface in salt solutions is a function of the position of the electrolyte in the lyotropic series, viz.

$$LiCl < NaCl < KCl < RbCl < CsCl$$

for monovalent ions. For divalent cations the sequence is:

$$CaCl_2 < SrCl_2 < BaCl_2$$

Experiments designed to remove adherent particles from a surface revealed that at pH 10 particles that adhered in the presence of Ca^{2+}

were more difficult to remove than particles deposited in the presence of Na^+. It was suggested that at pH 10 Ca^{2+} no longer acts to bridge negatively charged surface groups of the particle and the solid support, but rather neutralizes two adjacent moieties of negative charge on one surface, this makes it possible for the particle and the support to come much closer together resulting in a stronger adhesive force.

Gordon & Millero (1984) found that at low ionic strength (<0·1 M) the adsorption of the bacterium *Vibrio alginolyticus* to hydroxyapatite increased with increasing ionic strength, in accord with DLVO theory. At increased ionic strength (>0·2 M) bacterial adhesion decreased. Abbott *et al.* (1983) made similar observations using *Streptococcus mutans* adherent to glass. Lance & Gerba (1984) also demonstrated that the effect of ionic composition of suspending solutions on virus adsorption to soil was in accord with DLVO theory predictions. Mozes *et al.* (1987) assessed the adhesion of yeast and bacteria to solid surfaces in response to electrostatic and non-electrostatic interactions. It was found that the yeast *Saccharomyces cerevisiae* did not adhere to any of the test solid surfaces including glass and a range of polymeric materials unless the materials were treated with Fe^{3+}. With *Acetobacter aceti* (Mozes *et al.*, 1987) adhesion to one surface only was low at pH 3·5 in the absence of Fe^{3+} treatment. However, exposure to Fe^{3+} improved markedly the adsorption of bacteria to the test surfaces. In the case of *Moniliella pollinis*, a fungus with relatively hydrophobic cell wall characteristics, adhesion was favoured on hydrophobic surfaces. Adhesion increased significantly as the pH of the suspending liquid was decreased from 5 to 4 as would be expected. This was attributed to decreased cell/solid-support electrostatic repulsion caused by decreased surface charge as measured by electrophoretic mobility.

Similar results have been obtained for cultured plant cells adherent to a range of polymeric surfaces including fluorinated ethylenepropylene, polyethylene terephthalate, polystyrene, sulphonated polystyrene, and glass. The electrophoretic mobility of cultured cells of *C. roseus* decreases as expected in response to increased ionic strength of the ions Na^+, Ca^{2+} and Al^{3+}. Also as expected, the negative slope of the decrease increased with increasing cation valency, in accord with DLVO predictions. In 1·0 M $CaCl_2$ or 0·1 $AlCl_3$ the electrophoretic mobility approached zero; however up to 1·0 M NaCl decreased the electrophoretic mobility only slightly. In solutions containing concentrations of $AlCl_3$ greater than 0·1 M charge reversal from $-2·25 \times 10^{-8} m^2 V^{-1} s^{-1}$ to $1·5 \times 10^{-8} m^2 V^{-1} s^{-1}$ was observed. Cell adhesion experiments were performed on a range of supports of widely differing surface tensions (hydrophobicity) in solutions of various concentrations of NaCl, $CaCl_2$, or $AlCl_3$. Cells of

C. roseus adsorbed to the more hydrophobic surfaces preferentially when the surface tension of the liquid was less than that of the cells as predicted by the thermodynamic model, but also the maximum level of adhesion occurred in the presence of 0.1 M $AlCl_3$. This trend was apparent on all solid support surfaces. Adhesion also decreased however, when the concentration of $AlCl_3$ was either greater or less than 0.1 M. It will be recalled that in 0.1 M $AlCl_3$ solution the electrophoretic mobility of the cells was close to zero; at this point electrostatic repulsion between the cells and the solid surfaces was relatively low resulting in increased adhesion. With increasing concentrations of $AlCl_3$ above 0.1 M, up to saturating levels at pH near 5, adsorption decreased because adsorption of the ion to the cell and support surfaces would result in charge reversal from negative to positive sign of charge. It is expected that electrostatic repulsion between the cells and the support would then ensue. The decreased adhesion observed lends support to this view (Facchini *et al.*, 1989*a,b*). The presence of any of the salts at any concentration used did not alter the surface tension of the liquid by more than 0.5 mJ m^{-2}.

Thus, if a surface (e.g. glass) which carries a negative sign of charge is treated with colloidal alumina, a monolayer of alumina particles will be deposited onto the glass and the net charge will be neutralized; adsorption of increasing levels of alumina is unfavourable. Excess alumina may be rinsed away leaving a monolayer of alumina particles. Consequently, a layer of cells may be deposited on the surface of the alumina monolayer. The technique has been exploited with other positively charged particles (Mozes & Rouxhet, 1985; Mozes *et al.*, 1987; van Haecht *et al.*, 1985) to yield immobilized cells of various types. Shimada & Shimahara (1989) noted that *Escherichia coli* K-12 cells would adsorb efficiently onto the surfaces of ceramic powders, such as γ-alumina, silica, titania, zirconia or glass in either acidic or alkaline solution in the presence of the powerful cationic flocculant poly-(methacryl-oxyethyl trimethylammonium chloride). However, in the absence of the flocculant adsorption was poor at pH 9. An explanation for these results could be that at pH 9 both the cells and the support surfaces have relatively high negative zeta potentials, and consequently, electrostatic repulsion would not favour efficient adsorption.

The data presented clearly indicate that the adsorption of cells to surfaces may be either increased or decreased if the surface tensions of the cell, liquid or solid support are specifically modified, or if the support materials and cells are treated under appropriate conditions with agents that modify the electrostatic condition of the system. The following conclusions can be made regarding the adsorption of cells in suspension to various solid support carriers and several conditions must be fulfilled and optimized. It is clear that adsorption of cells to surfaces can indeed be predicted, controlled and modified. The technique is easily applicable to a wide range of cells under mild and physiological conditions. However, the process seems to be under-utilized, and certainly under-developed for industrial applications. The following observations may be of some use to those interested in applying the technique.

1. The initial phase of cell adsorption is dependent on thermodynamic and electrostatic phenomena prevalent in the system. Adhesion will be greatest to hydrophobic surfaces when the surface tension of the liquid is greater than the surface tension of the cells; however, when the surface tension of the suspending liquid is less than the surface tension of the cells, adhesion will be greatest to hydrophillic surfaces.

2. When the surface tension of the liquid, or of the solid-support is close to, or equals that of the cells, then adhesion may be strongly influenced by the electrostatic condition of the system.

3. The extent of adhesion to surfaces will be increased when the negative zeta potential of the cells and support are effectively neutralized. This can be accomplished by the use for cations such as Ca^{2+}, Fe^{3+}, Al^{3+}, or cationic flocculants such as alkyl-substituted quaternary ammonium salts.

4. The hydrophilic properties of cells are modulated throughout the growth cycle, it is therefore important to determine at what stage of the growth cycle the relative hydrophilicity of the cell is at the optimum level for the application.

5. Although cells adhere to surfaces under static conditions it is beneficial to provide some degree of mild stirring or agitation to improve the rate and extent of adhesion.

6. The initial inoculum-load is important in determining the rate of adhesion to surfaces.

Several benefits for the production of biotechnological products can be realized through the use of immobilized biocatalysts. The benefits include, high biocatalyst densities with high biocatalytic activity, possible reuse of the biocatalyst, continuous or semi-continuous operation, maintenance of the biocatalyst in the most active phase for extended periods, and improved downstream processing. An efficient, and inexpensive cell immobilization strategy will no doubt help to stimulate the development of commercial-scale processes for the production of high-value-added ~oduct. DiCosmo and co-workers investigated the fundamental mechanisms that influenced the process of cell adhesion to various polymeric materials with a view toward developing a plant cell immobilization process based on spontaneous cell adhesion (DiCosmo *et al.*, 1988; Facchini *et al.*, 1988*a,b*, 1989*a,b*); a similar strategy was used by Archambault *et al.* (1989). Based on thermodynamic and electrostatic considerations, it was found that indeed, adhesion of cultured plant cells to polymeric surfaces could be predicted, modified and controlled. Subsequently, the spontaneous adhesion of cultured plant cells to glass fibre mats having various surface treatments was evaluated. The glass fibre mat was chosen to provide a large surface area for efficient and high levels of cell immobilization, good flow-through characteristics, chemical inertness and inexpensive process development under a variety of culture conditions.

4.4 IMMOBILIZATION OF CULTURED PLANT CELLS TO GLASS FIBRE MATS

The predictive value of the thermodynamic model of cell adhesion was applied to the spontaneous adhesion of cultured plant cells to several, chemically modified glass fibres (Table 4.1). The thermodynamic model predicts that for relatively hydrophilic cells, such as those of *Catharanthus roseus* (Facchini *et al.*, 1988*a*), adhesion would be greatest to the glass fibre with the greatest surface tension, i.e. the untreated glass, when the cells are suspended in nutrient media. Indeed, Fig. 4.3 demonstrates the predictions of the model. The untreated glass fibre mat with a surface tension of 60·1 mJ m^{-2} demonstrated the greatest efficiency as a cell immobilization substrate. In essence, the data presented in Fig. 4.3 confirm the theoretical predictions of the thermodynamic model in a practical application. The importance of surface tensions in mediating cell adsorption and

Table 4.1

Surface coatings, water contact angles, and surface tensions of the fibreglass substrates used for immobilization experiments

Coating	Contact angle θH_2O (degrees)	Surface tension (mJ m^{-2})
PS = phenyl	85·7 ± 2·6	31·5 ± 1·7
PG = polyglycol	73·7 ± 7·9	36·8 ± 7·4
CHO = aldehyde	62·5 ± 3·8	45·7 ± 2·2
CTMS = alkyl	52·5 ± 1·6	51·5 ± 1·3
AW = silanol	48·4 ± 6·1	53·9 ± 3·6
HC = untreated glass	37·5 ± 4·4	60·1 ± 2·2

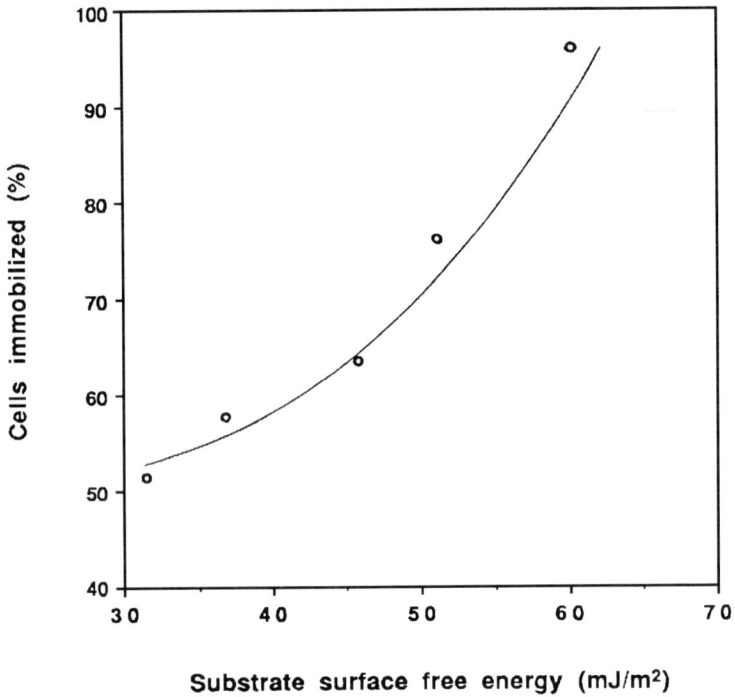

Fig. 4.3. Immobilization of *Catharanthus roseus* cultured cells to glass fibre mats with various surface coatings yielding a range of surface free energies.

immobilization is evident. Due to the morphology of the glass fibre support entrapment of cells also occurs. The two processes taken together allow for relatively high biomass loading onto the solid support.

Figure 4.4 illustrates the amount of plant cells immobilized onto a 1·3 g mass of glass fibre material as a function of the inoculum load (Facchini & DiCosmo, 1990). It is evident that over 90% of the cells were effectively immobilized when 2·5 g (fresh weight) of cells, or less were used as the inoculum load. However, when 3 g of cells were used the loading efficiency declined to approximately 55%. The loading efficiency for different cell masses is given in Fig. 4.4. The maximum loading efficiency in the system is approximately 90% with a ratio of 1·9 g cells per gram of fibre glass material.

The various surface modifications of the fibreglass materials significantly affected the loading efficiency as demonstrated in Fig. 4.5. The greatest practical immobilization efficiency was provided by the untreated glass fibre surface coating; for this reason the untreated glass fibre support was chosen. Two morphologies of the support material are

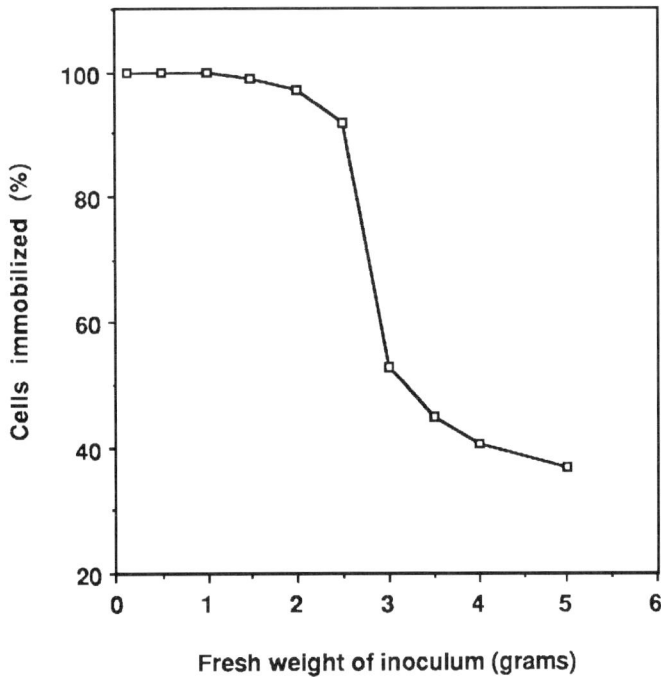

Fig. 4.4. Immobilization of *Catharanthus roseus* cells onto the untreated glass fibre substrate (HC). Immobilization efficiency was determined after four days in culture.

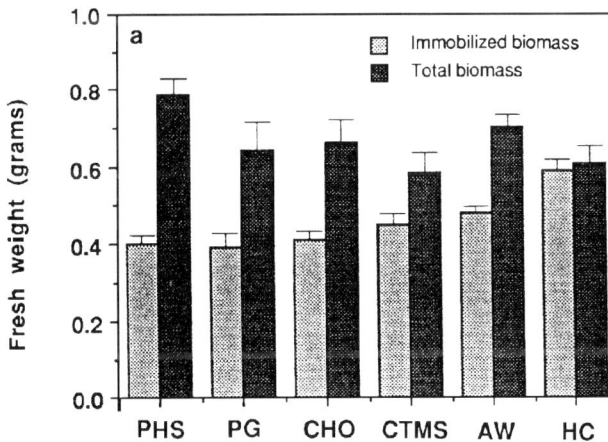

Fig. 4.5. Effect of the glass fibre surface coatings on total culture growth of immobilized and non-immobilized cells and the actual immobilized biomass of *Catharanthus roseus*. Surface coatings: PHS – phenyl; PG – polyglycol; CHO – aldehyde; CTMS – alkyl; AW – silanol; HC – untreated glass. Error bars are the standard deviation from the mean value.

A   B

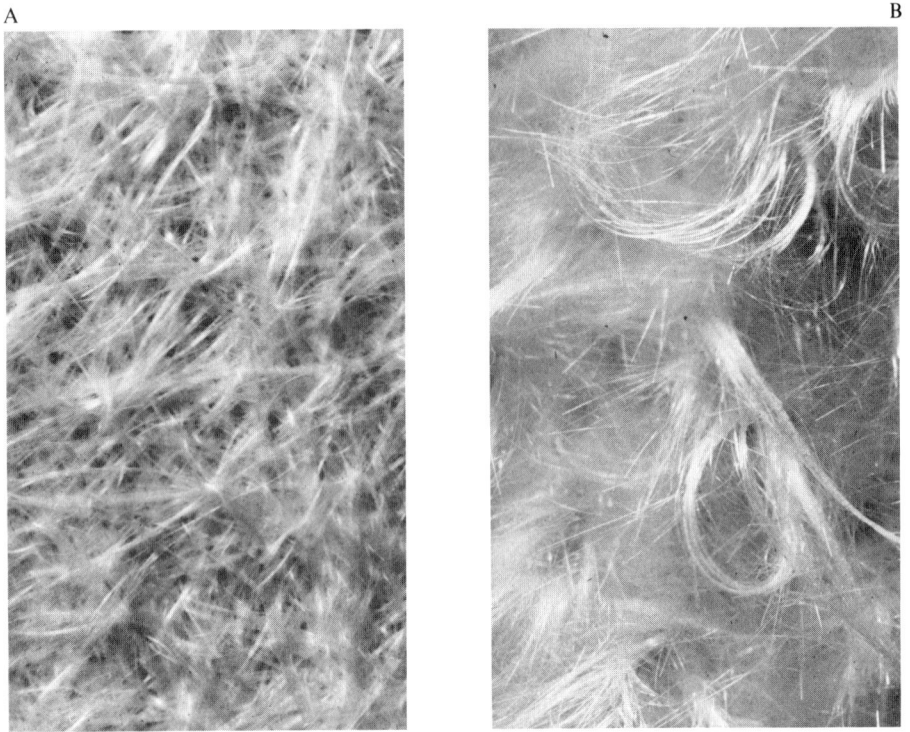

Fig. 4.6. Two morphologies of the glass fibre immobilization substrate.

illustrated in Fig. 4.6, and Fig. 4.7 illustrates the material in Fig. 4.6B with a full load of adherent *Catharanthus roseus* cells after 14 days in culture. The practical utility of the glass fibre mats for cell immobilization strategies was verified further by incorporating the material into a 5 l bioreactor (Fig. 4.8). Facchini & DiCosmo (1990, 1991*a,b*) examined the production of indole alkaloids by *C. roseus*, and berberine and columbamine by *Thalictrum rugosum* cells immobilized to the glass fibre mats. In Fig. 4.8A, *C. roseus* cells were inoculated into the system; agitation was provided by a stirring-bar impeller and by mild aeration. After 24 hr (Fig. 4.8B) a portion of the cells were immobilized as seen by a decrease in the turbidity of the culture. After a further 48 hr (Fig. 4.8C), the cells are completely immobilized onto the fibre mats. The immobilization was firm and irreversible, and the system was maintained for an extended period.

The efficiency of cell immobilization by the glass fibre mat material is a generally applicable phenomenon. The technique has been used to immobilize *C. roseus* (Facchini & DiCosmo, 1990) and *Thalictrum rugosum*

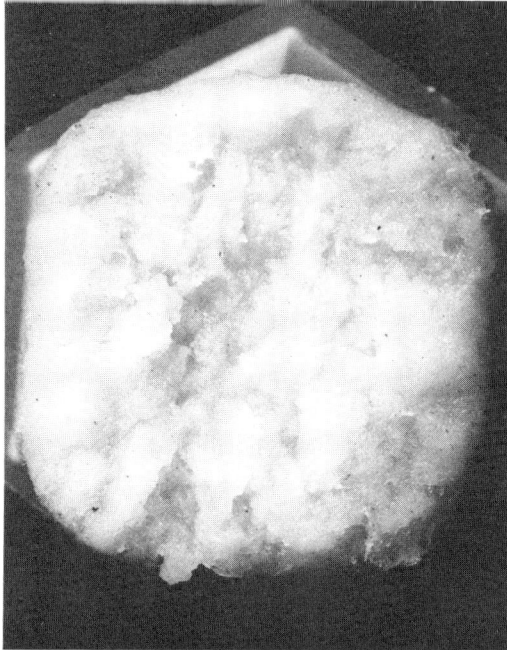

Fig. 4.7. Surface view of *Catharanthus roseus* cells immobilized onto the glass fibre material in Fig. 4.6B.

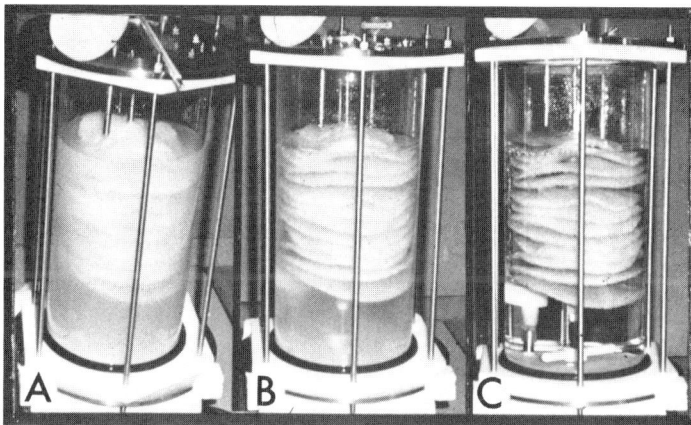

Fig. 4.8. The immobilization of *Catharanthus roseus* cells to glass fibre mats in a 5 l bioreactor; (A) At time of inoculation; (B) 24 hr post-inoculation; (C) 72 hr post-inoculation.

(Facchini & DiCosmo, 1991*a,b*). The immobilization process was shown to involve non-specific adhesion mechanisms because the immobilization efficiency was a function of the surface tension of the glass fibres. However, the woven configuration of the glass fibres may have resulted in the passive entrapment of cells also, particularly the larger cell aggregates. The distribution of cells within the glass fibre matrix, and a complete description of the interaction of adhesion and entrapment mechanisms during the immobilization process, remain to be fully defined. However, the success of the immobilization technique demonstrates the industrial potential of the strategy, and provides a system to answer questions regarding the physiology and metabolism of adherent plant cells.

The accumulation of indole alkaloids and tryptamine, an alkaloid precursor, by glass fibre-immobilized cells of *C. roseus* was lower than in suspension-cultured cells. However, the accumulation of berberine, a benzylisoquinoline alkaloid by immobilized and suspension cultured cells of *T. rugosum* was similar to that found in suspension. The growth rate of immobilized cells of *C. roseus* and *T. rugosum* was reduced in comparison to suspended cells. The effect of immobilization on the accumulation of metabolites by cultured plant cells is probably a function of the species, and the nature of the biosynthetic pathway. Subsequent research should determine the effect of the immobilization method on the accumulation of secondary metabolites in various other species. In addition, the factors that affect the physiology of biofilm formation, viz. increased aggregation, reduced oxygen transfer, and pH changes must be elucidated.

The reported increase in the accumulation of indole alkaloids by cells of *C. roseus* cells entrapped in alginate beads may occur because of phosphate limitation, rather than the process of immobilization. Glass fibre-immobilized *C. roseus* cells, like those immobilized using all other non-alginate techniques show decreased production of indole alkaloids. Continued investigation of the biosynthesis of secondary products in glass fibre-immobilized cells will provide further insight into the regulation of metabolism in cultured cells, and improve the potential for the successful commercial exploitation of immobilized plant cell systems.

The immobilization of cultured plant cells using a glass fibre support and a predictive model of cell adhesion offers advantages over other immobilization methods. It demonstrates that the strategy deserves further development. The relatively simple, but successful, incorporation of the immobilization strategy into a simple laboratory-scale bioreactor should provide encouragement to the plant cell fermentation industry. The limited success of suspension-cultured plant cells for the production of secondary products is, in many cases, primarily the result of the low product yields and high cost of batch-culture techniques. The potential for the development of a semi-continuous or continuous-culture system that use immobilized cultured plant cells on biochemically inert, potentially reusable, and inexpensive glass fibres was demonstrated by *T. rugosum* cultivated in a 4·5 l bioreactor with prolonged berberine accumulation (Facchini & DiCosmo, 1991). However, the selection of high producing cell lines and the optimization of culture conditions are necessary in order to develop an economically feasible system.

4.5 REFERENCES

Abbott, A., Rutter, P. R. & Berkeley, R. C. W. (1983). The influence of ionic strength, pH and a protein layer on the interaction between *Streptococcus mutans* and glass surfaces. *J. Gen. Microbiol.*, **129**, 439–45.

Absolom, D. R., Lamberti, F. V., Policova, Z., Zingg, W., van Oss, C. J. & Neumann, A. W. (1983). Surface thermodynamics of bacterial adhesion. *Appl. Environ. Microbiol.*, **46**, 90–7.

Absolom, D. R., Zingg, W. & Neumann, A. W. (1985). Surface thermodynamics of cellular and protein interactions. In *Comprehensive Biotechnology, Vol. 4* (Robinson, C. W. & Howell, J. A., eds). Pergamon Press, Oxford, pp. 433–46.

Adamson, A. W. (1982). *Physical Chemistry of Surfaces*. 4th edn. John Wiley & Sons. New York, 664 pp.

Archambault, J., Volesky, B. & Kurz, W. G. W. (1989). Surface immobilization of plant cells. *Biotechnol. Bioeng.*, **33**, 293–399.

Arcuri, E. J. (1982). Continuous ethanol production and cell growth in an immobilized-cell bioreactor employing *Zymomonas mobilis*. *Biotechnol. Bioeng.*, **24**, 595–604.

Betinec, D. L. & Mendelson, N. H. (1988). Immobilization of *Bacillus subtilis* macro–fibers by entanglement. *Eng. Microb. Technol.*, **10**, 418–25.

Black, G. M., Webb, C., Mathews, T. M. & Atkinson, B. (1984). Practical reactor systems for yeast cell immobilization using biomass support particles. *Biotechnol. Bioeng.*, **26**, 134–41.

Bornman, C. H. & Zachrisson, A. (1987). Immobilization of plant protoplasts using microcarriers. In *Methods in Enzymology, Vol. 135*, Academic Press, New York, pp. 421–33.

Buchs, J., Mozes, N., Wandrey, C. & Rouxhet, P. G. (1988). Cell adsorption control by culture conditions. *Appl. Microbiol. Biotechnol.*, **29**, 119–28.

Bucke, C. (1987). Cell immobilization in calcium alginate. In *Methods in Enzymology, Vol. 135* (Mosbach, K., ed.). Academic Press, New York, pp. 175–89.

Busscher, H. J., Weerkamp, A. H., van der Mei, H. C., van Pelt, A. W. J., de Jong, H. P. & Arends, J. (1984). Measurement of the surface free energy of bacterial cell surfaces and its relevance for adhesion. *Appl. Environ. Microbiol.*, **48**, 980–3.

Champluvier, B., Kamp, B. & Rouxhet, P. G. (1988). Immobilization of β-galactosidase retained in yeast: adhesion of the cells on a support. *Appl. Microbiol. Biotechnol.*, **27**, 464–9.

Changui, C., Doren, A., Stone, W. E. E., Mozes, N. & Rouxhet, P. G. (1987). Surface properties of polycarbonate and promotion of yeast cells adhesion. *J. Chim. Phys.*, **84**, 275–81.

Chibata, I., Tosa, T. & Sato, T. (1986). Biocatalysis: immobilized cells and enzyme. *J. Mol. Catal.*, **37**, 1–24.

Chibata, I., Tosa, T., Sato, T. & Mori, T. (1976). Production of L–amino acids by aminocyclase adsorbed on DEAE–Sephadex. In *Methods in Enzymology, Vol. XLIV* (Mosbach, K., ed.). Academic Press, New York, pp. 746–59.

Daly, J. G. & Stevenson, R. M. W. (1987). Hydrophobic and haemagglutonating properties of *Renibacterium salmoninarum*. *J. Gen. Microbiol.*, **13**, 3575–80.

DiCosmo, F., Facchini, P. J. & Neumann, A. W. (1988). Does the spontaneous adhesion of cultured plant cells to polymer surfaces have potential as an immobilization technique? *Trends Biotechnol.*, **6**, 137–40.

DiCosmo, F., Facchini, P. J. & Neumann, A. W. (1989). Plant cell adhesion to surfaces as predicted by a thermodynamic model and modified by electrostatic interaction. *Coll. Surfaces*, **42**, 255–69.

Doyle, R. J., Nesbitt, W. E. & Taylor, K. G. (1982). On the mechanisms of adherence of *Streptococcus sanguis* to hydroxyapatite. *FEMS Microbiol. Lett.*, **15**, 1–5.

Duddridge, J. E., Kent, C. A. & Lewis, J. F. (1985). Effect of surface shear stress on the attachment of *Pseudomonas fluorescens* to stainless steel under defined flow conditions. *Biotechnol. Bioeng.*, **24**, 153–64.

Dunn, I. J. (1987). Methods for selecting and growing mixed cultures in biofilm fluidized sand beds. In *Methods in Enzymology, Vol. 135* (Mosbach, K., ed.). Academic Press, New York, pp. 300–7.

Efremov, J. F. (1976). Periodic colloid structures. In *Surface and Colloid Science, Vol. 8* (Matijevic, E., ed.). John Wiley & Sons, New York, pp. 85–192.

Ehrlich, H. L. & Holmes, D. S. (eds). (1986). Biotechnology for the mining, metal-refining, and fossil fuel processing industries. *Biotechnol. Bioeng. Symp.*, **16**, 386 pp.

Facchini, P. J. & DiCosmo, F. (1990). Immobilization of cultured *Catharanthus roseus* cells using a fibreglass substratum. *Appl. Microbiol. Biotechnol.*, **33**, 36–42.

Facchini, P. J. & DiCosmo, F. (1991*a*). Plant cell bioreactor for the production of protoberberine alkaloids from immobilized *Thalictrum rugosum* cultures. *Biotechnol. Bioeng.*, **37**, 397–403

Facchini, P. J. & DiCosmo, F. (1991*b*). Secondary metabolite biosynthesis in cultured cells of *Catharanthus roseus* (L.) G. Don immobilized by adhesion to glass fibres. *Appl. Microbiol. Biotechnol.*, **35**, 382–92.

Facchini, P. J., DiCosmo, F., Radvanyi, L. C. & Giguere, Y. (1988*a*). Adhesion of *Catharanthus roseus* cells to surfaces: effects of substrate hydrophobicity. *Biotechnol. Bioeng.*, **32**, 935–8.

Facchini, P. J., Neumann, A. W. & DiCosmo, F. (1988*b*). Thermodynamic aspects of plant cell adhesion to polymer surfaces. *Appl. Microbiol. Biotechnol.*, **29**, 346–55.

Facchini, P. J., Neumann, A. W. & DiCosmo, F. (1989*a*). The effect of inoculum concentration on the kinetics and extent of cultured plant cell adhesion to polymer surfaces under static conditions. *Biotechnol. Appl. Biochem.*, **11**, 74–82.

Facchini, P. J., Neumann, A. W. & DiCosmo, F. (1989*b*). Adhesion of suspension-cultured *Catharanthus roseus* cells to surfaces: effect of pH, ionic strength, and cation valency *Biomaterials*, **10**, 318–24.

Flörsheimer, A., Kula, M.-R., Schutz, H. J. & Wandrey, C. (1989). Continuous production of kyotorphin. *Biotechnol. Bioeng.*, **33**, 1400–5.

Gbewonyo, K., Meier, J. & Wang, D. I. C. (1987). Immobilization of mycelial cells on celite. In *Methods in Enzymology, Vol. 135* (Mosbach. K., ed.). Academic Press, New York, pp. 318–33.

Gibbons, R. J. (1980). Adhesion of bacteria to the surfaces of the mouth. In *Microbial Adhesion to Surfaces* (Berkeley, R. C. W., Lynch J. M., Melling, J., Rutter, P. R. & Vincent, B., eds). Ellis Horwood, Chichester, pp. 351–78.

Gordon, A. S. & Millero, F. J. (1984). Electrolyte effects on attachment of an estuarine bacterium. *Appl. Environ. Microbiol.*, **47**, 495–9.

Hartmeir, W. (1988). *Immobilized Biocatalysts. An Introduction.* Springer-Verlag, New York, USA.

Hattori, R. & Hattori, T. (1985). Adsorptive phenomena involving bacterial cells and an ion exchange resin. *J. Gen. Appl. Microbiol.*, **31**, 147–63.

Hiemenz, P. C. (1986). *Principles of Colloid and Surface Chemistry*, 2nd Edn. Marcel Dekker Inc., New York, 815 pp.

Iamamoto, M. S. & Gushikem, Y. (1989). Adsorption of metal ions from aqueous and ethanol solutions by silica gel functionalized with pyridinium ions. *J. Coll. Interface Science*, **129**, 162–5.

Iler, R. K. (1973). Colloidal silica. In *Surface and Colloid Science, Vol. 6* (Matijevic, E., ed.). John Wiley & Sons, New York, pp. 1–100.

Keller, L. (1982). Acid–bacterial and ferric sulfate leaching of pyrite single crystals. *Biotechnol. Bioeng.*, **24**, 83–96.

Kennedy, J. F. & Cabral, J. M. S. (1987). Immobilization of enzymes on transition metal-activated supports. In *Methods in Enzymology, Vol. 135* (Mosbach, K., ed.). Academic Press, New York, pp. 117–30.

Keshavarz, T., Walker, E., Elgin, R., Lilly, G., Holt, G., Bull, A. T. & Lilly, M. D. (1989). Immobilization of *Penicillium chrysogenum*: spore growth on celite. *Appl. Microbiol Biotechnol.*, **30**, 487–91.

Kim, D. W. *et al.* (1988). Adsorption of cellulase from *Trichoderma viride* on microcrystalline cellulose. *Appl. Microbiol. Biotechnol.*, **28**, 148–54.

Klein, J. & Vorlop, K.-D. (1985). Immobilization techniques-cells. In *Comprehension Biotechnology, Vol. 2* (Cooney, C. L. & Humphrey, A. E., eds). Pergamon Press. New York, pp. 203–25.

Kolster, A. & Beiderbeck, R. (1987). Conditions for the attachment of *Calystegia* cells to solid substrates. *Ann. Bot.*, **59**, 471–3.

Lamptey, J. & Moo-Young, M. (1987). Immobilization of yeast cells by adsorption on wood chips. In *Methods in Enzymology, Vol. 135* (Mosbach, K., ed.). Academic Press, New York, pp. 333–40.

Lance, J. C. & Gerba, C. P. (1984). Effect of ionic composition of suspending solution on virus adsorption by a soil column. *Appl. Environ. Microbiol.*, **47**, 484–8.

Lewis, J. A., Black, G. M., Mavituna, F. & Wilkson, A. (1986). The application of biomass support particles to the immobilization of plant cell cultures. In *Process Engineering Aspects of Immobilized Cell Systems* (Webb, C., Black,

G. M., Atkinson, B., eds.). The Institution of Chemical Engineers, Rugby, U.K, pp. 290–94.

Li, D., Gaydos, J. & Neumann, A. W. (1989). The phase rule for systems containing surfaces and lines. I. Moderate curvature. *Langmuir*, **5**, 1133–40.

Lindsey, K. & Yeoman, M. M. (1987). Techniques for the immobilization of plant cells. In *Methods in Enzymology, Vol. 135* (Mosbach, K., ed.). Academic Press, New York, pp. 410–21.

Messing, R. A. (1969). Molecular inclusions. Adsorption of macromolecules on porous glass membranes. *J. Am. Chem. Soc.*, **91**, 2370–1.

Messing, R. A. (1974). A stannous bridge for coupling urease to controlled pore titanium *Biotechnol. Bioeng.*, **16**, 1419–23.

Messing, R. A. 1985. Immobilization techniques-enzymes. In *Comprehensive Biotechnology, Vol. 2* (Cooney, C. L. & Humphrey, A. E., eds). Pergamon Press, New York, pp. 191–201.

Mosbach, K. (ed.) (1976). *Methods in Enzymology, Vol. 44.* Academic Press, New York, p. 999.

Mosbach, K. (ed.) (1987*a*). *Methods in Enzymology, Vol. 135.* Academic Press, New York, p. 675.

Mosbach, K. (ed.) (1987*b*). *Methods in Enzymology, Vol. 136.* Academic Press, New York, p. 584.

Mosbach, K. (ed.) (1988). *Methods in Enzymology, Vol. 137.* Academic Press, New York, p. 767.

Mozes, N., Marchal, F., Hermesse, M. P., van Haecht, J. L., Reuliaux, L., Leonard, A. J. & Rouxhet, P. G. (1987). Immobilization of microorganisms by adhesion: interplay of electrostatic and non–electrostatic interactions. *Biotechnol. Bioeng.*, **30**, 439–50.

Mozes, N. & Rouxhet, P. G. (1985). Metabolic activity of yeast immobilized as support monolayer. *Appl. Microbiol. Biotechnol.*, **22**, 92–7.

Nilsson, K., Brodelius, P. & Mosbach, K. (1987). Entrapment of microbial and plant cells, in beaded polymers. In *Methods in Enzymology, Vol. 135.* Academic Press, New York, pp. 222–30.

Neumann, A. W., Absolom, D. R., Zingg, W. & van Oss, C. J. (1979). Surface thermodynamics of leukocytes and platelet adhesion to polymer surfaces. *Cell Biophysics*, **1**, 79–82.

Neumann, A. W., Absolom, D. R., Francis, D. W. & van Oss, C. J. (1980*a*). Conversion tables of contact angles surface tensions. *Sep. Purif. Methods.*, **9**, 69–163.

Neumann, A. W., Hum, O. S., Francis, D. W., Zingg, W. & van Oss, C. J. (1980*b*). Kinetic and thermodynamic aspects of platelet adhesion from suspension to various substrates. *J. Biomed. Mater. Res.*, **14**, 499–509.

Neumann, A. W., Good, R. J., Hope, C. J. & Sejpal, M. (1974). An equation-of-state approach to determine surface tensions of low-energy solids from contact angles. *J. Colloid Interface Sci.*, **49**, 291–304.

O'Driscoll, K. F. (1976). Techniques of enzyme entrapment in gels. In *Methods in Enzymology, Vol. XLIV* (Mosbach, K., ed.). Academic Press, New York, pp. 169–83.

Parnas, R. S., Chaimberg, M., Taepaisitphongse, V. & Cohen, Y. (1989). The adsorption of polyvinylpyrrolidone and polyethylene oxide onto chemically modified silica. *J. Colloid Interface Sci.*, **129**, 441–50.

Pratt-Tepstra, I. H., Weerkamp, A. H. & Busscher, H. J. (1987). Adhesion of oral streptococci from a flowing suspension to uncoated and albumin-coated surfaces. *J. Gen. Microbiol.*, **133**, 3199–206.

Pratt-Tepstra, H., Weerkamp, A. H. & Busscher, H. J. (1989). Microbial factors in a thermodynamic approach of oral streptococcal adhesion to solid substrata. *J. Colloid Interface Sci.*, **129**, 568–74.

Ridgeway, H. F., Kelly, A., Justice, C. & Olson, B. H. (1983). Microbial fouling of reverse osmosis membranes used in advanced waste-water treatment technology: chemical, bacteriological and ultrastructural analysis. *Appl. Environ. Microbiol.* **45**, 1066–84.

Rosenberg, M., Gutnick, D. & Rosenberg, E. (1980). Adherence of bacteria to hydrocarbons: a simple method for measuring cell surface hydrophobicity. *FEMS Microbiol. Lett.*, **2**, 29–33.

Sakurai, H., Lee, H. W., Sato, S., Mukataka, S. & Takahashi J. (1989). Gluconic acid production at high concentration by *Aspergillus niger* immobilized on a nonwoven fabric. *J. Fermentation Bioeng.*, **67**, 404–8.

Seiskari, P., Linko, Y.-Y. & Linko, P. (1985). Continuous production of gluconic acid by immobilized *Gluconobacter oxydans* cell bioreactor. *Appl. Microbiol. Biotechnol.*, **21**, 356–60.

Shimada, K. & Shimahara, K. (1989). Application of powdered ceramics support to the production of L–tryptophan by resting *Eschenchia coli* K-12 cells. *J. Ferm. Bioeng.*, **67**, 215–18.

Shiraishi, F., Kawakami, K., Kono, S., Tamura, A., Tsuruta, S. & Kusunoki, K. (1989). Characterization of free gluconic acid by *Gluconobacter suboxydans* adsorbed on ceramic honeycomb monolith. *Biotech. Bioeng.*, **33**, 1413–18.

Skryabin, G. K. & Koshcheenko, K. A. (1987). Immobilization of living microbial cells in polyacrylamide gel. In *Methods in Enzymology, Vol. 135* (Mosbach, K., ed.). Academic Press, New York, pp. 198–216.

Smith, R. P., Absolom, D. R., Spelt, J. R. & Neumann, A. W. (1986). Approaches to determine the surface tension of small particles: equation-of-state considerations. *J. Colloid Interface Sci.*, **110**, 521–32.

Thonart, P., Custinne, M. & Paquot, M. (1982). Zeta potential of yeast cells: application in cell immobilization. *Enzyme Microb. Technol.*, **4**, 191–94.

Vega, J. L., Clausen, E. C. & Gaddy, J. L. (1988). Biofilm reactors for ethanol production. *Enz. Microb. Technol.*, **10**, 390–402.

van Haecht, J. L., M'Bombo, B. & Rouxhet, P. G. (1985). Immobilization of *Saccharomyces cerevisiae* by adhesion: treatment of the cells by Al ions. *Biotechnol. Bioeng.*, **27**, 217–24.

Visser, J. (1970). Measurement of the force of adhesion between submicron carbon-black particles and a cellulose film in aqueous solution. *J. Colloid Interface Sci.*, **34**, 26–31.

Visser, J. (1976). Adhesion of colloidal particles. In *Surface and Colloid Science, Vol. 8* (Matijevic, E., ed.). John Wiley & Sons, New York, pp. 3–84.

Ward, C. A. & Neumann, A. W. (1974). On the surface thermodynamics of a two-component liquid-vapour-ideal solid system. *J. Colloid Interface Sci.*, **49**, 286–90.

Webb. C., Black, G. M. & Atkinson, B. (eds). (1986). *Process Engineering Aspects of Immobilized Cell Systems*. Institution of Chemical Engineers, Rugby, UK.

Chapter 5

BIOFILMS, NATURALLY OCCURRING COMMUNITIES OF IMMOBILIZED CELLS

ROBERT J. C. MCLEAN,[1] DOUGLAS E. CALDWELL[2] AND
J. WILLIAM COSTERTON[3]

[1]*Department of Biology, Southwest Texas State University,
San Marcos, USA*
[2]*Department of Applied Microbiology & Food Science,
University of Saskatchewan, Saskatoon, Canada*
[3]*Center for Biofilm Engineering, College of Engineering,
Montana State University, Bozeman, USA*

CONTENTS

5.1 INTRODUCTION

The concept of using immobilized cells (addressed in other chapters of this book) has greatly enhanced the industrial scale production of a number of secondary metabolites by microorganisms. Generally the organisms are confined within a supporting matrix such as carrageenin or agar. In this environment, cellular growth is greatly reduced and the production of secondary metabolites thereby enhanced. Separation of microbial products from the immobilized culture is also facilitated due to the ease with which the relatively large clusters of cells can be removed.

Artificial cellular immobilization actually mimics a natural process. Microorganisms such as bacteria often incorporate into biofilm communities which adhere to a wide variety of surfaces. In lieu of an artificial supporting matrix, these organisms produce a variety of extracellular polysaccharide or polypeptide capsule polymers often referred to as the glycocalyx (Costerton *et al.*, 1981*b*, 1987). When growing on surfaces, biofilm or sessile bacteria are physiologically quite different from their planktonic counterparts (Marshall, 1985; Geesey & White, 1990; van Loosdrecht *et al.*, 1990). Yet virtually all knowledge of microbial physiology is based essentially on planktonic cells grown at maximum growth rates in rich liquid culture!

Biofilms are very widespread in nature. In a number of environments such as oligotrophic (nutrient poor) mountain streams, virtually all microbial activity occurs within biofilms, and the very few planktonic organisms present act as pioneers in that they enable bacteria to colonize new surfaces and form biofilms (Marshall, 1985, 1988; Geesey *et al.*, 1978). In an industrial context, biofilms are of great importance being responsible for such diverse activities as anaerobic corrosion in water pipelines and the removal of heavy metals and organic matter mineralization from waste water in trickling filters (Costerton *et al.*, 1987; McLean & Beveridge, 1990; MacLeod *et al.*, 1990).

In this chapter, we describe the formation, distribution, ecology, and physiology of these naturally occurring biofilm communities of immobilized cells. Whenever possible, we will present applications of biofilm physiology and ecology in relationship to industry. This represents a relatively new and important facet of microbiology in that the biofilm mode often represents the preferred growth strategy of bacteria in nature (Costerton *et al.*, 1987). It is our hope that we will generate an appreciation of this natural mode of bacterial growth and stimulate others to conduct investigations in this area since much basic and applied research remains to be done.

5.2   COLONIZATION AND BIOFILM FORMATION

In their natural environment, bacteria are subjected to a variety of stresses of a physical, chemical, and biological nature. Rather than being given a rich laboratory medium and culture conditions in which to grow, these organisms are often in oligotrophic environments, and must compete for any available nutrients. In addition, conditions such as temperature, oxygen concentration, Eh, water activity, chemical composition, presence of harmful chemicals, and predation often complicates matters even further (Marshall, 1985, 1988; Brock & Madigan, 1988).

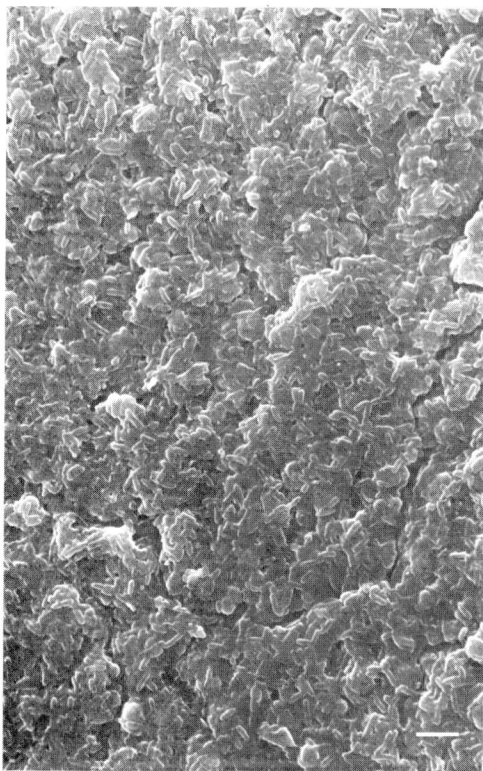

Fig. 5.1.   Bacterial colonization and growth on a surface often results in the formation of a biofilm. Biofilm growth allows the component organisms to persist in spite of many adverse environmental conditions (liquid flow and accompanying shear forces, antibacterial agents, low nutrient concentrations, etc.), and can quickly change the chemical and physical properties of the underlying surface. In this case, *Pseudomonas aeruginosa* was allowed to colonize a metal pipe from a heat exchanger for 24 hours. In this short time span, the underlying surface was completely occluded by bacterial growth. Bar represents 5 μm.

Bacteria have adopted a seemingly simple strategy for dealing with these environmental stresses in that they colonize surfaces and form encapsulated communities of cells often referred to as biofilms (Fig. 5.1) (Costerton *et al.*, 1987). This mode of growth enables bacteria to demonstrate remarkable resistance to physical disruption (Reid *et al.*, 1990), predation (Caldwell, 1984, 1987; Lawrence & Caldwell, 1987), antibiotics (Nickel *et al.*, 1985*a*; Stiekler *et al.*, 1987; Brown *et al.*, 1988; Nichols *et al.*, 1988), disinfectants (Costerton, 1984*a*; LeChevallier *et al.*, 1988*a,b*; Millward & Wilson, 1989), and the immune response (Costerton *et al.*, 1983; Costerton, 1984*b*). In addition, adhesion allows these organisms ready access to insoluble nutrients such as cellulose (Cheng & Costerton, 1980*b*; Cheng *et al.*, 1981*a*), and the ability to persist in spite of liquid shear forces such as those present in the urinary tract (McLean *et al.*, 1988) and fast-flowing aquatic environments (Geesey *et al.*, 1978).

5.2.1   Influence of Surface Properties on Colonization

Attachment is the first prerequisite for biofilm formation. Here, an organism must encounter a suitable surface, attach to it, and eventually grow in order to produce a mature biofilm. During the initial attachment process, the chemical and physical characteristics of the surface, and the surrounding medium can greatly influence the relative ease with which a particular bacterium can adhere. In some cases, bacteria are attracted to and attach to a specific target site. This specific adhesive process is quite evident in the medical field, and is one of the major reasons for the localization of some pathogens to particular tissues (Svanborg Edén *et al.*, 1985; Beachey *et al.*, 1988). It has also been shown to be responsible for the specific binding of *Rhizobium* sp. to the roots of legumes during nitrogen fixation (Dazzo & Truehet, 1983). The specificity of bacterial adhesion can be shown experimentally by removing or chemically modifying the putative target molecule(s) or by using a surface analogue with the appropriate molecular target(s). Quite often blood cells are used for this purpose as sugars on their cell surfaces are chemically well defined. Specific bacterial attraction to the erythrocyte cell surface is shown by an agglutination or clumping of the blood cells (haemagglutination) and represents a strong indication that the adhesions expressed by the organism possess a lectin-like form of activity. If no suitable blood cell type exists, other cell types or carbohydrate-coated latex beads may be substituted in the agglutination test. The role of lectins in specific binding can be further verified by a variety of hapten-inhibition studies during which a variety of sugar compounds (mimicking potential target molecules) are preabsorbed onto the bacterial cells in an effort to block adhesion (Dazzo & Truchet, 1983; Ofek & Perry, 1985; Svanborg Edén *et al.*, 1985; Beachey *et al.*, 1988). The lectins involved in specific bacterial adhesion are not always found on the bacterial cell. In the case of the *Rhizobium* legume symbiosis, lectins on the plant root hair cells recognize the carbohydrate capsule structures of the appropriate strain of *Rhizobium* (Dazzo & Truchet, 1983).

Non-specific adhesion of bacteria to surfaces is very common. As the name implies, this involves adhesion to a variety of surfaces, rather than a particular target site. Even so, a given organism may demonstrate a preference for a particular surface chemistry. The role of surface chemistry as it relates to bacterial adhesion has been an area of increasing study recently. Several investigations have shown that hydrophobic surfaces are more prone to colonization by dental and soil bacteria than are their hydrophilic counterparts (McEldowney & Fletcher, 1986; Pringle & Fletcher, 1986; Humphries *et al.*, 1987; van Loosdrecht *et al.*, 1987; Stenström, 1989). As might be expected, hydrophobic chemical interactions have been shown to play a major role in this type of bacterial colonization. In this case, adhesion can be diminished by the addition of detergents (Marshall *et al.*, 1989; Cowan *et al.*, 1987; Bryant *et al.*, 1984), or enhanced by the addition of cations (Fletcher, 1988) such as Na^+, Ca^{2+}, La^{3+}, or Fe^{3+}.

Adhesion is not a simple hydrophobic interaction. In an examination of *Streptococcus sanguis* adhesion to hydroxyapatite, Cowan *et al.* (1987) found that while hydrophobic interactions were evident, electrostatic bonding was even more significant. In this case, the addition of the electrolyte, NH_4SO_4, diminished binding. With the variety of surfaces which are colonized by bacteria, it is likely that the binding chemistry consists of a mixture of hydrophobic, electrostatic, hydrogen, van der Waals, and possibly even covalent bonds. The importance of each type of bond is influenced by the chemical environment of the surface and surrounding fluids, and the physiological status of the organisms.

The physical structure and location of the surface also plays a role in its ability to be colonized. Aquatic microorganisms tend to colonize surfaces more readily in streams than in a pond (Mills & Maubery, 1981). Presumably the higher water flow of the stream carried more organisms past these rocks and also created a stress factor in which these organisms were obliged to adhere in order to persist in this environment. Ferris *et al.* (1989*b*) established that colonization of rocks in a freshwater stream was inversely proportional to the hardness of the rock surface being colonized. They attributed this to an increased weathering of the softer rocks which in turn produced a rougher surface more prone to colonization. Baker (1984) demonstrated that colonization of materials in a natural environment such as a stream was enhanced by the roughness of the particular surface. Hydrophobicity in this environment was not as important. Certainly liquid flow rates which are reflected in very high surface shear forces are quite important in determining adhesion. At very high shear forces, the organisms showed an increasing tendency to either not attach or to be swept away. Adherent microorganisms in regions of high shear forces such as the plaque organisms within the mouth must develop very tenacious means of attachment (Christersson et al., 1988, 1989).

While there is ample evidence that physical and chemical properties of surfaces will greatly influence the rate of initial bacterial colonization, we have seen that over time virtually any surface will become colonized (Costerton et al., 1987). There are a number of explanations for this. Certainly, after long periods (24–72 h) of colonization by planktonic bacteria, a monolayer will be formed even in the slowest adhesion systems and subsequent colonization will be influenced by the surface of this monolayer and not that of the original surface. If one attempts to ambush colonizing bacteria by incorporating an antibacterial agent onto a surface, the initial colonizers will be killed. However, should these dead cells remain attached to the surface, future colonizers will only experience and attach to this layer of dead bacteria and not encounter the original surface (Chang & Merritt, 1991).

5.2.2 Bacterial Structures Involved in Adhesion

Bacteria possess a wide variety of mechanisms that allow them to attach to surfaces and persist in various environments. These include the more familiar structures such as fimbriae (pili), adhesins and capsules, as well as the less familiar structures including various holdfast structures, stalks, prostheca, cell wall components and slimes. From a mechanistic viewpoint, the role of these various structures in the adhesion of bacteria to a variety of surfaces has been the subject of a large number of studies recently. Of equal or greater importance is the concept of overall microbial strategies during colonization. We now turn our attention to these two ideas.

Microbial colonization requires that the organism(s) encounter an appropriate surface. The target may be a source of potential nutrition such as a food particle in the digestive tract of a ruminant (Cheng & Costerton, 1980b; Cheng et al., 1981a) or an inanimate object such as a polystyrene surface (McEldowney & Fletcher, 1986) which the organism

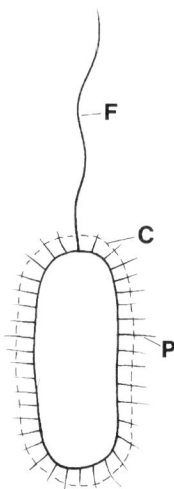

Fig. 5.2. Schematic representation of a bacterial cell showing the structures most responsible for initial surface adhesion. Motility usually arises from the propeller-like action of one or more flagella (F) (singular – flagellum). Adhesion generally occurs through the binding of pili (P) and/or the capsule (C) (also called the glycocalyx) to the surface. The capsule is a highly hydrated polysaccharide or polypeptide structure which can extend up to 10 μm from the cell surface. It represents the most significant structure in a mature biofilm.

would use as an anchor. Should some distance separate the organism from its particular target, then the ability to swim through a liquid to the target represents an obvious advantage. The most common bacterial structures involved in motility are flagella (singular – flagellum). These are proteinaceous whip-like structures, anchored in the cell membrane, which extend some distance from the cell surface (Fig. 5.2) (Beveridge, 1981). Motility arises when these structures are rotated by the cell. This overall process closely resembles the action of a propeller driving a boat (Larsen *et al.*, 1974).

In an elegant experiment, Korber *et al.* (1989) demonstrated that motility greatly enhanced the ability of *Pseudomonas fluorescens* to colonize glass surfaces in a laminar flow cell. In this experiment, motile and non-motile strains of *Ps. fluorescens* were added to a glass laminar flow cell and subjected to a flow velocity above (120 μm sec^{-1}) and below (8 μm sec^{-1}) the organism's maximum motility rate (85 μm sec^{-1}). In both cases, motility increased the ability of the organisms to adhere, enhanced dispersion over the surface, and increased the rate of colonization upstream, against the laminar flow. In other studies, positive chemotaxis, where the organisms are attracted to chemical(s) on a particular surface, has also been implicated in bacterial adhesion (reviewed in Finlay &

Falkow (1989)). Aside from playing a role in motility and chemotaxis, flagella themselves have also sometimes been directly implicated in adhesion (Marshall *et al.*, 1971; Belas & Colwell, 1982; Lawrence *et al.*, 1987). The long extension of flagella from the bacterial cell often enables these proteinaceous structures to come in contact with surfaces and to tether the cell to the surface.

The two major structures of bacteria most often associated with adhesion are fimbriae (also called pili), and the capsule (Duguid *et al.*, 1955; Fader & Davis, 1980; Costerton *et al.*, 1981*a,b*, 1987; Eighmy *et al.*, 1983; Doig *et al.*, 1987, 1988; Finlay & Falkow, 1989). Fimbriae are proteinaceous rods which extend out from the cell surface (Fig. 5.2). They are smaller structures than flagella having a typical dimension of 0·2–2 μm in length by 7 nm diameter (Eisenstein, 1987). The corresponding dimensions for flagella are 5–10 μm in length and 20 nm in diameter (MacNab, 1987). The fimbriae of *Escherichia coli* consist of repeating subunits of a single polypeptide of 17 kDa molecular weight. Biochemical analysis of this peptide subunit (reviewed in Eisenstein (1987)) shows it to be largely hydrophobic and to be acidic (pI, *ca.* 3·9). These two characteristics may help to explain the role it has in adhesion in that hydrophobicity and electrostatic bonding are major chemical components of this process (reviewed earlier). Many fimbriae of Gram negative organisms act as adhesins in that they bind very specifically to receptor molecules on the target surface often in a lectin-like manner. Indeed, pili and their associated adhesins have been shown to be a major virulence factor in many Gram negative pathogens such as *E. coli* in that these organisms bind specifically to tissues containing receptor molecules such as galactose and mannose containing residues in the urinary tract (Hagberg *et al.*, 1986; Eisenstein, 1987).

The other major structure of bacteria associated with adhesion is the capsule (Fig. 5.2). This is a highly hydrated extracellular polysaccharide or polypeptide polymer which may extend a considerable distance from the cell surface (Troy, 1973; Sutherland, 1977; Costerton *et al.*, 1981*b*). Capsules vary widely in their chemistry but quite often contain a number of components such as uronic acids, amino acids, phosphate, or sulphur groups, in addition to neutral sugars (Sutherland, 1977; Kenne & Lindberg, 1983). Most of these polymers are therefore anionic as evidenced by their staining with cationic ruthenium red and their ability to bind metals (McLean & Beveridge, 1990) (Fig. 5.3). As is the case with fimbriae, the anionic and to a lesser extent, hydrophobic character of these polymers suggests that electrostatic and hydrophobic bonding are involved in the attachment of these structures to surfaces. In addition one would expect a role for hydrogen bonding in some instances due to the carbohydrate nature of these polymers.

In a study using freshwater bacterial isolates, Allison and Sutherland (1987) found that initial adhesion to surfaces occurred independently of capsule production. However, it was required for subsequent growth of colonized cells into mature biofilms. There are numerous instances where capsule and/or fimbriae production have been exclusively or pre-dominantly implicated in adhesion (see for example Costerton *et al.* (1981*a*); Bryant *et al.* (1984); Costerton *et al.* (1987); Doig *et al.* (1987, 1988)).

A number of other cell surface components have been implicated in adhesion. These include Gram positive cell wall polymers such as lipo-teichoic acids (Kenne & Lindberg, 1983). These polymers are thought

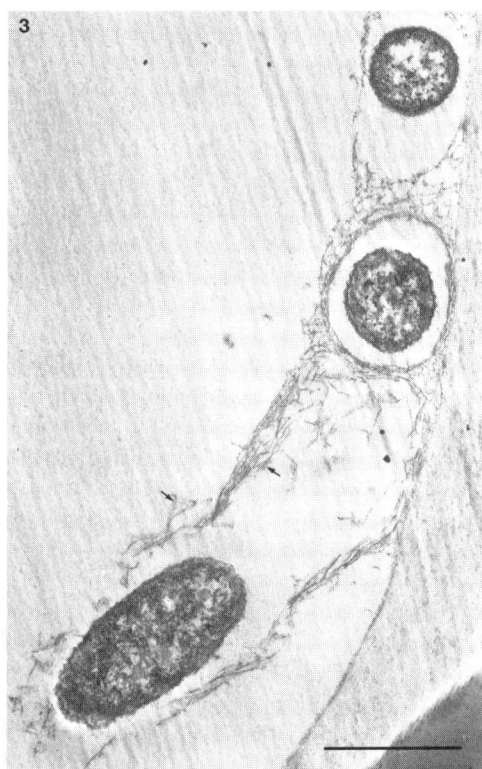

Fig. 5.3. Bacteria in an oligotrophic alpine stream adhere readily to available surfaces including experimental plastic disks. Dissolved minerals (iron, calcium, silicon, etc.) can precipitate onto bacterial structures such as the capsule. This results in the formation of minerals including iron hydroxides and silicates (arrows). Figure taken from Lappin-Scott & Costerton (1989) and reprinted with the permission of the copyright holder, Harwood Academic Publishers GmbH, Montreux, Switzerland. Bar represents 0·5 μm.

to act as adhesins in that they generally bind specifically to receptors on tissue surfaces (Op den Camp *et al.*, 1985; Aly & Levit, 1987). Other bacterial cell surface components such as outer membranes (Jacques *et al.*, 1988), regularly structured protein surface arrays, Streptococcal M proteins (Poirier *et al.*, 1989; Caparon *et al.*, 1991), etc. may be involved in adhesion in special instances. Other unusual stalked organisms such as *Caulobacter* sp. employ a holdfast mechanism for adhesion (Staley & Fuerst, 1989). With these exceptions, the capsule and fimbriae represent the major mechanisms of bacterial adhesion to surfaces.

5.2.3   Strategies and Mechanisms of Biofilm Formation

Biofilm formation can be essentially summarized into several distinct phases (Marshall *et al.*, 1971; Costerton *et al.*, 1987). Initially microorganisms encounter a target surface, bind reversibly, then irreversibly, and then grow into small clusters of cells (referred to as microcolonies). These microcolonies continue to grow and spread until a confluent biofilm forms over the surface. Once a biofilm reaches a certain stage of maturity, it begins to shed planktonic cells. These shed cells act as pioneers in that they are free to disperse and eventually colonize new surfaces.

   If we examine colonization of a surface by an organism such as *Pseudomonas aeruginosa* only at specific time points, this behaviour appears to be the case (Fig. 5.4) (Costerton *et al.*, 1987). The true complexity of colonization strategy is only realized when microorganisms are subjected

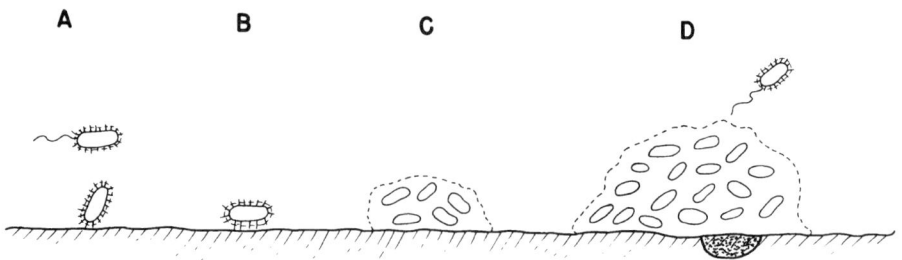

Fig. 5.4.   Generalized schematic of bacterial biofilm formation. Planktonic cells in the fluid phase attach reversibly (A), then irreversibly (B) to a suitable surface (Marshall *et al.*, 1971) by means of pili and/or capsule. Further growth (C) results in the formation of an encapsulated microcolony. As the microcolony grows into a biofilm (D) some cells are released into the fluid phase and can then colonize new surfaces. In spite of the surrounding fluid characteristics (such as the presence of dissolved O_2), biofilms and microcolonies often generate a chemically distinct microenvironment which may lead to activities such as anaerobic corrosion of the surface (darkened region).

to continuous scrutiny. For practical purposes, this can only be accomplished by light microscopy coupled with computer-enhanced image analysis (Caldwell & Germida, 1985; Caldwell, 1985; Caldwell & Lawrence, 1988). Light microscopy, especially phase contrast or dark field microscopy, allows the visualization of individual, living, unstained microorganisms. Another related technique is the use of interference reflection microscopy (IRM) (Marshall *et al.*, 1989). IRM allows cell-surface separation distances from 0 to 100 nm to be estimated since the approach of bacterial cells to a glass surface induces interference in reflected light leading to an apparent darkening of the cell as the distance from the cell to the glass approaches zero (Marshall *et al.*, 1989). It allows the direct examination of unstained living material, yet allows an estimation of the distance between the adhering organisms and the surface to which they are attached (Fletcher, 1988; Godwin *et al.*, 1989; Marshall *et al.*, 1989). The addition of computer enhanced image analysis to these optical techniques allows microbial activity such as adhesion, detachment, growth, and spacing of individual cells and clumps of cells to be evaluated (Caldwell & Lawrence, 1988). Confocal scanning laser microscopy (CSLM) is an exciting new development in microscopy which employs a laser light source and highly refined optics. CSLM enables investigators to obtain optical thin sections through specimens with a resolution of <1 μm (Brackenhoff *et al.*, 1988). Through the use of appropriate fluorescent probes (Haugland, 1992) and computerized image analysis, CSLM allows one to construct three-dimensional maps of chemical microenvironments within biological materials such as cells and biofilms (Caldwell *et al.*, 1992). The major limitations of these techniques are that studies are limited to materials that readily transmit light, such as glass, and that once the adherent cells form a thick biofilm, a reduction in light transmission drastically diminishes resolution. In spite of these difficulties, we are now beginning to realize the tremendous diversity of colonization strategies that bacteria exhibit.

A wide variety of other techniques are available for measuring bacterial adhesion (summarized in Costerton *et al.* (1987)). The most basic technique used involves standard plate counts whereby bacteria are dislodged from a surface by scraping with a sterile scalpel blade or gentle sonication, agitated to break up cell clumps, and plated onto a suitable medium. During this, one assumes that all organisms can be separated and easily grown. As these assumptions are never completely met, microbial numbers are generally underestimated by this technique. Other techniques involve counting adherent bacteria directly using a viable stain (Haugland, 1992) coupled with epifluorescence microscopy (Ladd *et al.*,

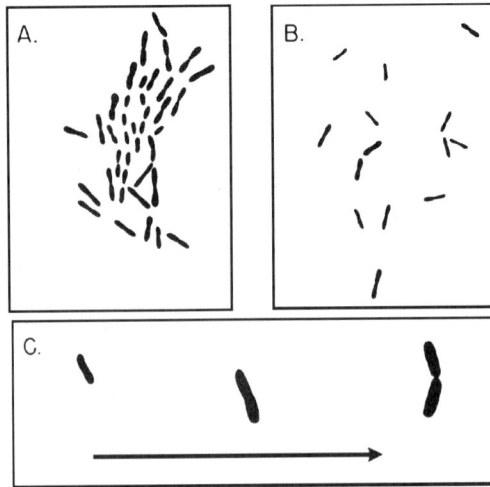

Fig. 5.5. Differences in colonization strategy can alter the appearance of microcolonies. Note the differences between packing (A) and shedding (B). Spreading of attached bacteria would result in a similar appearance to that of shedding (B). Rolling (C) involves loose attachment of an organism to a surface where it is moved along in the direction of flow (arrow). For details refer to text and also to references (Lawrence & Caldwell, 1987; McLean & Nickel, 1991). Figure reprinted from McLean & Nickel (1991) with permission of the copyright holder, Churchill Livingstone Medical Journals, Edinburgh, Scotland.

1985), measuring radiolabelled bacteria, measuring a biochemical component of the attaching organisms such as diaminopimellic acid (Beveridge, 1981), or measuring cellular metabolism using a radiolabelled substrate (Ladd *et al.*, 1987).

If bacteria adapted to colonization are allowed to colonize a glass laminar flow cell, several different types of cell movement strategies are seen (Fig. 5.5). Lawrence and Caldwell (1987) noted that once organisms from a freshwater stream colonized glass surfaces four distinct types of cell movement designated as 'packing', 'spreading', 'shedding', and 'rolling' were seen. Packing involved colonization by a mother cell. Upon growth and cell division, the daughter cells remained in tight proximity to the original colonization site thus forming a tightly packed microcolony. In the spreading manoeuvre, daughter cells remained attached, but moved some distance from each other producing a diffuse microcolony.

Colonies exhibiting shedding involved an attached mother cell which continuously shed daughter cells that then migrated to new locations. The fourth type of movement seen, rolling involved attached cells which did not firmly adhere, but rather rolled or tumbled along the surface in the direction of the flow. Each form of colonization behaviour appeared to be associated with and be a unique characteristic of a specific species. In other studies, *Pseudomonas fluorescens* has been shown to colonize exclusively via a packing strategy (Caldwell & Lawrence, 1986; Lawrence *et al.*, 1987), and *Proteus mirabilis* via a shedding strategy (McLean *et al.*, 1991; McLean & Nickel, 1991). One can speculate that an organism such as *Ps. fluorescens* would be able to colonize and rapidly cover a given surface in a natural environment thus giving it a competitive advantage over other organisms. In the human urinary tract, colonization of bladder epithelial cells induces them to slough into the urine (Aronson *et al.*, 1988), so the shedding strategy of *Pr. mirabilis* whereby progeny continuously colonize other surfaces would aid in the persistence of this organism in this environment (McLean & Nickel, 1991).

Colonization of surfaces is often enhanced as their chemistry is altered and when a number of other organisms are present. This is most evident in the mouth. Here mineral surfaces (teeth) are continuously coated by a pellicle of salivary proteins, which can affect colonization (Rosan *et al.*, 1985; Pratt-Tepstra *et al.*, 1989). Primary colonizers of freshly cleaned teeth are predominantly *Actinomyces viscosus, Streptococcus sanguis* and related bacteria (Ellen *et al.*, 1988; Kolenbrander, 1988). Other organisms do not colonize teeth as readily, however they often specifically recognize and bind quite readily to the primary colonizers and are thus able to colonize these surfaces (Ofek & Perry, 1985; Mergenhagen *et al.*, 1987; Liljemark *et al.*, 1988). This process is now referred to as coaggregation since mixing two complementary strains of bacteria such as sucrose-grown cells of *Streptococcus mutans* and *A. viscosus* induces a clumping reaction (Gibbons & Nygaard, 1970; Kolenbrander, 1988). A number of studies (Kolenbrander, 1988) have shown this reaction to consist of a series of lectin-like reactions between accessible surface constituents in bacterial partners, and may be responsible in part for the spatial distribution of microorganisms within the oral environment. Other physiological features such as capsule synthesis by *Strep. mutans* are also involved in dental plaque colonization (Koga *et al.*, 1988). Synergistic relationships between bacterial species has also been reported during adhesion in other environments (Chang & Merritt, 1991), however, the importance of this is quite variable, and the mechanisms are at present unknown (McEldowney & Fletcher, 1987).

From the aforementioned studies, it is evident that bacteria possess very diverse mechanisms and strategies that enable them to colonize virtually any type of surface. If a particular surface contains an antibacterial compound it can still be colonized by killed microorganisms, or chemically altered by the deposition of organic material of bacterial or other origin. Eventually incoming microorganisms will only encounter the chemically or bacterially modified surface, and be able to colonize. In this way the antibacterial nature will have been circumvented. From a practical viewpoint bacterial colonization and biofilm formation is ubiquitous – the problem is not so much inducing bacterial colonization, but how to prevent it. This is an area of active research, and will be addressed further.

5.3 DIVERSITY OF BIOFILMS IN NATURE

Surface adhesion and biofilm formation represent a fundamental strategy of bacterial growth in nature. Consequently it is not surprising to find that bacteria predominantly grow as biofilms in widely diverse environments so as to gain access to available nutrients, and protect themselves from predation and antimicrobial agents.

5.3.1 Natural Environments

In natural ecosystems, the sessile biofilm population of heterotrophic bacteria was found to exceed the planktonic population by 3–4 orders of magnitude in oligotrophic alpine streams (Geesey *et al.*, 1978) (Fig. 5.3) and 200-fold in sewage effluent (Costerton *et al.*, 1987). The only natural community of microorganisms which are predominately freeswimming or planktonic are the photosynthetic, unicellular microorganisms in lakes (Caldwell & Tiedje, 1975a,b; Caldwell, 1977). By not stacking themselves as layered communities upon submerged surfaces, these organisms gain optimum access to light at the water surface (Caldwell, 1977). Significant numbers of particle-associated and planktonic organisms are also found in the open ocean and in deep terrestrial aquifers (Ghiorse & Wilson, 1988). In these environments, organisms are under advanced stages of starvation as readily metabolizable carbon is virtually non-existent. Consequently these organisms survive by drastically altering and slowing down their metabolism (Kjelleberg *et al.*, 1987), and many energy-requiring biosynthetic reactions, notably capsule production (essential for biofilm growth), are stopped. The capacity of these organisms to survive without metabolizable carbon is phenomenal with estimates ranging up to 1000–10000 years (W. C. Ghiorse, personal communication). Yet even here many organisms adhere readily to available surfaces (Bright & Fletcher, 1983; Eighmy *et al.*, 1983; Pringle & Fletcher, 1986; Kjelleberg *et al.*, 1987; Iriberri *et al.*, 1990).

One would expect that heterotrophic organisms would be the predominant benefactors from a strategy of biofilm formation in that by adhesion, they would gain access to insoluble carbon sources such as cellulose. It is therefore quite interesting that biofilm formation has been described in at least one instance involving an autotrophic organism (Kieft & Caldwell, 1984). One often sees other autotrophic organisms most notably algae readily adhering to surfaces.

5.3.2   Industrial Environments

The nutrient-rich aquatic environments and large surface areas common to many industrial processes stimulate the formation of biofilms. In many instances this has proved to be quite costly in that biofilm formation causes such problems as microbial plugging in oil injection wells (Shaw *et al.*, 1985; Cusack *et al.*, 1987; McKinley *et al.*, 1988; Lappin-Scott & Costerton, 1989), reverse osmosis membranes (Ridgway *et al.*, 1983; Payment, 1989), industrial filters (Costerton, 1984a; Camper *et al.*, 1985; Lytle *et al.*, 1989), heat exchangers (Costerton *et al.*, 1987) and water treatment plants (LeChevallier *et al.*, 1987, 1988a). Biofilm formation often represents a major nuisance in that filtration capacity, efficiency of oil injection wells, and heat exchange capability may be greatly diminished or even eliminated. When conditions are suitable, anaerobic microenvironments develop which allow the growth of sulphate-reducing bacteria and the initiation of associated anaerobic corrosion (Costerton, 1984a; Hamilton, 1985). In addition, biofilms in water and waste-water treatment plants usually represent reservoirs of potentially pathogenic bacteria such as *Klebsiella, Salmonella, Yersinia,* and other enteric pathogens (Camper *et al.*, 1985; LeChevallier *et al.*, 1987, 1988a). The industrial cost of biofilm fouling is astronomical in that millions of dollars are spent yearly on biocides, drilling of new injection wells, and equipment replacement.

Traditionally biocides have been somewhat indiscriminately applied in industrial pipelines, and the bacterial population has been monitored through the use of standard plate counting techniques of the fluid phase. While planktonic organisms are rapidly eradicated by this treatment, very little effect is seen against biofilms. The usual result is that anaerobic corrosion and/or microbial plugging continues in supposedly 'sterilized' industrial systems. This led to the development of the Robbins biofilm sampling device (Ruseska *et al.*, 1982). The Robbins device is a length of pipe (usually 4 feet long, made from brass). Its distinguishing feature is that sections of the pipe wall, mounted on studs, can be easily removed and replaced. Biofilms which colonize the inside of pipes (including the Robbins device) can be easily sampled by removing the sampling plugs, dislodging the biofilm by scraping with a razor blade or scalpel, and plating on appropriate microbiological media. Incorporation of the Robbins device into industrial aquatic pipelines has allowed the continuous monitoring of microbial corrosion and the assessment of the efficacy of biocide treatment (Ruseska et al., 1982; Costerton, 1984a). Other treatments which have been successfully employed to alleviate biofilm associated biofouling include the use of a slow freezing cycle to induce large ice crystals within the biofilms of heat exchangers (Costerton, 1990). Thawing of these crystals then leads to the complete removal of the biofilm. In oil formations, microbial plugging is most successfully controlled through the administration of sodium hypochlorite (bleach) to kill and dissolve organic biofilm matter, followed by acid to dissolve small particulate inorganic minerals or fines (Cusack et al., 1987). The resistance of biofilms to antimicrobial agents is well-documented. Recently it has been shown that this resistance can be overcome if a biofilm is exposed to an antimicrobial agent in the presence of a small electrical field (Blenkinsopp et al., 1992). This new discovery is highly significant, given the enormous expenditure in the industrial and medical fields in controlling biofilms.

There are many positive aspects of industrial biofilm formation. Waste-water treatment systems have made extensive use of biofilms in trickling filters, and other treatment protocols. In trickling filters, sewage is sprayed onto beds of loose gravel. Biofilms rapidly form on these surfaces, and due to the circulation of air through the gravel and the large surface area, they can aerobically degrade large amounts of organic matter (Hamilton, 1987). Biofilms under more anaerobic conditions have been used for methane production (Eighmy et al., 1983; Eighmy & Bishop, 1984; LeChevallier et al., 1987) and sulphate reduction (Nielsen, 1987). In addition to carbon mineralization, biofilms and their anionic bacterial cell wall and capsular components are also able to assist in the recovery of trace metals from mining and other industrial wastes (Olson, 1986; Ferris et al., 1989a). The use of biofilms and bacterial surface components has tremendous potential for use in heavy metal pollution control and biologically catalyzed mineral formation (biomineralization) (Ferris et al., 1986; Mann, 1988; Beveridge, 1989; McLean & Beveridge, 1990).

When an oil well is first drilled, gas pressure within the reservoir results in approximately 8–30% of the total oil being forced to the surface. In order to increase oil recovery when the gas pressure is depleted, water is pumped into the reservoir through injection wells and thus displaces the oil to the production wells. During water injection, 'fingering' occurs as the injection water will preferentially be driven through high permeability zones in the reservoir bypassing the residual oil in place (MacLeod *et al.*, 1988). The application of biofilm technology to this problem shows great promise in enhancing oil recovery from these reservoirs by enabling selective plugging of these high permeability zones. In a series of experiments, a strain of *Klebsiella pneumoniae* was isolated from the produced water in an oil well. Injection of the vegetative cells into model rock cores results in the formation of a thin biofilm 'skin' which forms predominantly in areas of high permeability. Penetration of these organisms into the rock cores is enhanced if the cells are first starved (and thereby inhibited in capsule production). The resulting ultramicrobacteria (UMB) (Fig. 5.6) can then be pumped into the formation where

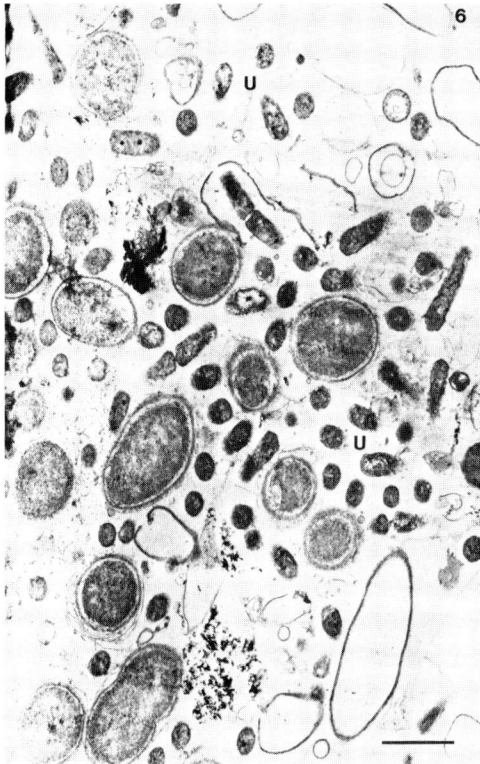

Fig. 5.6. Starvation of *Klebsiella* sp. isolated from produced water in oil fields results in the formation of ultramicrobacteria (U). Ultramicrobacteria do not produce capsule nor form biofilms under starvation conditions, yet upon nutrient resuscitation recover both of these capabilities (Lappin-Scott *et al.*, 1988a). Bar represents 1 μm.

they follow the injected water into regions of high permeability. After resuscitation with nutrients (Lappin-Scott *et al.*, 1988*a*), these starved cells rapidly recover and produce a relatively thick viscous biofilm (Lappin-Scott *et al.*, 1988*b*; MacLeod *et al.*, 1988) (Fig. 5.7) that plugs the high permeability water breakthrough zones and facilitates additional oil recovery.

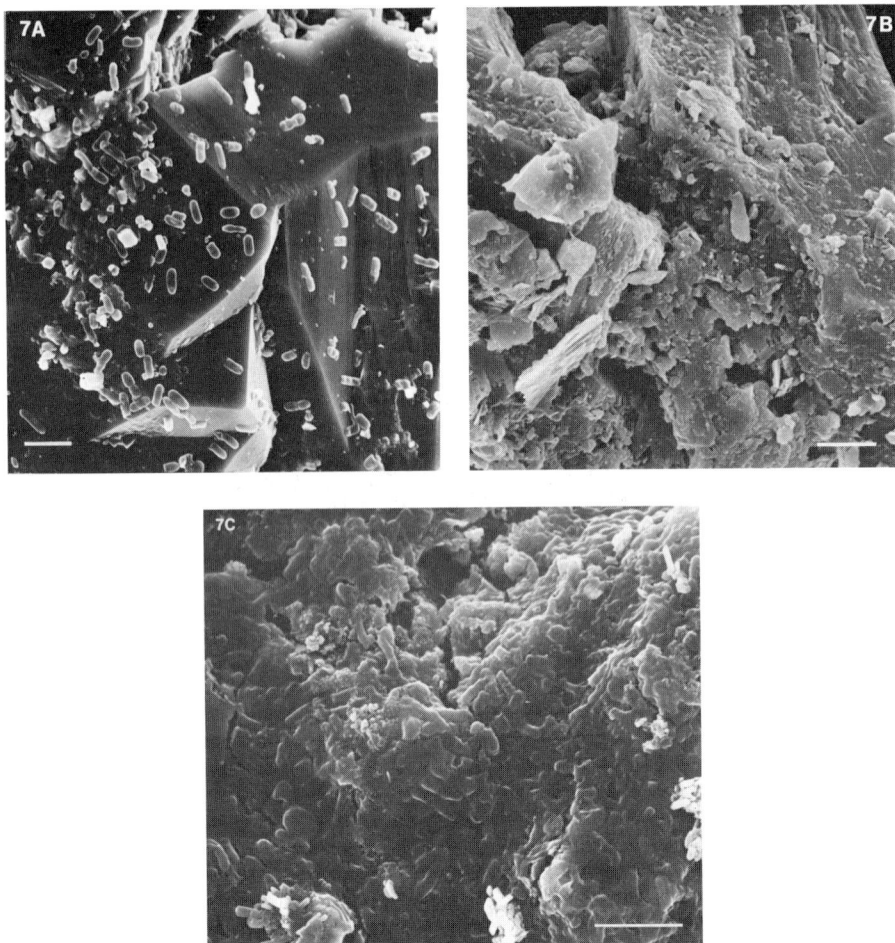

Fig. 5.7. Plugging of model sandstone rock cores using ultramicrobacteria (UMB). Introduction of UMB allows their penetration deep into the sandstone (7A). Exposure to nutrients results in rapid production of capsule (7B) and eventual biofilm formation (7C) which can then plug the high permeability zone of the oil-bearing rock formation. Bar represents 5 μm.

5.3.3   Medical Importance

Biofilm formation is an important bacterial strategy in the development of many chronic infections (see for detailed review Costerton (1984*b*); Gristina & Costerton (1985); Costerton *et al.* (1987); Jacques *et al.* (1987); Dart & Seal (1988)). Although this environment is obviously different from the industrial, aquatic and terrestrial habitats addressed above, organisms still face such basic stresses as predation (by leucocytes), competition with the host and other flora for nutrients, adverse chemical conditions (such as antibiotics and disinfectants) and nutrient limitation (notably iron). It is therefore not surprising that bacteria often employ the same basic strategy of biofilm growth in order to persist in

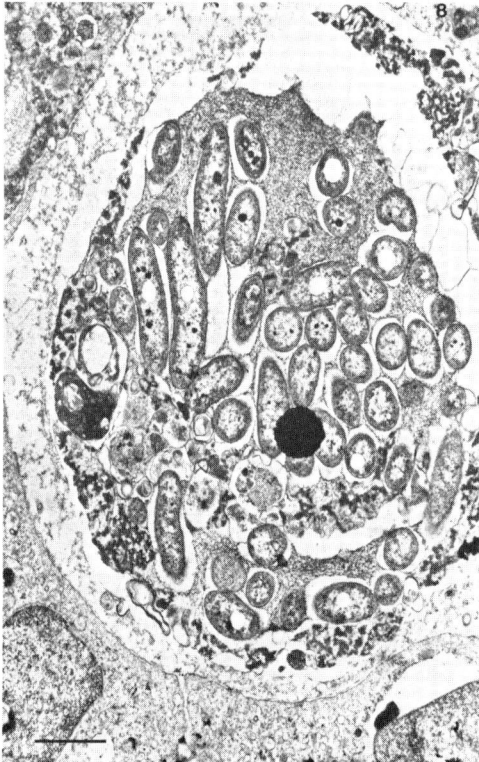

Fig. 5.8.   Microcolony of *Pseudomonas aeruginosa* within the lung of a rat used in the study of cystic fibrosis. As a consequence of their microcolony mode of growth, these organisms can easily withstand the host immune system and large dosages of antibiotics. See Cochrane *et al.* (1988) for details. Bar represents 1 μm.

these pathogenic environments (Fig. 5.8). As in the case of the industrial environment, biofilm growth presents both its advantages and disadvantages.

Advances in modern medicine have resulted in the development of a wide variety of prosthetic devices and instrumentation which are implanted into various regions of the body. Although these developments have resulted in considerable reduction of morbidity and mortality, they do present an 'immunologically inert' surface that can be readily colonized. Biofilms which form on these surfaces act as a nidus of infection which may ultimately develop into a systemic, possibly life-threatening infection (for review see Jacques *et al.* (1987)). Probably the most prevalent nosocomial or hospital-acquired infections in North America are urinary tract infections associated with the use of urinary catheters. Using a modification of the Robbins device, Curtis Nickel and his colleagues have shown that the antibiotic resistance of catheter associated biofilm bacteria is considerably higher (up to 1000 fold) than their planktonic counterparts (Nickel *et al.*, 1985*b*). The cost of biofilm-associated infections is considerable, and has led to the development of active

Fig. 5.9. Mineralization within biofilm-associated infections results in the formation of several types of medically important calculi. Two examples shown here are struvite ($NH_4MgPO_4 \cdot 6H_2O$) calculi (9A) which form in the urinary tract from *Proteus mirabilis* infection, and calcium bilirubinate calculi (9B) that form in the gall-bladder from β-glucuronidase-producing *Escherichia coli* infection. Crystals are indicated by arrows. Bar represents 2·5 μm in 9A, 1 μm in 9B. Figure 5.9B was provided courtesy of Dr Joe Sung, University of Calgary.

research programs by a number of companies to develop polymers which are able to resist colonization. We have successfully used the modified Robbins device to examine the resistance of many experimental polymers to bacterial colonization and biofilm formation (Nickel and Costerton, unpublished observations; McLean and Nickel. unpublished observations) and no material has yet shown a total inherent resistance to this inexorable natural process.

Biomineralization involving biofilms also has its parallel in the medical field in the formation of brown pigment gall stones (Leung *et al.*, 1989), dental calculi (Driessens *et al.*, 1985), and struvite ($NH_4MgPO_4 \cdot 6H_2O$) urinary calculi (McLean *et al.*, 1988). In all three cases, mineralization is preceded by biofilm formation. The physiological activities of the bacteria within these biofilms alter the chemistry of the surrounding urine, saliva or bile resulting in the formation of precipitates. Formation of these crystals within the gel matrix of the biofilm ensures that they become cemented together and ultimately grow to become clinically significant calculi (Fig. 5.9). In the case of struvite calculi, we have shown that alkaline microenvironments created within *Pr. mirabilis* biofilms enable acid-labile struvite crystals to form in acidic conditions (Clapham *et al.*, 1990; McLean *et al.*, 1991).

The most striking beneficial aspect of biofilm formation in the medical and agricultural environments is perhaps the role of bacteria in digestive processes. In the absence of bacteria, animals are unable to digest complex carbohydrates such as cellulose. Symbiotic bacteria are intimately associated with all animals including man, and particularly in herbivores, play an integral role in the digestion of feed, most notably cellulose (Cheng & Costerton, 1980a). K.-J. Cheng and colleagues have shown that microcolonies and consortia enable digestive tract microorganisms to metabolize feed more efficiently in that the organisms gain access to their insoluble substrate and essential physiological interactions between component organisms are facilitated (Cheng & Costerton, 1980b; Cheng *et al.*, 1981a) (Fig. 5.10).

Examinations of human and animal tissues have also shown that autochthonous bacterial populations develop on the tissue surfaces of many organs (Cheng *et al.*, 1979, 1981b; Cheng & Costerton, 1980a; Costerton *et al.*, 1983; Jacques *et al.*, 1986) that are exposed to environmental sources of microorganisms (mouth, oropharynx, vagina, etc.). Natural biofilm populations often play a pivotal role in the resistance of these organs to invasion by pathogens (Chan *et al.*, 1985; Reid *et al.*, 1988) and in many instances, their maintenance in good order is assiduously cultivated (Cheng *et al.*, 1979; Reid *et al.*, 1988).

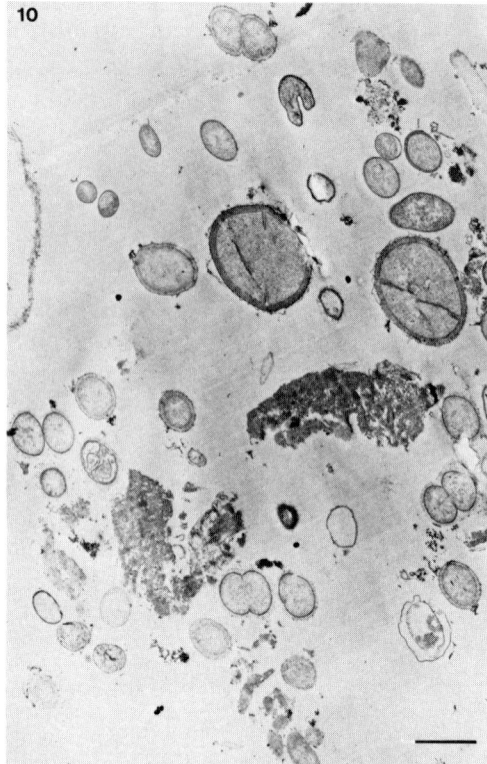

Fig. 5.10. Metabolic activities by natural populations of bacteria are often a result of cooperative (synergistic) metabolic interactions between several different organisms. In this micrograph are shown examples of the diverse microorganisms responsible for the digestion of cellulose and eventual production of volatile fatty acids in the digestive tract of cattle. During cellulose digestion, these mixed cultures adhere to feed particles and to each other, forming consortia in order to facilitate interbacterial metabolic reactions. See text and Cheng & Costerton (1980b), Cheng et al. (1981a) for details. Bar represents 1 μm. Micrograph was provided courtesy of Dr K.-J. Cheng, Agriculture Canada Research Station, Lethbridge, Alberta, Canada.

5.4   BIOFILM COMPOSITION

The predominant feature of any biofilm community is the large quantity of extracellular polysaccharide or polypeptide capsular material within which the organisms are encased (Costerton et al., 1987). The quantity and composition of these polymers reflect the physiology and composition of the organisms within the community. Capsule synthesis and composition varies quite widely among bacteria and other organisms (Sutherland, 1977, 1985, 1988; Kenne & Lindberg, 1983; Whitfield,

1988). Various strains of a given bacterial species generally show a wide diversity in the quantity and composition of polymers produced. Recently, White and coworkers showed that analytical techniques such as infrared spectroscopy could be used to analyse biofilm polymers and in so doing help to determine the taxonomic composition of these bacterial communities (Nichols *et al.*, 1985).

We are only beginning to realize that the vast majority of bacteria in nature produce extracellular polymers. Early observations that some types of bacteria are non-encapsulated have been shown to be artifacts of the growth of these organisms in rich laboratory culture (Geesey *et al.*, 1978; Costerton *et al.*, 1981*b*, 1987; Chan *et al.*, 1984; Costerton, 1984*a*; Beynon *et al.*, 1992). Bacterial capsule production is a highly regulated physiological activity, especially in that it requires a considerable quantity of energy. If a particular organism is in the presence of a rich supply of nutrients such as a laboratory growth medium, capsule production will not be necessary for the survival of the organism. Switching off capsule production will therefore enable the metabolism of the cell to be more directed toward cell growth and division. The net result is that these non-encapsulated organisms are able to quickly overgrow their capsule-producing counterparts (Costerton *et al.*, 1987).

The prevalence of encapsulated bacteria in nature suggests that extracellular polymer production is vital to the survival of these organisms. As this often represents the outermost layer of the cell, it is with this region that bacteria first make contact with their environment. The anionic nature of many capsules is reflected in their affinity for cations such as metals (McLean & Beveridge, 1990; McLean *et al.*, 1990, 1992) and antibiotics such as tobramycin (Nichols *et al.*, 1988; Anwar *et al.*, 1989). In this aspect they serve as an ion exchange resin. One would also expect that anionic capsules might repel penetration by uncharged or anionic compounds. As a consequence, the bacterial cell surface will therefore only experience a greatly reduced concentration of a given toxic cationic compound. Conversely, should a cationic compound such as Mg^{2+} represent a nutrient, the capsule can act as a reservoir whereby Mg^{2+} is absorbed during periods of abundance and released to high affinity transport systems of the cell during periods of starvation.

Extracellular polymers represent the most common feature of biofilms. In its simplest form a biofilm will consist of a surface-adherent community of one or more species of prokaryotic and/or eucaryotic cells which is largely immobilized in an amorphous glycocalyx matrix. Three-dimensional examination of living biofilms by confocal scanning laser microscopy has shown that extracellular material can account for as much as 73–98% of the biofilm volume. Bacterial cells generally grow as discrete microcolonies within the biofilm matrix (Lawrence *et al.*, 1991). At any time there will be a number of large and small molecules present in the matrix introduced from the resident population through cell metabolism or cell death and lysis, or introduced from the liquid phase. Interestingly many biofilms found in nature become encrusted with a wide variety of minerals that are not found in the liquid phase or within the organisms. These minerals may arise as a consequence of biofilm metabolism or non-biological chemical reactions of bound metals. Examples of metabolically induced biofilm mineralization are the production of struvite (Fig. 5.9A) in urinary tract biofilms (McLean *et al.*, 1988) and calcium bilirubinate in gall-bladder biofilms (Fig. 5.9B) (Leung *et al.*, 1989). Conversely, metabolic processes are not involved in the nucleation of several silicates and other types of minerals by bacteria in aquatic environments (reviewed in Beveridge (1989)). In a study of mine tailings ponds, Grant Ferris and Terry Beveridge showed that biofilm bacteria growing on surfaces suspended into these ponds could bind a wide variety of metals (Ferris *et al.*, 1989*a*). The net result of this metal binding was the formation of a number of minerals including iron ferrihydrite ($Fe_2O_3 \cdot H_2O$) (an example of which is shown in Fig. 5.3). The most commonly seen example of metal binding by biofilms is the slimy rust coating that builds up along surfaces in a wide assortment of water containers such as toilet tanks. This is due to the sorption of dissolved iron from water by the bacterial biofilm and its oxidation into rust-coloured ferric hydroxide. The passive sorption of metals from solution by bacteria has implications far beyond the scope of microbiology. In a large number of studies Ferris, Beveridge and Fyfe reviewed in Beveridge (1989) have shown that various cellular constituents of bacteria and bacterial biofilms may in fact be largely responsible for the formation of several clays and minerals of geological importance. Future developments in this area may eventually lead to the use of bacteria and bacterial biofilms in the commercial reclamation of metals from mines and industrial wastes, and to the production of industrially important minerals (McLean & Beveridge, 1990).

5.5   BIOFILM ECOLOGY

For a particular species to survive, members must be able to obtain sufficient nutrition, grow, and reproduce. In addition, there must be mechanisms whereby at least some organisms can survive periods of stress (including starvation, adverse environmental conditions and predation). The obvious price of a species' failure to meet these demands is extinction. Bacteria are no different in these basic requirements from other organisms. As a group bacteria have been enormously successful in that they represent the earliest and most ubiquitous organisms to have colonized the earth. Their range of habitats is truly impressive in that they survive and often grow in regions of extreme acidity, alkalinity, temperature, pressure, radiation, salinity, osmotic stress, and harmful chemicals (Alexander, 1971). Bacteria are also found in association with higher forms of life where they can exhibit a pathogenic, commensal, or symbiotic relationship with their host. In many of these environments, adequate nutrition is often not readily accessible and so microorganisms have developed elaborate strategies including adhesion to surfaces and growth as biofilms to circumvent these problems as well as to deal with the ecological stresses of competition and predation.

5.5.1 Role in Persistence and Nutrition

If we examine oligotrophic ecosystems such as fast-flowing streams, adherence to surfaces allows organisms to persist. Dissolved substances in aqueous environments tend to concentrate on surfaces – a process which is often called the interfacial effect (McLean & Beveridge, 1990). Because of this, adhesion to surfaces therefore places organisms in regions of higher nutrient concentrations (Hamilton, 1987).

Adhesion also permits microorganisms ready access to insoluble substrates such as cellulose, or to surfaces which exude nutrients. These two situations are quite commonly encountered within the gastrointestinal tract of animals and have been particularly well studied in the rumen and caecum of herbivores. The rumen represents the largest organ within the digestive tract of ruminant animals (cattle, sheep, goats, etc.) and is unique in that all of its digestive functions are performed by the bacteria and other microorganisms (Hungate, 1966; Cheng & Costerton, 1980*a*). Within this organ, cellulose-containing plant material is digested by a number of microorganisms (principally bacteria) into several simpler compounds such as acetate, propionate and butyrate which are then absorbed and used by the host animal. No one species of microorganism is capable of the complete digestion of cellulose. Instead several species of bacteria usually act as consortia in performing this task (Cheng & Costerton, 1980*a,b*; Cheng *et al.*, 1981*a*). Cellulolytic organisms such as *Ruminococcus albus*, *R. flavefaciens*, and *Bacteroides succinogenes* generally adhere to insoluble substrates such as crystalline cellulose (Hungate,

1966; Kudo *et al.*, 1987). Other bacteria which may utilize the metabolic end products of cellulose digestion are seen in close proximity to and often adherent to the initial layer of cellulose degrading organisms (Cheng & Costerton, 1980*a*; Cheng *et al.*, 1981*a*; Costerton *et al.*, 1983). The ecological advantages to these organisms in adhering to cellulose and to each other in microcolonies (Fig. 5.10) (Cheng *et al.*, 1981*a*) become quite apparent in terms of ready access to an insoluble substrate, and transfer of metabolic end products. In the rumen environment and probably in other environments the formation of microcolonies and biofilms will enhance synergistic metabolic activities among their component microorganisms. In addition the proximity of microorganisms within a biofilm community has been shown to greatly facilitate the exchange of genetic material between members (Bale *et al.*, 1988).

5.5.2   Resistance to Adverse Environmental Conditions

Biofilms offer many organisms an opportunity to dwell in what would otherwise be hostile environments. If we consider a typical single species biofilm of 10 or more layers of cells in thickness on an inert surface in an aerobic environment (Fig. 5.11A), any chemicals which reach the interior cells must first pass through the surface layer. In the case of nutrients, one would therefore anticipate that the surface microorganisms would demonstrate the fastest growth in that they would be exposed to the highest concentration of nutrients and their metabolic activity would largely deplete nutrients and oxygen available for the interior cells. Should the fluid phase contain any toxic materials (Fig. 5.11B), then we would expect the surface organisms to be inhibited or even killed, and growth or survival demonstrated by interior cells. Now if we consider a biofilm containing several different species on this same inert surface (Fig. 5.11C), the surface cells would be expected to again show the fastest growth, but in addition to depleting nutrients they would tend also to deplete oxygen. Should the biofilm be sufficiently thick, oxygen would be totally depleted allowing the growth of obligately anaerobic organisms within the interior. The growth of interior organisms whether anaerobic or not, would be enhanced if they require some metabolic end product(s) produced by the surface layer. If the surface on which the biofilm resides contains some form of nutrition such as cellulose (Fig. 5.11D), we could expect those organisms in intimate contact with the surface to utilize those nutrients and subsequent layers of other organisms to utilize preferentially their metabolic end products. These strategies nicely explain how anaerobic microorganisms can grow in otherwise

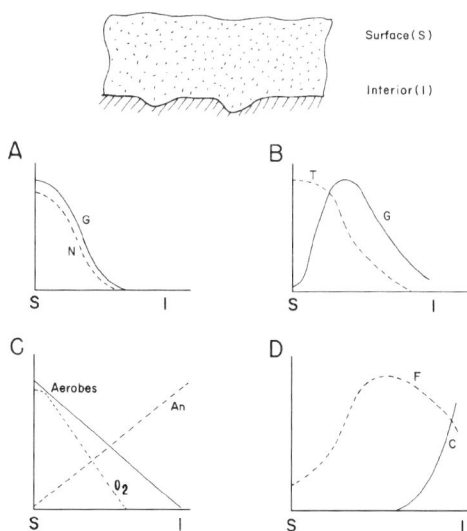

Fig. 5.11. Physiological profile within a model biofilm from the surface (S) into the interior (I). If the organisms are adherent onto an inert surface (example A), growth (G) will be most rapid near the surface since nutrients (N) will be rapidly metabolized by the surface layers of organisms. If a toxin (T) is present (example B), surface growth will be inhibited. Absorption or inactivation of the toxin by the first few layers of organisms allows the growth of underlying organisms. In example C, growth of aerobes is highest at the surface. Depletion of oxygen by these surface aerobes allows growth of anaerobes (An) in the interior. If adhesion is to an insoluble, metabolizable substrate such as cellulose (example D), cellulolytic activity (C) will be highest at the point of adhesion. Fermentation (F) of end-products of cellulolytic activity by other organisms will be highest in the immediate vicinity of the primary cellulose degraders.

hostile aerobic environments such as the skin (Costerton *et al.*, 1983, 1987), the mouth (Gibbons, 1989), and industrial water pipelines (Costerton, 1984*a*; Hamilton, 1987). Similar arguments can be used to explain the resistance of biofilms to other adverse environmental conditions such as antibiotics or heavy metals (Dean-Ross & Mills, 1989).

Many studies have shown that the biofilm mode of growth confers a considerable degree of protection to the component organisms against a wide variety of antimicrobial agents including antibiotics (Costerton, 1984*a*; Nickel *et al.*, 1985*b*; Ladd *et al.*, 1987; LeChevallier *et al.*, 1988*a,b*; Lytle *et al.*, 1989). In many cases this protection is quite significant in that biofilms can be up to several orders of magnitude more resistant than their planktonic counterparts. Several studies have been conducted to ascertain the mechanisms of this resistance. Anwar *et al.* (1989) showed that *Pseudomonas aeruginosa* biofilm resistance to the antibiotic tobramycin disappeared if the organisms were resuspended. In addition older biofilms were most resistant. The authors concluded that the adhesion process was essential for antibiotic resistance. Although the capsule itself mediates some protection due to its binding of antibacterial compounds and reducing their penetration into the biofilm (Nichols *et al.*, 1988), the chief mechanism of this resistance is now considered to be related to reduced cellular metabolism (Ladd *et al.*, 1987; Brown *et al.*, 1988; Gilbert *et al.*, 1990). When compared to planktonic cells grown in conventional rich laboratory media, biofilm cells show a greatly reduced growth rate (expressed in terms of metabolic activity per cell or cell doublings per unit time). This is due to the energy required for biosynthetic production of capsular material as well as the necessity for diffusion of nutrients or other compounds (such as O_2) into the matrix. Since antibiotics largely act as inhibitors in many biosynthetic processes, the naturally lower growth rate within biofilms would reduce the effectiveness of these compounds. The surprising finding by Tim Ladd and Curtis Nickel (Ladd *et al.*, 1987) was that the metabolism of *Ps. aeruginosa* within biofilms (as expressed by ^{14}C-glutamate uptake) is lowered even more upon exposure to the antibiotic tobramycin. This metabolic slowdown, which was not noted in planktonic cells, increases antibiotic resistance even further.

5.5.3 Relation of Biofilms to Other Bacterial Growth Strategies

As attested by their ubiquitous nature, biofilms represent an extremely successful ecological strategy for microorganisms in nature. However, they do not represent the only mode of growth available. As mentioned previously, introducing bacteria into rich laboratory media often induces them to forgo capsule production and biofilm growth and to grow as planktonic cells. These planktonic cells then quickly overgrow their sessile biofilm counterparts in that they are able to devote a greater fraction of their cellular metabolism to cell growth and reproduction (Costerton *et al.*, 1987). The other instance in which biofilm growth is rarely found is during conditions of long term starvation.

Under natural conditions, bacteria like other forms of life usually experience times of feast or famine. During periods of plenty, cellular physiology is directed towards growth and reproduction. Conversely during famine conditions, physiological strategy is geared mainly to survival. In higher forms of life such as animals and plants, dormancy often represents a common strategy. This is illustrated in seed production by annual plants and hibernation in many animals. In either case, these organisms survive periods of low nutrient availability during winter months in a dormant state and provide themselves with sufficient nutrients for their basal metabolism from stored energy reserves such as starch, glycogen, and fat. Bacteria are certainly no different in their basic strategy from these other forms of life. Upon depletion of readily available nutrients, bacteria initially synthesize high affinity alternative transport systems and metabolic pathways to scavenge what nutrients are available. Eventually endogenous metabolism begins to predominate as the organisms utilize internal sources of nutrition such as non-essential proteins, and high energy compounds such as poly β hydroxybutyrate (Kjelleberg *et al.*, 1987; Matin *et al.*, 1989). Some bacteria notably members of the genera *Bacillus* and *Clostridium* produce endospores, while other microorganisms including yeasts, protozoa and fungi produce cysts or spores (Brock & Madigan, 1988).

There are several environments wherein bacteria encounter virtually no easily metabolizable carbon. These include the open ocean (Morita, 1982, 1988; Kjelleberg *et al.*, 1987; Roszak & Colwell, 1987; Matin *et al.*, 1989; Moyer & Morita, 1989) and porous rock aquifers deep below the terrestrial subsurface (Elliot *et al.*, 1984; Balkwill & Ghiorse, 1985; Balkwill *et al.*, 1989). The persistence and survival of some of these aquifer organisms is truly remarkable in that they may have been in their particular environments for several centuries or even millennia (W. C. Ghiorse, personal communication). Nevertheless the distinguishing feature of most of these organisms is that they are usually found as individual cells – not as biofilms. In the case of starved marine organisms, Morita and others have established that some organisms undergo a series of cell divisions without a net increase in overall population mass (Morita, 1982, 1988; Roszak & Colwell, 1987; Moyer & Morita, 1989). The net result of this process is the formation of spherical ultramicrobacteria, so named because they may only be 0.1–0.2 μm in diameter (in contrast to 0.5–2 μm size of many bacteria). In this case the strategy of these organisms is to make as many copies of their genome as possible so that when nutrients eventually become available, some representatives of the species will have survived. Other aspects of bacterial physiology during starvation stress are altered. A number of starvation-induced proteins are produced (Tormo *et al.*, 1990; Nyström *et al.*, 1992; Siegele & Kolter, 1992) and in at least one instance, starved microorganisms become much more resistant to heavy metals and ultraviolet radiation (Nyström *et al.*, 1992).

Very little experimental work to date has been done to study the effects of extreme and prolonged starvation upon biofilms. In a study with a marine *Pseudomonas* sp.; Wrangstadh *et al.* (1989) found that starvation induced the formation of capsular material which in turn promoted detachment from hydrophobic surfaces (Wrangstadh *et al.*, 1986, 1989).

Presumably these starved organisms went off in search of 'greener pastures' or at least more nutrition. Other investigators have found that starved microorganisms adhere more readily to surfaces (Marshall, 1988). This whole concept of starvation survival and its relevance to microbial and biofilm ecology and physiology represents an area of much needed future study.

5.6   BIOFILM PHYSIOLOGY

When compared to their planktonic counterparts, biofilm organisms would be expected to individually demonstrate slower growth (as expressed in terms of cell divisions and increased cell mass). This is due to the physiological requirements of capsule production, and possibly poor diffusion of soluble nutrients into and waste products out of the biofilm matrix (Rittmann *et al.*, 1986). This is more than offset by the many advantages of biofilms outlined above.

Biofilm physiology as an area of research is still basically in its infancy. The majority of basic physiological studies with bacteria have been performed using planktonic populations often under conditions that induce maximum growth. While this approach has yielded much in the fields of bacterial physiology and biochemistry, it must be realized that we are studying organisms under the stress of maximum growth. As most microorganisms in nature are starved or at least nutrient limited, it is possible that at least some of these discoveries will be shown to be largely irrelevant to the organisms as they exist in their natural environment (Brown & Williams, 1985). Most if not all the biochemical techniques that are available to us today require at the least ng to μg quantities of material and so we obtain values that reflect an average for a given population rather than individual cell values. In spite of these limitations, David White and his colleagues have adapted a number of sensitive biochemical tests to study the bacterial population characteristics of natural biofilm communities (White & Findlay, 1988). Using these and other techniques, we are able to detect specific chemical signatures of individual members within biofilm communities. Analysis of components such as membrane lipids, storage products, etc., unique to individual species and/or growth conditions allow us to estimate the overall physiological condition and constitution of biofilm communities. Other investigators, notably Ronald Atlas, Norman Pace, David Stahl, Betty Olson, and others are developing molecular techniques which enable specific gene sequences and viable organisms to be isolated, identified and differentiated from natural populations *in situ* (Stahl *et al.*, 1988; Steffan *et al.*, 1989; Amann *et al.*, 1990*a,b*, 1992; Bej *et al.*, 1991*a,b,c*; Voordouw *et al.*, 1991; Tsai & Olson, 1992). Direct observations by light and electron microscopy greatly complement these other techniques in that they allow detailed examinations of individual cells (Costerton, 1979; Holt & Beveridge, 1982), and may often suggest appropriate future experimental strategies. Specific localization of enzymes within a population by either immuno-cytochemical or histochemical techniques can enhance morphological techniques in that it will illustrate the distribution of various metabolic functions among members of a biofilm microbial community (Costerton & Marks, 1977; McLean *et al.*, 1986).

Bright and Fletcher compared the metabolic activity of adherent and free living bacteria using a marine *Pseudomonas* sp. (Bright & Fletcher, 1983). In these experiments, the organisms were allowed to attach to a variety of surfaces, and their metabolic activity was evaluated by uptake of radiolabelled amino acids as shown by microautoradiography, and electron transport activity measured by staining with a tetrazolium salt. Activity by adherent organisms was compared to that of detached organisms, and to free living organisms. The authors found that as the hydrophobicity of the surface increased, there was an increase in the proportion of adherent cells showing metabolic activity (^{3}H-leucine uptake) and electron transport system activity but overall this relationship of activity between adherent and non-adherent cells depended upon the substratum to which they were attached. Their only firm conclusion was that metabolic activity appeared to promote firm attachment, and that cells that detached from the surface substrata often assimilated fewer amino acids. Overall they found no general effect of surfaces on the activity of attached bacteria, and that the relative activity of attached versus free-living bacteria depended upon the amino acid used, the concentration of the amino acid, and the properties of the particular surface involved. Several studies have shown the adhesion process itself to have some influence on microbial physiological activity (Pedersen *et al.*, 1986; Geesey & White, 1990). Other investigators have found that metabolic activities of the dental microorganisms, *Streptococcus mutans*, and *Veillonella* sp. (as expressed in DNA and acid production) were enhanced when these organisms were growing as a mixed biofilm or plaque (Noorda *et al.*, 1988).

Cells within the biofilm matrix would not be expected to have equal access to substrates due to problems of diffusion and possibly competition. If a particular substrate is necessary for growth, one would expect bacterial cells to possess several mechanisms by which the substrate in question might be assimilated. In order to economize biosynthetic requirements, individual cells normally only express one particular system at any given time. However due to the aforementioned problems of diffusion and competition within a biofilm, both low and high affinity transport mechanisms for a given substrate might be expected within this microbial community. This has been shown in the concurrent expression of both low and high affinity aspartate transport systems within wastewater biofilms (Eighmy & Bishop, 1984). Indeed physical disruption of a biofilm has been shown to augment its overall collective metabolism in that the access of all cells to nutrients is enhanced (Murray *et al.*, 1987).

Metabolic activities unique to biofilms have been described in the literature. Many animals are dependent upon microbial fermentation in their digestive processes. In ruminants, fermentation occurs predominantly in the rumen and caecum (Hungate, 1966). Cheng and coworkers have repeatedly observed that microbial digestion of cellulose within the rumen occurs as a result of colonization of this insoluble substrate by consortia of mixed microbial biofilm populations (Cheng & Costerton, 1980b; Cheng *et al.*, 1981a). As described previously, metabolic waste products (such as volatile fatty acids and H_2) of the primary cellulose digesting organisms represent substrates for other organisms (such as methanogens) (Wolin & Miller, 1983). Biofilms, in this instance, increase the efficiency of cellulose digestion in that synergistic microbial interactions are greatly enhanced. The other example of metabolism unique to biofilms is the anaerobic corrosion processes within pipelines. Under appropriate environmental conditions, consortia of bacteria colonize pipes and through their metabolic processes actually generate an anaerobic reducing microenvironment in the interior of the biofilm. Quite frequently these anaerobic microenvironments possess a sufficiently low Eh to support the growth of sulphate reducing bacteria. Due to localized differences in metabolism, bacteria in one region may scavenge electrons from the metal itself, thereby converting the metal to soluble cations. In another region, bacteria may transfer electrons to an electron acceptor such as sulphate. The resulting sulphide ions then are able to react with the metal to form a metal sulphide. The net consequence is the generation of a biological electrochemical cell and corrosion of the metal (reviewed in Hamilton (1985); Costerton *et al.* (1987); Hamilton (1987)).

5.7   PRACTICAL ASPECTS OF BIOFILM FORMATION

The ubiquitous nature of bacterial growth as biofilms ensures that they will influence many aspects of industry. Detrimental aspects of biofilms include their ability to induce microbial fouling and corrosion in industrial pipelines and other aquatic systems. In addition, the biofilm mode of growth is very instrumental in the ability of bacteria to persist. Consequently they can act as a reservoir for potentially harmful or pathogenic bacteria in many environments such as waste-water treatment systems, and infections. These topics have been addressed previously.

Many industrial processes employ the fermentation or biosynthetic capabilities of microbes. In a relatively simple process, ethanol is produced by the anaerobic fermentation of glucose by a monoculture of *Saccharomyces cerevisiae* or *Zymomonas* sp. These organisms are provided with appropriate levels of substrate (often a saccharified grain mash), and as a result of their growth and primary anaerobic metabolism, produce ethanol. In more complicated microbial biosyntheses, the desired product (such as an antibiotic) may only occur at the cessation of cell growth as a result of secondary metabolism. The efficiency of these biosynthetic processes is often enhanced by the use of immobilized cells (addressed elsewhere in this book) in that the immobilization process induces the formation of secondary metabolites. In either of these examples, the microbes in question are provided with appropriate levels of substrate, and other growth requirements, and the fermentation endproducts removed through centrifugation, filtration, distillation, or other procedures. More difficult problems arise when poor substrate quality and quantity, and adverse environmental conditions are present.

Many industrial and agricultural wastes contain high concentrations of organic matter. In anaerobic digesters, these substances are converted to methane through microbial fermentation (MacLeod *et al.*, 1990). No single species of organism is responsible for all aspects of anaerobic digestion. Rather these fermentation processes follow several metabolic pathways with one or more species being responsible for and deriving benefit from particular reactions. During microbial cometabolism, each organism adapts to its particular niche, and relies upon its neighbours to remove inhibitory metabolic waste products. Such cell–cell interactions are enhanced when these organisms coexist within biofilms (Eighmy *et al.*, 1983; Carey *et al.*, 1984; Eighmy & Bishop, 1984; Camper *et al.*, 1985; Nielsen, 1987). As substrate levels in waste-water are rarely constant, and small or significant levels of harmful compounds may be present, biofilm growth becomes very advantageous to the organisms in terms of survival.

Recent technological developments in molecular biology have enabled us to get a very detailed understanding about many physiological processes of bacteria. Through selection, amplification, appropriate regulation, and expression of particular genes or gene sequences we are able to develop 'super organisms' which in the laboratory are able to perform all aspects of a desired metabolic function, produce huge yields of product and/or consume huge quantities of substrate. Often, introduction of foreign genes, or alteration of naturally occurring genes, disrupt overall cell biology in that the loss of some other essential cell function can result. The outcome of these metabolic imbalances may not become apparent until these engineered organisms are moved into less ideal environments outside the laboratory and are exhibited when laboratory scale processes fail miserably at the pilot plant or field testing stage.

Biofilm growth offers several possible strategies for the development of new industrial microbial processes. Aspects particularly evident in wastewater treatment are that interspecies metabolic processes are enhanced and persistence during times of nutrient depletion and/or the presence of harmful compounds is enhanced. Rather than develop one super organism to perform all aspects of a desired process, it might be prudent to employ more naturally efficient metabolic processes involving several types of organisms growing as consortia within biofilms.

Another potential of biofilm technology would be preliminary testing of genetically engineered organisms as to their suitability and survival capabilities in the environment. Biofilms containing one or more species of microorganism are relatively easily produced in the laboratory (Costerton, 1984a; Costerton et al., 1987). This being the case, one could conceivably produce biofilms containing the test organism and any number of other organisms. Due to the close proximity of the individual members within biofilms, cell–cell interactions are maximized and such processes as microbial synergy or competition, cometabolism (such as interspecies hydrogen transfer), and even intercellular genetic transfer would be enhanced. The potential for biofilm technology is therefore quite significant.

ACKNOWLEDGEMENTS

Work in the authors' laboratories has been funded through grants from the Kidney Foundation of Canada (RJCM), Alberta Heritage Foundation for Medical Research (RJCM and JWC), Medical Research Council of Canada (JWC), Natural Sciences and Engineering Council of Canada (RJCM, DEC, and JWC), Alberta Oil Sands and Technology Research Authority (JWC), Alberta Farming for the Future Program (RJCM and JWC), Office of Naval Research (DEC), numerous industrial contracts

(JWC), and the Ontario Ministry of Health Career Scientist Program (RJCM). J. W. Costerton was the holder of a Natural Sciences and Engineering Research Council of Canada (NSERC) Industrial Research Chair in Biofilm Microbiology and is now the Director of the Center for Biofilm Engineering, Montana State University, Bozeman, MT.

These basic concepts and developments of biofilm technology would not have been possible without the enthusiasm and help of a large number of our colleagues and collaborators through the years.

We dedicate this chapter to Kevin C. Marshall, a gentleman and a scholar, and a pioneer in biofilm microbiology, on the occasion of his retirement.

5.8 REFERENCES

Alexander, M. (1971). Biochemical ecology of microorganisms. *Annual Review of Microbiology*, **25**, 361–92.

Allison, D. G. & Sutherland I. W. (1987). The role of exopolysaccharides in adhesion of freshwater bacteria. *Journal of General Microbiology*, **133**, 1319–28.

Aly, R. & Levit, S. (1987). Adherence of *Staphylococcus aureus* to squamous epithelium: role of fibronectin and teichoic acid. *Reviews of Infectious Diseases*, **9**, S341–50.

Amann, R. I., Krumholz, L. & Stahl, D. A. (1990a). Fluorescent oligonucleotide probing of whole cells for determinative, phylogenetic, and environmental studies in microbiology. *Journal of Bacteriology*, **172**, 762–70.

Amann, R. I., Binder, B. J., Olson, R. J., Chisholm, S. W., Devereux, R. & Stahl, D. A. (1990b). Combination of 16S rRNA–targeted oligonucleotide probes with flow cytometry for analyzing mixed microbial populations. *Applied and Environmental Microbiology*, **56**, 1919–25.

Amann, R. I., Stromley, J., Devereux, R., Key, R. & Stahl, D. A. (1992). Molecular and microscopic identification of sulfate-reducing bacteria in multi-species biofilms. *Applied and Environmental Microbiology*, **58**, 614–23.

Anwar, H., Van Biesen, T., Dasgupta, M., Lam, K. & Costerton, J. W. (1989). Interaction of biofilm bacteria with antibiotics in a novel *in vitro* chemostat system. *Antimicrobial Agents and Chemotherapy*, **33**, 1824–6.

Aronson, M., Medalia, O., Amichay, D. & Nativ, O. (1988). Endotoxin-induced shedding of viable uroepithelial cells is an antimicrobial defense mechanism. *Infection and Immunity*, **56**, 1615–17.

Baker, J. H. (1984). Factors affecting the bacterial colonization of various surfaces in a river. *Canadian Journal of Microbiology*, **30**, 511–15.

Bale, M. J., Day, M. J. & Fry, J. C. (1988). Novel method for studying plasmid transfer in undisturbed river epilithon. *Applied and Environmental Microbiology*, **54**, 2756–8.

Balkwill, D. L., Fredrickson, J. K. & Thomas, J. M. (1989). Vertical and horizontal variations in the physiological diversity of the aerobic chemoheterotrophic bacterial microflora in deep southeast coastal plain subsurface sediments. *Applied and Environmental Microbiology*, **55**, 1058–65.

Balkwill, D. L. & Ghiorse, W. C. (1985). Characterization of subsurface bacteria associated with two shallow aquifers in Oklahoma. *Applied and Environmental Microbiology*, **50**, 580–8.

Beachey, E. H., Giampapa, C. S. & Abraham, S. N. (1988). Bacterial adherence. Adhesin receptor-mediated attachment of pathogenic bacteria to mucosal surfaces. *American Review of Respiratory Diseases*, **138**, S45–48.

Bej, A. K., Mahbubani, M. H. & Atlas, R. M. (1991*a*). Detection of viable *Legionella pneumophila* in water by polymerase chain reaction and gene probe methods. *Applied and Environmental Microbiology*, **57**, 597–600.

Bej, A. K., DiCesare, J. L., Haff, L. & Atlas, R. M. (1991*b*). Detection of *Escherichia coli* and *Shigella* spp. in water by using the polymerase chain reaction and gene probes for *uid*. *Applied and Environmental Microbiology*, **57**, 1013–17.

Bej, A. K., McCarty, S. C. & Atlas, R. M. (1991*c*). Detection of coliform bacteria and *Escherichia coli* by multiplex polymerase chain reaction: comparison with defined substrate and plating methods for water quality monitoring. *Applied and Environmental Microbiology*, **57**, 2429–32.

Belas, M. R. & Colwell, R. R. (1982). Adsorption kinetics of laterally and polarly flagellated *Vibrio*. *Journal of Bacteriology*, **151**, 1568–80.

Beveridge, T. J. (1981). Ultrastructure, chemistry, and function of the bacterial wall. *International Review of Cytology*, **72**, 229–317.

Beveridge, T. J. (1989). Role of cellular design in bacterial metal accumulation and mineralization. *Annual Review of Microbiology*, **43**, 147–71.

Beynon, L. M., Dumanski, A. J., McLean, R. J. C., MacLean, L. L., Richards, J. C. & Perry, M. B. (1992). Capsule structure of *Proteus mirabilis* (ATCC 49565). *Journal of Bacteriology*, **174**, 2172–7.

Blenkinsopp, S. A., Khoury, A. E. & Costerton, J. W. (1992). Electrical enhancement of biocide efficacy against *Pseudomonas aeruginosa* biofilms. *Applied and Environmental Microbiology*, **58**, 3770–3.

Brackenhoff, G. J., van der Voort, H. T. M., van Spronsen, E. A. & Nanninga, N. (1988). 3-Dimensional imaging of biological structures by high resolution confocal microscopy. *Scanning Microscopy*, **2**, 33.

Bright, J. J. & Fletcher, M. (1983). Amino acid assimilation and electron transport system activity in attached and free–living marine bacteria. *Applied and Environmental Microbiology*, **45**, 818–25.

Brock, T. D. & Madigan, M. T. (1988). *Biology of microorganisms*. Prentice Hall, Englewood Cliffs, NJ.

Brown, M. R. W., Allison, D. G. & Gilbert, P. (1988). Resistance of bacterial biofilms to antibiotics: a growth-rate related effect. *Journal of Antimicrobial Chemotherapy*, **22**, 777–80.

Brown, M. R. W. & Williams, P. (1985). The influence of environment on envelope properties affecting survival of bacteria in infections. *Annual Review of Microbiology*, **39**, 527–56.

Bryant, R. D., Costerton, J. W. & Laishley, E. J. (1984). The role of *Thiobacillus albertis* glycocalyx in the adhesion of cells to elemental sulfur. *Canadian Journal of Microbiology*, **30**, 81–90.

Caldwell, D. E. (1977). The planktonic microflora of lakes. *Critical Reviews in Microbiology*, **5**, 306–70.

Caldwell, D. E. (1984). Surface colonization parameters from cell density and distribution. In *Microbial adhesion and aggregation*, pp. 125–136. Ed. Marshall, K. C. Springer-Verlag Berlin.

Caldwell, D. E. (1985). New developments in computer-enhanced microscopy (CEM). *Journal of Microbiological Methods*, **4**, 117–25.

Caldwell, D. E. & Germida, J. J. (1985). Evaluation of difference imagery for visualizing and quantitating microbial growth. *Canadian Journal of Microbiology*, **31**, 35–44.

Caldwell, D. E. & Lawrence, J. R. (1986). Growth kinetics of *Pseudomonas fluorescens* microcolonies within the hydrodynamic boundary layers of surface microenvironments. *Microbial Ecology*, **12**, 299–312.

Caldwell, D. E. (1987). Microbial colonization of solid-liquid interfaces. In *Biochemical Engineering V*, pp. 274–280. New York Academy of Sciences, New York.

Caldwell, D. E. & Lawrence, J. R. (1988). Study of attached cells in continuous–flow slide culture. In *CRC handbook of laboratory model system for microbial ecosystems, volume 1*, pp. 117–138. Ed. Wimpenny, J. W. T. CRC Press, Inc., Boca Raton, FL.

Caldwell, D. E., Korber, D. R. & Lawrence, J. R. (1992). Imaging of bacterial cells by fluorescence exclusion using scanning confocal laser microscopy. *Journal of Microbiological Methods*, **15**, 249–61.

Caldwell, D. E. & Tiedje, J. M. (1975*a*). A morphological study of bacteria in the hypolimnia of several Michigan lakes. *Canadian Journal of Microbiology*, **21**, 362–76.

Caldwell, D. E. & Tiedje, J. M. (1975*b*). The structure of anaerobic bacterial communities in the hypolimnia of several lakes in Michigan. *Canadian Journal of Microbiology*, **21**, 377–85.

Camper, A. K., LeChevallier, M. W., Broadaway, S. C. & McFeters, G. A. (1985). Growth and persistence of pathogens on granular activated carbon filters. *Applied and Environmental Microbiology*, **50**, 1378–82.

Caparon, M. G., Stephens, D. S., Olsen, A. & Scott, J. R. (1991). Role of M protein in adherence of group A streptococci. *Infection and Immunity*, **59**, 1811–17.

Carey, J. H., Fox, M. E., Brownlee, B. G., Metcalfe, J. L. & Platford, R. F. (1984). Disappearance kinetics of 2,4- and 3,4-dichlorophenol in a fluvial system. *Canadian Journal of Physiology and Pharmacology*, **62**, 971–5.

Chan, R., Lam, J. S., Lam, K. & Costerton, J. W. (1984). Influence of culture conditions on expression of the mucoid mode of growth of *Pseudomonas aeruginosa*. *Journal of Clinical Microbiology*, **19**, 8–16.

Chan, R. C. Y., Reid, G., Irvin, R. T., Bruce, A. W. & Costerton, J. W. (1985). Competitive exclusion of uropathogens from human uroepithelial cells by *Lactobacillus* whole cells and cell wall fragments. *Infection and Immunity*, **47**, 84–89.

Chang, C. C. & Merritt, K. (1991). Effect of *Staphylococcus epidermidis* on adherence of *Pseudomonas aeruginosa* and *Proteus mirabilis* to polymethyl methacrylate (PMMA) and gentamicin-containing PMMA. *Journal of Orthopedic Research*, **9**, 284–8.

Cheng, K.-J. & Costerton, J. W. (1980*a*). Adherent rumen bacteria – their role in the digestion of plant material urea and epithelial cells. In *Digestive Physiology and Metabolism in Ruminants*, pp. 227–250. Eds Ruckebusch, Y. & Thivend, P. MTP Press, Lancaster, England.

Cheng, K.-J. & Costerton, J. W. (1980*b*). The formation of microcolonies by rumen bacteria. *Canadian Journal of Microbiology*, **26**, 1104–13.

Cheng, K.-J., McCowan, R. P. & Costerton, J. W. (1979). Adherent epithelial bacteria in ruminants and their roles in digestive tract function. *American Journal of Clinical Nutrition*, **32**, 139–48.

Cheng, K.-J., Fay, J. P., Coleman, R. N., Milligan, L. P. & Costerton, J. W. (1981*a*). Formation of bacterial microcolonies of feed particles in the rumen. *Applied and Environmental Microbiology*, **41**, 298–305.

Cheng, K.-J., Irvin, R. T. & Costerton, J. W. (1981*b*). Autochthonous and pathogenic colonization of animal tissues by bacteria. *Canadian Journal of Microbiology*, **27**, 461–90.

Christersson, C. E., Dunford, R. G., Glantz, P. O. & Baier, R. E. (1989). Effect of critical surface tension on retention of oral microorganisms. *Scandinavian Journal of Dental Research*, **97**, 247–56.

Christersson, C. E., Glantz, P. O. & Baier, R. E. (1988). Role of temperature and shear forces on microbial detachment. *Scandinavian Journal of Dental Research*, **96**, 91–98.

Clapham, L., McLean, R. J. C., Nickel, J. C., Downey, J. & Costerton, J. W. (1990). The influence of bacteria on struvite crystal habit and its importance in urinary stone formation. *Journal of Crystal Growth*, **104**, 475–84.

Cochrane, D. M. G., Brown, M. R. W., Anwar, H., Weller, P. H., Lam, K. & Costerton, J. W. (1988). Antibody response to *Pseudomonas aeruginosa* surface protein antigens in a rat model of chronic lung infection. *Journal of Medical Microbiology*, **27**, 255–61.

Costerton, J. W. (1979). The role of electron microscopy in the elucidation of bacterial structure and function. *Annual Review of Microbiology*, **33**, 459–79.

Costerton, J. W. (1990). *The removal of microbial fouling by ice nucleation*. US Patent number 4,419,248.

Costerton, J. W., Irvin, R. T. & Cheng, K.-J. (1981*a*). The role of bacterial surface structures in pathogenesis. *Critical Reviews in Microbiology*, **8**, 303–38.

Costerton, J. W., Irvin, R. T. & Cheng, K.-J. (1981*b*). The bacterial glycocalyx in nature and disease. *Annual Review of Microbiology*, **35**, 299–324.

Costerton, J. W., Lam, J., Lam, K. & Chan, R. (1983). The role of the microcolony mode of growth in the pathogenesis of *Pseudomonas aeruginosa* infections. *Reviews of Infectious Diseases*, **5**, S867–73.

Costerton, J. W. & Marks, I. (1977). Localization of enzymes in procaryotic cells. In *Electron Microscopy of Enzymes: Principles and Methods*, pp. 98–134. Ed. Hayat, M. A. Van Nostrand Reinhold Co., New York.

Costerton, J. W., Rozee, K. R. & Cheng, K.-J. (1983). Colonization of particulates, mucous and intestinal tissue. *Progress in Food and Nutritional Science*, **7**, 91–105.

Costerton, J. W. (1984a). The formation of biocide-resistant biofilms in industrial, natural, and medical systems. *Developments in Industrial Microbiology*, **25**, 363–72.

Costerton, J. W. (1984b). The etiology and persistence of cryptic bacterial infections: a hypothesis. *Reviews of Infectious Diseases*, **6**, S608–16.

Costerton, J. W., Cheng, K.-J., Geesey, G. G., Ladd, T. I., Nickel, J. C., Dasgupta, M. & Marrie, T. J. (1987). Bacterial biofilms in nature and disease. *Annual Review of Microbiology*, **41**, 435–64.

Cowan, M. M., Taylor, K. G. & Doyle, R. J. (1987). Energetics of the initial phase of adhesion of *Streptococcus sanguis* to hydroxylapatite. *Journal of Bacteriology*, **169**, 2995–3000.

Cusack, F., Brown, D. R., Costerton, J. W. & Clementz, D. M. (1987). Field and laboratory studies of microbial/fines plugging of water injection wells: mechanism, diagnosis and removal. *Journal of Petroleum Science and Engineering*, **1**, 39–50.

Dart, J. K. & Seal, D. V. (1988). Pathogenesis and therapy of *Pseudomonas aeruginosa* keratitis. *Eye*, **2**, S46–55.

Dazzo, F. B. & Truchet, G. L. (1983). Interactions of lectins and their saccharide receptors in the *Rhizobium*–legume symbiosis. *Journal of Membrane Biology*, **73**, 1–16.

Dean-Ross, D. & Mills, A. L. (1989). Bacterial community structure and function along a heavy metal gradient. *Applied and Environmental Microbiology*, **55**, 2002–9.

Doig, P., Smith, N. R., Todd, T. & Irvin, R. T. (1987). Characterization of the binding of *Pseudomonas aeruginosa* alginate to human epithelial cells. *Infection and Immunity*, **55**, 1517–22.

Doig, P., Todd, T., Sastry, P. A., Lee, K. K., Hodges, R. S., Paranchych, W. & Irvin, R. T. (1988). Role of pili in adhesion of *Pseudomonas aeruginosa* to human respiratory epithelial cells. *Infection and Immunity*, **56**, 1641–6.

Driessens, F. C. M., Borggreven, J. M. P. M., Verbeeck, R. M. H., van Dijk, J. W. E. & Feagin, F. F. (1985). On the physicochemistry of plaque calcification and the phase composition of dental calculus. *Journal of Periodontal Research*, **20**, 329–36.

Duguid, J. P., Smith, I. W., Demster, G. & Edwards, P. N. (1955). Non-flagellar filamentous appendages ('fimbriae') and haemagglutinating activity in *Bacterium coli. Journal of Pathology and Bacteriology*, **70**, 335–48.

Eighmy, T. T., Maratea, D. & Bishop, P. L. (1983). Electron microscopic examination of wastewater biofilm formation and structural components. *Applied and Environmental Microbiology*, **45**, 1921–31.

Eighmy, T. T. & Bishop, P. L. (1984). Multiplicity of aspartate transport in thin wastewater biofilms. *Applied and Environmental Microbiology*, **48**, 1151–8.

Eisenstein, B. I. (1987). Fimbriae. In *Escherichia coli* and *Salmonella typhimurium Cellular and Molecular Biology, volume 1*, pp. 84–90. Eds Neidhardt, F. C., Ingraham, J. L., Low, K. B., Magasanik, B., Schaechter, M. & Umbarger, H. E. American Society for Microbiology, Washington DC.

Ellen, R. P., Schwarz-Faulkner, S. & Grove, D. A. (1988). Coaggregation among periodontal pathogens, emphasizing *Bacteroides gingivalis–Actinomyces viscosus* cohesion on a saliva-coated mineral surface. *Canadian Journal of Microbiology*, **34**, 299–306.

Elliot, E. T., Coleman, D. C., Ingham, R. E. & Trofymow, J. A. (1984). Carbon and energy flow through microflora and microfauna in the soil subsystem of terrestrial ecosystems. In *Current Perspectives in Microbial Ecology*, pp. 424–33. Eds. Klug, M. J. & Reddy, C. A. American Society for Microbiology, Washington.

Fader, R. C. & Davis, C. P. (1980). Effect of piliation on *Klebsiella pneumoniae* infection in rat bladders. *Infection and Immunity*, **30**, 554–61.

Ferris, F. G., Beveridge, T. J. & Fyfe, W. S. (1986). Iron-silica crystallite nucleation by bacteria in a geothermal sediment. *Nature*, **320**, 609–11.

Ferris, F. G., Schultze, S., Witten, T. C., Fyfe, W. S. & Beveridge, T. J. (1989*a*). Metal interactions with microbial biofilms in acidic and neutral pH environments. *Applied and Environmental Microbiology*, **55**, 1249–57.

Ferris, F. G., Fyfe, W. S., Witten, T., Schultze, S. & Beveridge, T. J. (1989*b*). Effect of mineral substrate hardness on the population density of epilithic microorganisms in two Ontario rivers. *Canadian Journal of Microbiology*, **35**, 744–7.

Finlay, B. B. & Falkow, S. (1989). Common themes in microbial pathogenicity. *Microbiological Reviews*, **53**, 210–30.

Fletcher, M. (1988). Attachment of *Pseudomonas fluorescens* to glass and influence of electrolytes on bacterium–substratum separation distance. *Journal of Bacteriology*, **170**, 2027–30.

Geesey, G. G., Mutch, R., Costerton, J. W. & Green, R. B. (1978). Sessile bacteria: an important component of the microbial population in small mountain streams. *Limnology and Oceanography*, **23**, 1214–23.

Geesey, G. G. & White, D. C. (1990). Determination of bacterial growth and activity at solid–liquid interfaces. *Annual Review of Microbiology*, **44**, 579–602.

Ghiorse, W. C. & Wilson, J. T. (1988). Microbial ecology of the terrestrial subsurface. *Advances in Applied Microbiology*, **33**, 107–72.

Gibbons, R. J. (1989). Bacterial adhesion to oral tissues: a model for infectious diseases. *Journal of Dental Research*, **68**, 750–60.

Gibbons, R. J. & Nygaard, M. (1970). Interbacterial aggregation of plaque bacteria. *Archives of Oral Biology*, **15**, 1397–400.

Gilbert, P., Collier, P. J. & Brown, M. R. W. (1990). Influence of growth rate on susceptibility to antimicrobial agents: Biofilms, cell cycle, dormancy, and stringent response. *Antimicrobial Agents and Chemotherapy*, **34**, 1865–8.

Godwin, S. L., Fletcher, M. & Burchard, R. P. (1989). Interference reflection microscopic study of sites of association between gliding bacteria and glass substrata. *Journal of Bacteriology*, **171**, 4589–94.

Gristina, A. G. & Costerton, J. W. (1985). Bacterial adherence to biomaterials and tissue. The significance of its role in clinical sepsis. *American Journal of Bone and Joint Surgery*, **67**, 264–73.

Hagberg, L., Lam, J., Svanborg Edén, C. & Costerton, J. W. (1986). Interaction of a pyelonephritogenic *Escherichia coli* strain with the tissue components of the mouse urinary tract. *Journal of Urology*, **136**, 165–72.

Hamilton, W. A. (1985). Sulphate-reducing bacteria and anaerobic corrosion. *Annual Review of Microbiology*, **39**, 195–217.

Hamilton, W. A. (1987). Biofilms: microbial interactions and metabolic activities. In *Ecology of microbial communities*, pp. 361–385. Ed. Hamilton, W. A. University Press, Cambridge, UK.

Haugland, R. P. (1992). *Handbook of Fluorescent Probes and Research Chemicals*. Molecular Probes, Inc., Eugene, OR.

Holt, S. C. & Beveridge, T. J. (1982). Electron microscopy: its development and application to microbiology. *Canadian Journal of Microbiology*, **28**, 1–53.

Humphries, M., Nemcek, J., Cantwell, J. B. & Gerrard, J. J. (1987). The use of graft copolymers to inhibit the adhesion of bacteria to solid surfaces. *FEMS Microbial Ecology*, **45**, 297–304.

Hungate, R. E. (1966). *The Rumen and its Microbes*. Academic Press, New York.

Iriberri, J., Unanue, M., Ayo, B., Barcina, I. & Egea, L. (1990). Bacterial production and growth rate estimation from [^{3}H]thymidine incorporation for attached and free-living bacteria in aquatic systems. *Applied and Environmental Microbiology*, **56**, 483–7.

Jacques, M., Marrie, T. J. & Costerton, J. W. (1987). Review: Microbial colonization of prosthetic devices. *Microbial Ecology*, **13**, 173–91.

Jacques, M., Olson, M. E., Crichlow, A. M., Osborne, A. D. & Costerton, J. W. (1986). The normal microflora of the female rabbit's genital tract. *Canadian Journal of Comparative Medicine*, **50**, 272–4.

Jacques, M., Roy, G. & Mittal, K. R. (1988). Hemagglutinating properties of *Actinobacillus pleuropneumoniae*. *Canadian Journal of Microbiology*, **34**, 1046–9.

Kenne, L. & Lindberg, B. (1983). Bacterial polysaccharides. In *The polysaccharides, volume 2*, pp. 287–363. Ed. Aspinall, G. O. Academic Press, New York.

Kieft, T. L. & Caldwell, D. E. (1984). Chemostat and in-situ colonization kinetics of *Thermothrix thiopara* on calcite and pyrite surfaces. *Geomicrobiology Journal*, **3**, 217–29.

Kjelleberg, S., Hermansson, M., Marden, P. & Jones, G. W. (1987). The transient phase between growth and nongrowth of heterotrophic bacteria, with emphasis on the marine environment. *Annual Review of Microbiology*, **41**, 25–49.

Koga, T., Toda, Y., Moro, I. & Hamada, S. (1988). Electron-microscopic observation of adherence of serotype c *Streptococcus mutans* to the enamel surface due to glucan synthesis. *Zentralblatt für Bakteriologie, Mikrobiologie, und Hygiene, Series A*, **269**, 492–500.

Kolenbrander, P. E. (1988). Intergeneric coaggregation among human oral bacteria and ecology of dental plaque. *Annual Review of Microbiology*, **42**, 627–56.

Korber, D. R., Lawrence, J. R., Sutton, B. & Caldwell, D. E. (1989). Effect of laminar flow velocity on the kinetics of surface recolonization by Mot$^+$ and Mot$^-$ *Pseudomonas fluorescens*. *Microbial Ecology*, **18**, 1–19.

Kudo, H., Cheng, K.-J. & Costerton, J. W. (1987). Electron microscopic study of the methylcellulose-mediated detachment of cellulolytic rumen bacteria from cellulose fibers. *Canadian Journal of Microbiology*, **33**, 267–72.

Ladd, T. I., Schmiel, D., Nickel, J. C. & Costerton, J. W. (1985). Rapid method for detection of adherent bacteria on Foley urinary catheters. *Journal of Clinical Microbiology*, **21**, 1004–6.

Ladd, T. I., Schmiel, D., Nickel, J. C. & Costerton, J. W. (1987). The use of a radiorespirometric assay for testing the antibiotic sensitivity of catheter-associated bacteria. *Journal of Urology*, **138**, 1451–6.

Lappin-Scott, H. M., Cusack, F., MacLeod, A. & Costerton. J. W. (1988*a*). Starvation and nutrient resuscitation of *Klebsiella pneumoniae* isolated from oil well waters. *Journal of Applied Bacteriology*, **64**, 541–9.

Lappin-Scott, H. M., Cusack, F. & Costerton, J. W. (1988*b*). Nutrient resuscitation and growth of starved cells in sandstone cores: a novel approach to enhanced oil recovery. *Applied and Environmental Microbiology*, **54**, 1373–82.

Lappin-Scott, H. M. & Costerton, J. W. (1989). Bacterial biofilms and surface fouling. *Biofouling*, **1**, 323–42.

Larsen, S. H., Reader, R. W., Kort, E. N., Tso, W.-W. & Adler, J. (1974). Change in direction of flagellar rotation is the basis of the chemotactic response in *Escherichia coli*. *Nature*, **249**, 74–7.

Lawrence, J. R. & Caldwell, D. E. (1987). Behavior of bacterial stream populations within the hydrodynamic boundary layers of surface microenvironments. *Microbial Ecology*, **14**, 15–27.

Lawrence, J. R., Delaquis, P. J., Korber, D. & Caldwell, D. E. (1987). Behavior of *Pseudomonas fluorescens* within the hydrodynamic boundary layers of surface microenvironments. *Microbial Ecology*, **14**, 1–14.

Lawrence, J. R., Korber, D. R., Hoyle, B. D., Costerton, J. W. & Caldwell, D. E. (1991). Optical sectioning of microbial biofilms. *Journal of Bacteriology*, **173**, 6558–67.

LeChevallier, M. W., Babcock, T. M. & Lee, R. G. (1987). Examination and characterization of distribution system biofilms. *Applied and Environmental Microbiology*, **53**, 2714–24.

LeChevallier, M. W., Cawthon, C. D. & Lee, R. G. (1988*a*). Factors promoting survival of bacteria in chlorinated water supplies. *Applied and Environmental Microbiology*, **54**, 649–54.

LeChevallier, M. W., Cawthon, C. D. & Lee, R. G. (1988*b*). Inactivation of biofilm bacteria. *Applied and Environmental Microbiology*, **54**, 2492–9.

Leung, J. W. C., Sung, J. Y. & Costerton, J. W. (1989). Bacteriological and electron microscopy examination of brown pigment stones. *Journal of Clinical Microbiology*, **27**, 915–21.

Liljemark, W. F., Bloomquist, C. G., Coulter, M. C., Fenner, L. J., Skopek, R. J. & Schachtele, C. F. (1988). Utilization of a continuous streptococcal surface to measure interbacterial adherence *in vitro* and *in vivo*. *Journal of Dental Research*, **67**, 1455–60.

Lytle, M. S., Adams, J. C., Dickman, D. G. & Bressler, W. R. (1989). Use of nutrient response techniques to assess the effectiveness of chlorination of rapid sand filter gravel. *Applied and Environmental Microbiology*, **55**, 29–32.

MacLeod, F. A., Lappin-Scott, H. M. & Costerton, J. W. (1988). Plugging of a model rock system by using starved bacteria. *Applied and Environmental Microbiology*, **54**, 1365–72.

MacLeod, F. A., Guiot, S. R. & Costerton, J. W. (1990). Layered structure of bacterial aggregates produced in an upflow anaerobic sludge bed and filter reactor. *Applied and Environmental Biology*, **56**, 1598–1607.

MacNab, R. M. (1987). Flagella. In *Escherichia coli* and *Salmonella typhimurium cellular and molecular Biology, volume 1*, pp. 70–83. Eds Neidhardt, F. C., Ingraham, J. L., Low, K. B., Magasanik, B., Schaechter, M. & Umbarger, H. E. American Society for Microbiology, Washington DC.

Mann, S. (1988). Molecular recognition in biomineralization. *Nature*, **332**, 119–24.

Marshall, K. C., Stout, R. & Mitchell, R. (1971). Mechanisms of the initial events in the sorption of marine bacteria to solid surfaces. *Journal of General Microbiology*, **68**, 337–48.

Marshall, K. C. (1985). Bacterial adhesion in oligotrophic habitats. *Microbiological Science*, **2**, 321–6.

Marshall, K. C. (1988). Adhesion and growth of bacteria at surfaces in oligotrophic habitats. *Canadian Journal of Microbiology*, **34**, 503–6.

Marshall, P. A., Loeb, G. L., Cowan, M. M. & Fletcher, M. (1989). Response of microbial adhesives and biofilm matrix polymers to chemical treatments as determined by interference reflection microscopy and light section microscopy. *Applied and Environmental Microbiology*, **55**, 2827–31.

Matin, A., Auger, E. A., Blum, P. H. & Schultz, J. E. (1989). Genetic basis of starvation survival in nondifferentiating bacteria. *Annual Review of Microbiology*, **43**, 293–316.

McEldowney, S. & Fletcher, M. (1986). Variability of the influence of physicochemical factors affecting bacterial adhesion to polystyrene substrata. *Applied and Environmental Microbiology*, **52**, 460–5.

McEldowney, S. & Fletcher, M. (1987). Adhesion of bacteria from mixed cell suspension to solid surfaces. *Archives of Microbiology*, **148**, 57–62.

McKinley, V. L., Costerton, J. W. & White, D. C. (1988). Microbial biomass, activity, and community structure of water and particulates retrieved by backflow from a waterflood injection well. *Applied and Environmental Microbiology*, **54**, 1383–93.

McLean, R. J. C., Beauchemin, D. & Beveridge, T. J. (1992). Influence of oxidation state on iron binding by *Bacillus licheniformis* capsule. *Applied and Environmental Microbiology*, **58**, 405–8.

McLean, R. J. C., Beauchemin, D., Clapham, L. & Beveridge, T. J. (1990). Metal binding characteristics of the gamma-glutamyl capsule polymer of *Bacillus licheniformis* ATCC 9945. *Applied and Environmental Microbiology*, **56**, 3671–7.

McLean, R. J. C. & Beveridge, T. J. (1990). Metal binding capacity of bacterial surfaces and their ability to form mineralized aggregates. In *Microbial Mineral Recovery*, pp. 185–222. Eds Ehrlich, H. L. & Brierley, C. L. McGraw-Hill, New York.

McLean, R. J. C., Cheng, K.-J., Gould, W. D., Nickel, J. C. & Costerton, J. W. (1986). Histochemical and biochemical urease localization in the periplasm and outer membrane of two *Proteus mirabilis* strains. *Canadian Journal of Microbiology*, **32**, 772–8.

McLean, R. J. C., Lawrence, J. R., Korber, D. R. & Caldwell, D. E. (1991). *Proteus mirabilis* biofilm protection against struvite crystal dissolution and its implications in struvite urolithiasis. *Journal of Urology*, **146**, 1138–42.

McLean, R. J. C. & Nickel, J. C. (1991). Bacterial colonization behaviour: a new virulence strategy in urinary infections? *Medical Hypotheses*, **36**, 269–72.

McLean, R. J. C., Nickel, J. C., Cheng, K.-J. & Costerton, J. W. (1988). The ecology and pathogenicity of urease-producing bacteria in the urinary tract. *Critical Reviews in Microbiology*, **16**, 37–79.

Mergenhagen, S. E., Sandberg, A. L., Chassy, B. M., Brennan, M. J., Yeung, M. K., Donkersloot, J. A. & Cisar, J. O. (1987). Molecular basis of bacterial adhesion in the oral cavity. *Reviews of Infectious Diseases*, **9**, S467–74.

Mills, A. L. & Maubery, R. (1981). Effect of mineral composition on bacterial attachment to submerged rock surfaces. *Microbial Ecology*, **7**, 315–22.

Millward, T. A. & Wilson, M. (1989). The effect of chlorhexidine on *Streptococcus sanguis* biofilms. *Microbios*, **58**, 155–64.

Morita, R. Y. (1982). Starvation-survival of heterotrophs in the marine environment. *Advances in Microbial Ecology*, **6**, 171–98.

Morita, R. Y. (1988). Bioavailability of energy and its relationship to growth and starvation survival in nature. *Canadian Journal of Microbiology*, **34**, 436–41.

Moyer, C. L. & Morita, R. Y. (1989). Effect of growth rate and starvation–survival on the viability and stability of a psychrophilic marine bacterium. *Applied and Environmental Microbiology*, **55**, 1122–7.

Murray, R. E., Cooksey, K. E. & Priscu, J. C. (1987). Influence of physical disruption on growth of attached bacteria. *Applied and Environmental Microbiology*, **53**, 2997–9.

Nichols, P. D., Henson, J. M., Guckert, J. B., Nivens, D. E. & White, D. C. (1985). Fourier transform–IR spectroscopic methods for microbial ecology analysis of bacteria, bacteria-polymer mixtures and biofilms. *Journal of Microbiology Methods*, **4**, 79–94.

Nichols, W. W., Dorrington, S. M., Slack, M. P. & Walmsley, H. L. (1988). Inhibition of tobramycin diffusion by binding to alginate. *Antimicrobial Agents and Chemotherapy*, **32**, 518–23.

Nickel, J. C., Wright, J. B., Ruseska, I., Marrie, T. J., Whitfield, C. & Costerton, J. W. (1985*a*). Antibiotic resistance of *Pseudomonas aeruginosa* colonizing a urinary catheter *in vitro*. *European Journal of Clinical Microbiology*, **4**, 213–18.

Nickel, J. C., Ruseska, I., Wright, J. B. & Costerton, J. W. (1985*b*). Tobramycin resistance of *Pseudomonas aeruginosa* cells growing as a biofilm on urinary catheter material. *Antimicrobial Agents and Chemotherapy*, **27**, 619–24.

Nielsen, P. H. (1987). Biofilm dynamics and kinetics during high-rate sulfate reduction under anaerobic conditions. *Applied and Environmental Microbiology*, **53**, 27–32.

Noorda, W. D., Purdell-Lewis, D. J., van Montfort, A. M. & Weerkamp, A. H. (1988). Monobacterial and mixed bacterial plaques of *Streptococcus mutans* and *Veillonella alcalescens* in an artificial mouth: development, metabolism, and effect on human dental enamel. *Caries Research*, **22**, 342–7.

Nyström, T., Olsson, R. M. & Kjelleberg, S. (1992). Survival, stress resistance, and alterations in protein expression in the marine *Vibrio* sp. strain S14 during starvation for different individual nutrients. *Applied and Environmental Microbiology*, **58**, 55–65.

Ofek, I. & Perry, A. (1985). Molecular basis of bacterial adherence to tissues. In *Molecular Basis Oral Microbial adhesion*, pp. 7–13. Eds. Mergenhagen, S. E. & Rosan, B. American Society for Microbiology, Washington, DC.

Olson, G. J. (1986). Microbial intervention in trace element-containing industrial process streams and waste products. In *The Importance of Chemical 'speciation' in Environmental Processes*, pp. 493–512. Eds. Bernhard, M., Brinckman, F. E. & Sadler, P. J. Springer–Verlag, Berlin.

Op den Camp, H. J. M., Oosterhof, A. & Veerkamp, J. H. (1985). Interaction of Bifidobacterial lipoteichoic acid with human intestinal epithelial cells. *Infection and Immunity*, **47**, 332–4.

Payment, P. (1989). Bacterial colonization of domestic reverse-osmosis water filtration units. *Canadian Journal of Microbiology*, **35**, 1065–7.

Pedersen, K., Holmstrom, C., Olsson, A.-K. & Pedersen, A. (1986). Statistical evaluation of the influence of species variation, culture conditions, surface wettability, and fluid shear on attachment and biofilm development of marine bacteria. *Archives of Microbiology*, **145**, 1–8.

Poirier, T. P., Kehoe, M. A., Whitnack, E., Dockter, M. E. & Beachey, E. H. (1989). Fibrinogen binding and resistance to phagocytosis of *Streptococcus sanguis* expressing cloned M protein of *Streptococcus pyogenes*. *Infection and Immunity*, **57**, 29–35.

Pratt-Tepstra, I. H., Weerkamp, A. H. & Busscher, H. J. (1989). The effects of pellicle formation on streptococcal adhesion to human enamel and artificial substrata with various surface free-energies. *Journal of Dental Research*, **68**, 463–7.

Pringle, J. H. & Fletcher, M. (1986). Influence of substratum hydration and adsorbed macromolecules on bacterial attachment to surfaces. *Applied and Environmental Microbiology*, **51**, 1321–5.

Reid, G., Jewett, M. A. S., Nickel, J. C., McLean, R. J. C. & Bruce, A. W. (1990). Effect of extra corporeal shock wave lithotripsy on bacterial viability: relationship to the treatment of struvite stones. *Urological Research*, **18**, 425–7.

Reid, G., McGroarty, J. A., Angotti, R. & Cook, R. L. (1988). *Lactobacillus* inhibitor production against *Escherichia coli* and coaggregation ability with uropathogens. *Canadian Journal of Microbiology*, **34**, 344–51.

Ridgway, H. F., Kelly, A., Justice, C. & Olson, B. H. (1983). Microbial fouling of reverse-osmosis membranes used in advanced wastewater treatment technology: chemical, bacteriological, and ultrastructural analyses. *Applied and Environmental Microbiology*, **45**, 1066–84.

Rittmann, B. E., Crawford, L., Tuck, C. K. & Namkung, E. (1986). In-situ determination of kinetic parameters for biofilms isolation and characterization of oligotrophic biofilms. *Biotechnology and Bioengineering*, **28**, 1753–60.

Rosan, B., Eifert, R. & Golub, E. (1985). Bacterial surfaces, salivary pellicles, and plaque formation. In *Molecular Basis of Oral Microbial Adhesion*, pp. 69–76. Eds Mergenhagen, S. E. & Rosan, B. American Society for Microbiology, Washington, DC.

Roszak, D. B. & Colwell, R. R. (1987). Survival strategies of bacteria in the natural environment. *Microbiological Reviews*, **51**, 365–79.

Ruseska, I., Robbins, J., Lashen, E. S. & Costerton, J. W. (1982). Biocide testing against corrosion-causing oilfield bacteria helps control plugging. *Oil and Gas Journal*, **3**, 253–64.

Shaw, J. C., Bramhill, B., Wardlaw, N. C. & Costerton, J. W. (1985). Bacterial fouling in a model core system. *Applied and Environmental Microbiology*, **49**, 693–701.

Siegele, D. A. & Kolter, R. (1992). Life after log. *Journal of Bacteriology*, **174**, 345–8.

Stahl, D. A., Flesher, B., Mansfield, H. R. & Montgomery, L. (1988). Use of phytogenetically based hybridization probes for studies of ruminal microbial ecology. *Applied and Environmental Microbiology*, **54**, 1079–84.

Staley, J. T. & Fuerst, J. A. (1989). Budding and/or appendaged bacteria. In *Bergey's Manual of Systematic Bacteriology, volume 3*, pp. 1890–1993. Eds Staley, J. T., Bryant, M. P., Pfennig, N. & Holt, J. G. Williams and Wilkins, Baltimore.

Steffan, R. J., Breen, A., Atlas, R. M. & Sayler, G. S. (1989). Application of gene probe methods for monitoring specific microbial populations in freshwater ecosystems. *Canadian Journal of Microbiology*, **35**, 681–5.

Steffan, R. J., Goksoyr, J., Bej, A. K. & Atlas, R. M. (1988). Recovery of DNA from soils and sediments. *Applied and Environmental Microbiology*, **54**, 2908–15.

Stenström, T. A. (1989). Bacterial hydrophobicity, an overall parameter for the measurement of adhesion potential to soil particles. *Applied and Environmental Microbiology*, **55**, 142–7.

Stickler, D. J., Clayton, C. L. & Chawla, J. C. (1987). Assessment of antiseptic bladder washout procedures using a physical model of the catheterised bladder. *British Journal of Urology*, **60**, 413–8.

Sutherland, I. W. (1977). Bacterial exopolysaccharides – their nature and production. In Surface Carbohydrates of the Prokaryotic Cell, pp. 27–96. Ed. Sutherland, I. W. Academic Press, London.

Sutherland, I. W. (1985). Biosynthesis and composition of gram-negative bacterial extracellular and wall polysaccharides. *Annual Review of Microbiology*, **39**, 243–70.

Sutherland, I. W. (1988). Bacterial surface polysaccharides: structure and function. *International Review of Cytology*, **113**, 187–231.

Svanborg Edén, C., Andersson, B., LeMer, H., Magnusson, G. & Noori, G. (1985). Specificity for glycolipid receptors of adhering bacteria: Experimental differences between uropathogenic *Escherichia coli* and *Streptococcus pneumoniae*. In *Molecular Basis of Oral Microbial Adhesion*, pp. 14–17. Eds Mergenhagen, S. E. & Rosan B. American Society for Microbiology, Washington, DC.

Tormo, A., AlmirQorn, M. & Kolter, R. (1990). *sur* A, an *Escherichia coli* gene essential for survival in stationary phase. *Journal of Bacteriology*, **172**, 4339–47.

Troy, F. A. (1973). Chemistry and biosynthesis of the poly (γ-D-glutamyl) capsule in *Bacillus licheniformis*. *Journal of Biological Chemistry*, **248**, 305–15.

Tsai, Y.-L. & Olson, B. H. (1992). Detection of low numbers of bacterial cells in soils and sediments by polymerase chain reaction. *Applied and Environmental Microbiology*, **58**, 754–7.

van Loosdrecht, M. C., Lyklema, J., Norde, W., Schraa, G. & Zehnder, A. J. (1987). Electrophoretic mobility and hydrophobicity as a measure to predict the initial steps of bacterial adhesion. *Applied and Environmental Microbiology*, **53**, 1898–1901.

van Loosdrecht, M. C., Lyklema, J., Norde, W. & Zehnder, A. J. (1990). Influence of interfaces on microbial activity. *Microbiological Reviews*, **54**, 75–87.

Voordouw, G., Voordouw, J. K., Karkhoff-Schweizer, R. R., Fedorak, P. M. & Westlake, D. W. S. (1991). Reverse sample genome probing, a new technique for identification of bacteria in environmental samples by DNA hybridization, and its application to the identification of sulfate–reducing bacteria in oil field samples. *Applied and Environmental Microbiology*, **57**, 3070–8.

White, D. C. & Findlay, R. H. (1988). Biochemical markers tor measurement of predation effects on the biomass, community structure, nutritional status, and metabolic activity of microbial biofilms. *Hydrobiologia*, **159**, 119–32.

Whitfield, C. (1988). Bacterial extracellular polysaccharides. *Canadian Journal of Microbiology*, **34**, 415–20.

Wolin, M. J. & Miller, T. L. (1983). Carbohydrate fermentation. In *Human Intestinal Microflora in Health and Disease*, pp. 147–165. Ed. Hentges, D. J. Academic Press, New York.

Wrangstadh, M., Conway, P. L. & Kjelleberg, S. (1986). The production and release of an extracellular polysaccharide during starvation of a marine *Pseudomonas* sp. and the effect thereof on adhesion. *Archives of Microbiology*, **145**, 220–7.

Wrangstadh, M., Conway, P. L. & Kjelleberg, S. (1989). The role of an extracellular polysaccharide produced by the marine *Pseudomonas* sp. S9 in cellular detachment during starvation. *Canadian Journal of Microbiology*, **35**, 309–12.

INDEX

337